Mechanisms and Machine Science

Volume 114

This book series establishes a well-defined forum for monographs, edited Books, and proceedings on mechanical engineering with particular emphasis on MMS (Mechanism and Machine Science). The final goal is the publication of research that shows the development of mechanical engineering and particularly MMS in all technical aspects, even in very recent assessments. Published works share an approach by which technical details and formulation are discussed, and discuss modern formalisms with the aim to circulate research and technical achievements for use in professional, research, academic, and teaching activities.

This technical approach is an essential characteristic of the series. By discussing technical details and formulations in terms of modern formalisms, the possibility is created not only to show technical developments but also to explain achievements for technical teaching and research activity today and for the future.

The book series is intended to collect technical views on developments of the broad field of MMS in a unique frame that can be seen in its totality as an Encyclopaedia of MMS but with the additional purpose of archiving and teaching MMS achievements. Therefore, the book series will be of use not only for researchers and teachers in Mechanical Engineering but also for professionals and students for their formation and future work.

The series is promoted under the auspices of International Federation for the Promotion of Mechanism and Machine Science (IFToMM).

Prospective authors and editors can contact Mr. Pierpaolo Riva (publishing editor, Springer) at: pierpaolo.riva@springer.com

Indexed by SCOPUS and Google Scholar.

More information about this series at https://link.springer.com/bookseries/8779

Tigran Parikyan
Editor

Advances in Engine and Powertrain Research and Technology

Design · Simulation · Testing · Manufacturing

Editor
Tigran Parikyan ⓘ
Advanced Simulation Technologies
AVL List GmbH
Graz, Austria

ISSN 2211-0984   ISSN 2211-0992   (electronic)
Mechanisms and Machine Science
ISBN 978-3-030-91871-2   ISBN 978-3-030-91869-9   (eBook)
https://doi.org/10.1007/978-3-030-91869-9

This Springer imprint is published by the registered company Springer Nature Switzerland AG
The registered company address is: Gewerbestrasse 11, 6330 Cham, Switzerland

Preface

The stringent legislation concerning the emissions of vehicles has become one of the strongest factors affecting engine and powertrain manufacturers during the last decade. Consequently, we witness a massive drift toward electrification of power-trains, especially in the automotive industry—a trend which will hold further and promises to get even stronger in the years and even decades to come. In parallel, research and development on much cleaner internal combustion (IC) engines based not only on improved design and control strategy of the combustion process, but also on new and alternative fuels, continues as well. Besides that, various types of hybrid powertrains including both IC engine and electric motor take a considerable share in many powertrains produced worldwide. One should not forget the power turbines, used both in transportation machinery and in power generation.

The aim of this book is to cover a wide range of applied research on engines and powertrains compactly presented in one volume, by showing innovative engineering solutions, which already found their place in the development process of improved engines and powertrains used in modern vehicles and power plants.

Addressing research and development topics in the area of engines and power-trains, the present collection of works will be of interest primarily for scientists, engineers and educators working in automotive, marine, aviation, and power gener-ation areas. At the same time, it can also be useful for graduate students and equally for all those who are relatively new to the subject and are looking for a single source with a good overview of the state-of-the-art as well as an up-to-date information on governing physical theories, numerical methods, and their application in design, simulation, testing, and manufacturing. The reader will find here a rich mixture of approaches, software tools, and case studies used to investigate and optimize diverse powertrains, their functional units, and separate machine parts based on different physical phenomena, their adequate mathematical representation, efficient solution algorithms, and experimental validation.

The diversification of the powertrain types mentioned above does not mean aban-doning the topics of traditional importance, like vibration and durability, tribology, NVH, CFD, control strategy, etc., which can all be found in the book, too.

According to the topics, the chapters are grouped into the following parts:

Part One: Dynamics and Vibrations/Tribology and Lubrication,
Part Two: Combustion/CFD/Emissions/Fuels,
Part Three: Hybrid and Electrified Powertrains,
Part Four: Testing/Calibration/Monitoring/Diagnostics, and
Part Five: Manufacturing.

The *Advances in Engine and Powertrain Research and Technology* is the first book of the Springer series *Mechanisms and Machine Science* completely dedicated to engines and powertrains. It is published under the auspices of the *International Federation for the Promotion of Mechanism and Machine Science* (IFToMM). All the chapters are authored (or co-authored) by members of the following IFToMM Technical Committees:

- *Engines and Powertrains* (in the majority of chapters),
- *Transportation Machinery,* and
- *Rotordynamics.*

The 15 chapters of the book are written by internationally recognized experts working at universities, research centers, and industrial companies which results in a combination of in-depth knowledge of the theory with an emphasized practical focus.

We hope that the readers will find the book worth reading and will use it as a reference in their work.

Graz, Austria   Dr. Tigran Parikyan
October 2021   Chairman, Technical Committee for
                                                          Engines and Powertrains (2019–23)

                                                          International Federation for the
                                                    Promotion of Mechanism and Machine
                                                                          Science (IFToMM)

Acknowledgements

I would like to thank Dr. Hubert Herbst (Scania, Sweden), Dr. Vladimir Ivanovic (Ford, USA), my IFToMM colleagues Prof. José R. Serrano (CMT, Universitat Politècnica de València, Spain), Prof. Avinash K. Agarwal (Institute of Technology Kanpur, India), Prof. Andrei Lipatnikov (Chalmers University of Technology, Sweden), Assoc. Prof. Pavel Novotný (Brno University of Technology, Czech Republic), Dipl.-Ing. Ilias Papadimitriou (GF Casting Solutions, Switzerland), Dr. Madhu Raghavan (General Motors R&D, USA), Prof. Didier Rémond (INSA Lyon, France), Prof. Thales F. Peixoto (University of Campinas, Brazil), as well as my colleagues from AVL List GmbH (Graz, Austria)—Dr. Robert Fairbrother, Dr. Herwig Ofner, Dr. Peter Priesching, Dr. Alexander Schenk, Dipl.-Ing. Henrik Schuemie, Dipl.-Ing. Martin Sopouch, Dr. Reinhard Tatschl—and also Dr. Ming-Tang Ma (AVL Shanghai TC, China), who all substantially contributed to improving the quality of the book by reviewing the chapters and making valuable comments and suggestions.

Tigran Parikyan

Contents

Dynamics and Vibrations/Tribology and Lubrication

Efficient Modeling of Engine Parts and Design Analysis Tasks in Simulation of Powertrain Dynamics: An Overview

Tigran Parikyan ⓘ

Abstract Simulation of dynamics is one of the important tasks in engineering calculation projects related to internal combustion engine-based powertrains. It allows to evaluate strength and durability of engine parts, reliability and wear of their lubricated contacts, noise, and vibratory interaction of the powertrain with the vehicle body. Depending on the modeling depth and on the scope and accuracy of results, the software tools used for simulation of powertrain dynamics, can be sub-divided into fast *basic tools* and precise *high-end tools*. It is shown that in order to reach the expected quality and to be timely efficient at the same time, a complete project should apply a combined dynamic simulation workflow consisting of two main phases: (1) *early phase*, with system check, basic analysis and layout, as well as coarse optimization using basic tools, and (2) *final phase*, with fine tuning and verification of critical parameters of the system using high-end tools. The present chapter gives an extended overview of the typical design analysis tasks and related software tools used in the early phase of powertrain simulation workflow. It should be noted that efficient dynamic models of engine parts—first of all crankshaft—significantly contribute to shorter preparation for high-performance calculations, and to smooth transition between different tasks and phases of the project.

Keywords Internal combustion engine · Powertrain dynamics · Simulation software · Design analysis · Early phase · Crankshaft model

1 Introduction

To better understand the reasons for different ways of dynamic modeling of engine parts and performing various simulation tasks, it makes sense to start with a background information on potential problems related to powertrain design and their solution methods, as well as to consider two main types of calculation projects.

T. Parikyan (✉)
Advanced Simulation Technologies, AVL List GmbH, Hans-List-Platz 1, 8020 Graz, Austria
e-mail: tigran.parikyan@avl.com

© The Author(s), under exclusive license to Springer Nature Switzerland AG 2022
T. Parikyan (ed.), *Advances in Engine and Powertrain Research and Technology*,
Mechanisms and Machine Science 114,
https://doi.org/10.1007/978-3-030-91869-9_1

Table 1 Usual problems in IC engines and respective counter-measures

	Problems and their causes ⟶ Consequences	Counter-measures/design analysis tasks
A	Increased overall vibration of engine block due to unbalance in cranktrain ⟶ excessive load on the vehicle body, reduction of driving comfort	• Balancing the crankshaft with masses of the counterweights • Additional balancing with balancer shafts • Layout of engine mounts
B	Increased dynamic stresses in transmission due to cyclic fluctuations of crankshaft rotation ⟶ noise and shorter service life of the gears	• Limitation of the rotational irregularity by designing a flywheel with sufficient inertia
C	Increased stresses in crankshaft due to torsional resonances ⟶ failure of the crankshaft	• Minimizing the resonance amplitudes by using torsional vibration damper
D	Increased stresses in crankshaft due to bending resonances ⟶ failure of the crankshaft	• Design of the counterweights • Design of the main bearings • Optimization of the fillet radii
E	Increased stresses in slider bearings due to journal inclination ⟶ intensive wear of the contact surfaces of the journal and the shell	• Balancing separate crank throws by a correct distribution of counterweights • Design of hydrodynamic sliding bearings

1.1 Usual Problems in Internal Combustion Engines and Design Analysis Tasks

When designing a new internal combustion (IC) engine or analyzing an existing one, application engineers are usually confronted with one or more problems (Table 1). The objective of *design analysis* is to take necessary counter-measures which would eliminate or mitigate the undesirable consequences caused by immature or inappropriate design.

1.2 Two Types of Powertrain Dynamics Calculation Projects

In engineering practice related to calculation of engine or powertrain dynamics, there are two main types of projects which are shortly described below.

1st Project type: new development

The engine (or powertrain) does not exist and has to be designed and manufactured from scratch. There are only few design parameters known and in some cases the engine parts might not even be shaped as CAD models at the beginning.

In such a project, the conventional approach is to start with the general layout of the whole powertrain in order to find out the most important missing design parameters. At the same time, the problems which may arise during the layout process itself, are being resolved using engineering know-how or by some short-loop iterations which employ variation of parameters or optimization with the help of basic simulation tools.

Only after that, the project enters the 2nd phase where the quality of the basic layout will be verified with more sophisticated simulation tools and the existing parameters can be adjusted to meet the ultimate design targets.

2nd Project type: fixing a problem

The engine (or powertrain) has already been designed and manufactured. However, during its testing on the test-bed or even during its operation within the vehicle, one or more problems occur (like breaking of engine parts, increased wear in bearings, high-level vibrations and noise, etc.) which cannot be resolved by the usual service measures.

In such projects, the conventional approach is to calculate the powertrain with the existing parameters (with possibly accurate simulation tools) to find out the root cause of the problem and to suggest counter-measures resulting in some—usually minor—parameter modifications. These changes to the design must proof efficient after virtual verification by computer simulation, and real validation on test bed afterwards.

In what follows, we consider the criteria for the correct choice of the simulation tools for each of the two project phases.

2   The Right Simulation Tool for Each Project Phase and Task

Here we will consider the modeling depth and the levels of complexity of the simulation tasks and will relate these to the phases of the project to be used in.

2.1   Two Levels of Software Tools to Simulate Powertrain Dynamics

Depending on the complexity of the problem to resolve, one or the other simulation tool can be chosen. According to the modeling depth and scope of results, these tools—and related simulation levels—can be divided into two groups:

(1)   *basic* simulation tools/levels,
(2)   *high-end* simulation tools/levels.

The basic tools (Fig. 1) are usually represented by a collection of software programs—each one resolving a separate design analysis task. Characteristically, they allow to consider the mechanical system in a decomposed or simplified way, which makes it possible for an application engineer to model quickly and to concentrate on a specific task. Accordingly, there is a restricted set of results generated

Fig. 1 Basic level simulation tools: simplified modeling, highest performance

Fig. 2 High-end level simulation tools: fine modeling, highest accuracy

in each of these tasks, related to either a single powertrain part—like crankshaft or engine block, or employing a single dimension—like in 1D torsional vibration analysis, or 2D hydrodynamic (HD) analysis of slider bearings.

Due to linear modeling of connections between rigid and flexible bodies representing engine parts, most of these dynamic simulation tools can solve in frequency domain that ensures a very high performance.

On the other hand, with the basic level simulation tools some physical phenomena can only be partly revealed, like in case of low-frequency NVH.[1]

In some cases, these tools can be used for optimization based on some relative criteria only—like in case of relative durability factors in the crankshaft strength evaluation task.

On the contrary, the high-end tools (Fig. 2) can employ all the possibilities of finite-element (FE) modeling of the elastic bodies, and by solving in time domain, they can apply the most complex non-linear force functions available for the connections between the parts, that assure the highest attainable accuracy of results. Here one has no restrictions concerning the scope of the generated results either—a full 3D simulation of the whole powertrain is possible. Usually, such a dynamic simulation

---

[1] Noise and vibration harshness.

is performed with a single high-end tool capable to cover all the physical phenomena of any specific application—calculation of absolute stress levels in engine parts, acoustics and NVH of the whole powertrain, and tribological behavior of slider bearings based on their elasto-hydrodynamic (EHD) modeling.

Regarding the crankshaft, it can be modeled in two different ways, and the structured model can be applied in most of the cases, except for some specific applications where a surface-to-surface contact within slider bearings is needed, and therefore the condensed model has to be used.

2.2 Simulation Levels Versus Project Phases

Each powertrain dynamics calculation project can consist of two phases—early concept phase and final design phase, or shortly—*early phase* and *final phase.*

As already mentioned before, based on the objectives, there exist two different types of projects. While in case of a new powertrain development it is a usual practice to go through both the phases, the "problem fixing" projects often make use of the final phase only, trying to fix the problem by fine tuning the parameters of the system and performing high-end simulations. It is advisable, however, not to skip the early phase in such "problem fixing" projects, either. The reason is that the basic calculations offer a relatively fast check of the whole system, so that one can make sure not having overlooked anything in the concept, and to move further by making finer adjustments on a verified basis.

At earlier times, as the simulation tools were not yet that sophisticated, basic tools were used in all calculations [1, 2], but the related modeling was mostly restricted and final results obtained this way were too coarse, so nowadays this approach is no longer practiced.

In some cases, high-end tools are still used for the whole project, including even the early phase, which can only make it difficult to optimize the powertrain subsystems (e.g. layout of the torsional vibration damper) and slow down the whole workflow.

One can conclude that in order to reach both quality and performance targets, one has to give preference to basic tools in the early phase followed by high-end tools in the final phase (Table 2).

Simulation tool level\project phase	I—early phase	II—final phase
A. Basic level simulation tools	Suitable	Mostly too coarse
B. High-end level simulation tools	Mostly too slow	Suitable

Table 2 Level of simulation tools versus calculation project phase

To additionally benefit from the positive sides of the both project phases and to provide a good usability of the overall solution, respective software tools may be integrated into the same simulation environment featuring common GUI and post-processing, like in case of *AVL EXCITE*™ [3, 4]. Besides the shared input data, it is also important to assure the equivalence of modeling for a smooth transition between two phases [5, 6].

3 Efficient Dynamic Modeling of Engine Parts

In this section we mainly consider the modeling of the crankshaft. For the design analysis tasks considered, the other parts like connecting rod (conrod), piston and piston pin can be modeled much simpler being represented either by a single mass point or by several points with correctly distributed masses and moments of inertia.

3.1 Two Types of Dynamic Models of Crankshaft

Having the initial geometry and material data of crankshaft on input (Fig. 3), one can generate two similar, but differently defined dynamic models, in terms of the number and location of nodes and of the information on distribution of mass and stiffness properties contained therein.

Condensed model of crankshaft is generated from its full volumetric FE model (Fig. 4) using a dynamic reduction method called condensation [7, 8].

Structured model of crankshaft (Fig. 5) applies a different dynamic reduction method which results in a set of mass lumps associated with nodes that are inter-connected by elastic elements in pairs (Fig. 6) [9]. It is a generalization of bar-mass

Fig. 3 Initial CAD model of an in-line 4-cylinder 1.9 L engine crankshaft in STL format

Fig. 4 Full volumetric FE model of the 4-cylinder engine crankshaft (left), and the nodes retained in its condensed model, shown on a single crank throw (right)

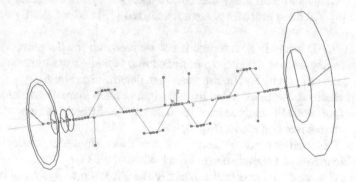

Fig. 5 Structured model of the 4-cylinder engine crankshaft

Fig. 6 Topology of the structured model shown for a single crank throw

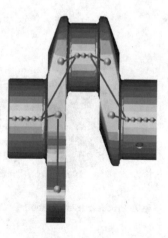

model which was earlier in use [10, 11].

While the condensed model of crankshaft is primarily used at high-end level simulation, the structured model can be used both at basic and at high-end levels [9, 12].

3.2 Generating Structured Model of Crankshaft

To transform the crankshaft CAD surface model (Fig. 3) into structured model, some intermediate steps are needed, based on mass partitioning and static analysis of FE meshes of complex-shape parts of the crankshaft—webs and disks (flywheel, pulley, etc.)—in combination with analytical calculations for cylindrical parts (journals, pins, etc.) [9]. All these procedures are automatized [13] and are shortly described below.

First, the CAD model in STL format is sub-divided into smaller parts (Fig. 7).

After meshing the webs, they are partitioned into separate mass lumps, for which the mass, center of gravity and inertia tensor are determined (Fig. 8).

The FE-meshes of webs extended by the neighbor main journal and crank pin are used to perform static FE analysis in order to define the stiffness matrices of the web elements within structured model (Fig. 9).

Similar FE-based procedures apply for the disks (flywheel, pulley) having considerable inertia and compliance to be sub-divided [9, 13].

As the FE meshing process and especially the FE static analysis can be rather time consuming, these tasks have been recently parallelized [14].

Fig.7 Processed surface model of the 4-cylinder engine crankshaft

Fig. 8 Mass partitioning of the webs: with counterweight (left), without counterweight (right)

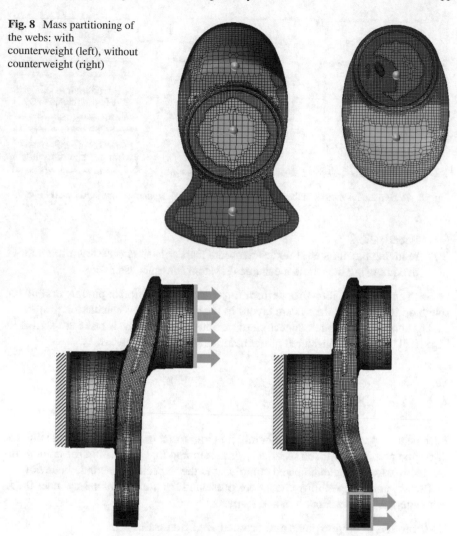

Fig. 9 Extended webs fixed at the main journal and deformed by loading the free end: at the crank pin (left); at the counterweight (right)

4 Design Analysis Tasks

4.1 Early Phase Workflow

The software tools used in the early phase united under the simulation suite called *AVL EXCITE™ Designer* [3] are very fast and simulate mostly in frequency domain. This becomes possible due to simplification of the system through:

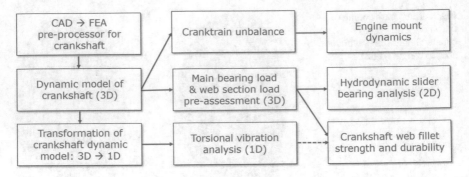

Fig. 10 Design analysis tasks in the early phase of powertrain dynamic simulation workflow

(a)   linearization,
(b)   reducing the number of nodes/elements/DOFs[2] to the necessary minimum,
(c)   decomposing the whole mechanical system into subsystems.

Such an approach allows for performing extensive variation of parameters and for rough optimization of the system layout in the early phase of calculation project.

The sequence and interconnection of various design analysis tasks are shown in Fig. 10 [15] and are considered more in detail in the next sections.

4.2   Preliminary Checks

Before starting any dynamic simulation, it is important to make sure that the models of engine parts, either taken separately or assembled together within the cranktrain, are geometrically compatible and comply with the expected kinematic behavior.

Below such compatibility checks are presented for the case of a 3-cylinder 0.8 L passenger car engine cranktrain and its parts.

CAD models of engine parts and cranktrain kinematics

Usually the workflow starts by putting together the main engine parts—crankshaft, conrods, pistons and piston pins—assembled within engine block, which at this stage can be simply represented by cylinder liners and main bearing shells. Using CAD models of the engine parts makes it possible to visually check their positions and geometric compatibility and to verify the correctness of their motion as well as the firing order by animating the cranktrain (Fig. 11).

---

[2] Degrees of freedom.

Fig. 11 CAD based model of a 3-cylinder 0.8 L passenger car engine cranktrain

Compatibility check of CAD and structured models of engine parts

Before using dynamic structured models these can be checked against their original CAD shapes by superimposing the two model types of the respective engine parts, an example for crankshaft and conrod is shown here in Fig. 12.

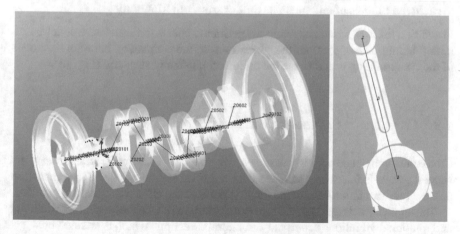

Fig. 12 Structured models of crankshaft and conrod together with their CAD models

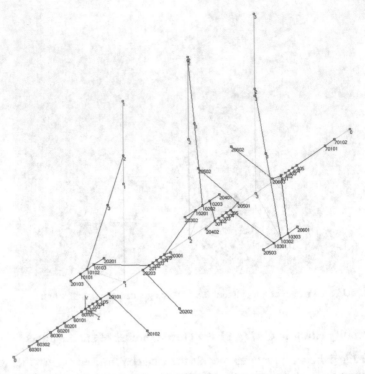

Fig. 13 Cranktrain kinematics and connections between the nodes of the structured models

Structured models of engine parts and cranktrain kinematics

In addition to CAD-based kinematic and geometry check, the structured model-based representation allows to proof the correctness of the connection between the nodes of the cranktrain parts and to see the animated kinematic structure (Fig. 13).

4.3 Cranktrain Unbalance Analysis

Cranktrain unbalance analysis assumes the crankshaft and other cranktrain parts rigid. It results in free forces and moments acting on engine block due to combustion process (gas excitation) and to the moving parts of the cranktrain (mass excitation).

The application target here is to analyze and try to reduce the level of translational and rotational vibrations in vertical and lateral directions which is attained by:

(a)   proper selection or modification of counterweights,
(b)   eventual use of balancer shafts, and
(c)   resilient mounting of the engine or powertrain.

The cranktrain unbalance analysis is illustrated by an example of a V6 4.0 L gasoline SUV engine shown in 3D (Fig. 14) and in projections (Fig. 15). The free forces and free moments at 3000 rpm –as functions both vs. crank angle and vs. harmonics—are shown in Figs. 16 and 17. Additionally, the reaction torque with which the cranktrain acts on the driveline, is shown in Fig. 18 as a total torque and its components—gas and mass excitations.

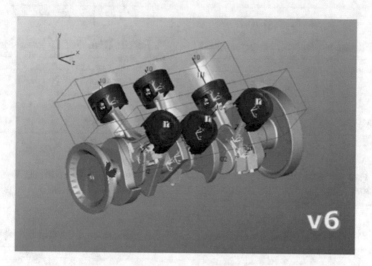

Fig. 14 CAD based cranktrain model of a V6-90° engine

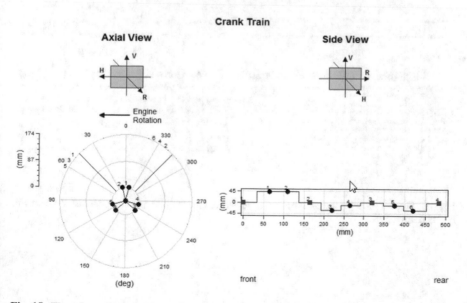

Fig. 15 The schematic projections of cranktrain model of a V6-90° engine

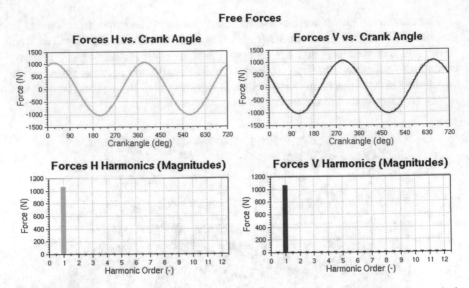

Fig. 16 Free forces of the cranktrain of a V6-90° engine at 3000 rpm (components: V—vertical, H—horizontal)

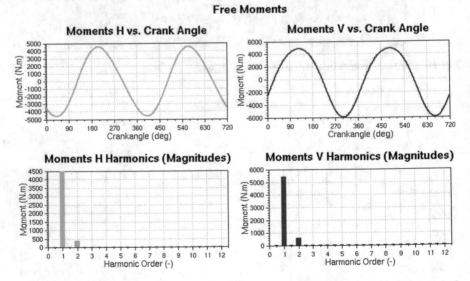

Fig. 17 Free moments of the cranktrain of a V6-90° engine at 3000 rpm components: V—vertical, H—horizontal)

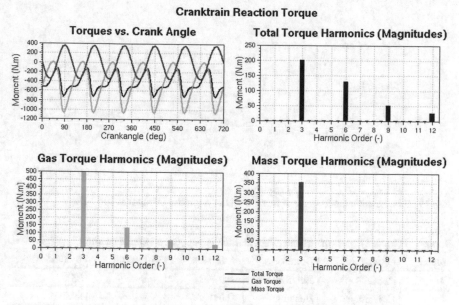

Fig. 18 Reaction torque of the cranktrain of a V6-90° engine at 3000 rpm

Table 3 Frequency spectrum of a free crankshaft

Mode	Freq. [Hz]	Dominant	% Torsion	% Tension	% Bending
7	**136.10**	**Bending**	**0.0**	**0.3**	**99.7**
8	188.64	Bending	0.0	0.0	100.0
9	267.05	Torsion	99.8	0.0	0.2
10	306.12	Tension	0.0	98.0	2.0
11	328.34	Bending	0.0	3.1	96.9
12	356.24	Bending	0.2	0.0	99.8

4.4 Crankshaft Modal Analysis

To determine the spectral properties of the structured model of crankshaft before using it in forced response analysis, modal analysis is performed.

First, a free crankshaft is considered. The spectrum of natural frequencies is shown in Table 3 starting with the 7th mode.[3] The dominant vibration type in each mode is specified based on detected shares of oscillatory motion components (torsion, tension, and bending) and can be visually checked by animating the mode.

The mode shape of the mentioned first bending mode is shown in Fig. 19.

---

[3] The first six *rigid-body modes* all give 0.0 Hz due to 6 DOFs of the free unconstrained and unsupported body motion.

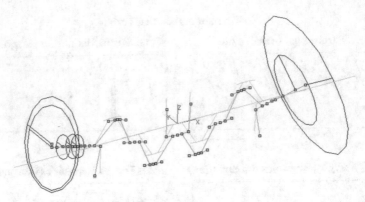

Fig. 19 First bending mode shape of free 4-cylinder engine crankshaft at 136.1 Hz

Table 4 Frequency spectrum of crankshaft elastically supported in main bearings

Mode	Freq. [Hz]	Dominant	% Torsion	% Tension	% Bending
3	**246.69**	**Bending**	**0.0**	**6.8**	**93.2**
4	257.83	Bending	0.0	0.8	99.2
5	262.98	Bending	8.1	0.0	91.9
6	270.31	Torsion	71.8	0.0	28.2
7	292.78	Bending	19.4	0.0	80.6
8	317.48	Tension	0.0	94.6	5.4

Both the spectrum of natural frequencies and the mode shapes can be used to verify the degree of correlation of spectral properties of the structured model of crankshaft with the corresponding volumetric FE model, e.g. using modal assurance criterion (MAC) [9].

After analyzing the modal properties of a free crankshaft, it can be useful to have a look at its behavior within the cranktrain, too. To do that, the stiffness of main journals supporting the crankshaft, and the additional masses (rotating mass of the conrod big end plus the effective share of the oscillating mass of the conrod small end, piston and piston pin) must be taken into account.[4]

The spectrum of natural frequencies of the supported crankshaft is shown in Table 4. Due to stiffness of additional supports, there remain only two rigid-body modes in the system,[5] and the 3rd mode becomes the first elastic one. The corresponding mode shape is shown in Fig. 20.

If effects of rotation of the crankshaft are also considered, then we speak of *gyroscopic modal analysis*. The natural frequencies in this case depend on the rotation speed. Table 5 shows the spectrum of gyroscopic frequencies at 3000 rpm. Each mode

---

[4] Actually, after modifying the dynamic properties of the crankshaft in such a way, one can speak of the modal analysis of the cranktrain rather than that of the crankshaft.

[5] In case no axial thrust stiffness is considered.

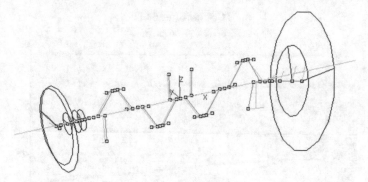

Fig. 20 First bending mode shape of supported 4-cylinder engine crankshaft at 246.7 Hz

Table 5 Frequency spectrum of crankshaft in gyroscopic modal analysis at 3000 rpm

Mode	Freq. [Hz]	Dominant/whirl	% Torsion	% Tension	% Bending
3	211.46	Bending/FW	0.1	0.4	99.5
4	**237.70**	**Bending/FW**	**0.9**	**4.5**	**94.5**
5	272.60	Torsion/BW	89.3	0.4	10.3
6	300.21	Bending/BW	7.0	12.5	80.5
7	308.66	Bending/BW	1.6	9.1	89.4
8	320.29	Tension/BW	0.0	69.9	30.0

additionally features a *whirl* (FW for forward whirl, and BW for backward whirl).

The mode shape of the highlighted mode #4 having forward whirl is shown in Fig. 21. It has to be noted that unlike the previous two types of modal analysis, the nodes in gyroscopic modal analysis trace elliptical orbits in the direction of crankshaft rotation (for the forward whirl) or in the opposite direction (for the backward whirl).

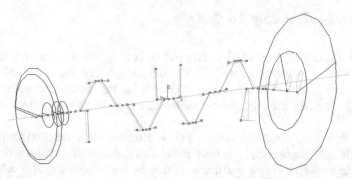

Fig. 21 Gyroscopic mode shape of 4-cylinder engine crankshaft (2nd forward-whirl bending-dominated 237.7 Hz mode at 3000 rpm)

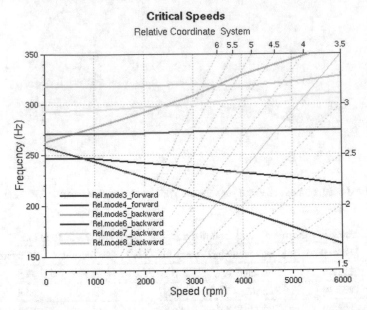

Fig. 22 Critical speeds diagram of gyroscopic modal analysis of a 4-cylinder engine crankshaft

The critical speeds diagram (Fig. 22) gives a good overview over the dependency of the natural frequencies on engine speed. The colored lines correspond to the modes keeping the same whirl direction. The inclined numbered dashed lines correspond to the orders of engine speed.

If performed for a large number of engine speeds, the gyroscopic modal analysis can take considerable time. To improve the performance, this calculation task has been recently parallelized [14].

4.5 Engine Mounting Analysis

For low-frequency NVH evaluation, it is sufficient to consider the engine as a rigid body supported by elastic mounts [16]. The mounts are modeled as hinged (single-axis) or fixed (3-axis) spring-damper elements having non-linear stiffness, and viscous or viscoelastic damping—for rubber or hydraulic mounts correspondingly (Fig. 23).

Under the action of gravity load, as well as driveline mean reaction torque, non-linear static analysis is performed first. In the static equilibrium position the mount springs attain a higher effective stiffness used in the linearized model for subsequent modal and dynamic forced response analyses.

The modal analysis is performed for each engine speed. Table 6 shows an example of spectrum of undamped natural frequencies at 1000 rpm, and Table 7—the spectrum

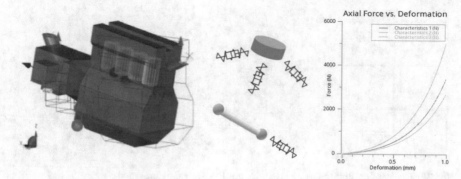

Fig. 23 Powertrain (rigid body) supported by mounts modeled as spring-dampers having non-linear stiffness

Table 6 Natural frequencies of a mounted 4-cylinder engine-based powertrain at 1000 rpm

Speed (rpm)	Mode	Frequency (Hz)	Dominant DOF	Kin. energy (%)	Mode shape
1000	1	11.2	Tx1	99.1	Transversal_1
	2	12.4	Ty1	59.6	Longitudinal_1
	3	20.1	Rz1	91.8	Yaw_1
	4	30.8	Tz1	95.3	Vertical_1
	5	46.7	Ry1	99.7	Roll_1
	6	96.6	Rx1	56.7	Pitch_1

Table 7 Damped frequencies of a mounted 4-cylinder engine-based powertrain at 1000 rpm

Speed (rpm)	Mode	Frequency (Hz)	Decay rate (−)	Dominant DOF	Kin. Energy (%)	Mode shape
1000	1	11.3	1.018e−01	Tx1	99.1	Transversal_1
	2	12.5	1.018e−01	Ty1	59.6	Longitudinal_1
	3	20.2	1.018e−01	Rz1	82.4	Yaw_1
	4	31.0	1.018e−01	Tz1	93.2	Vertical_1
	5	47.0	1.018e−01	Ry1	99.5	Roll_1
	6	97.1	1.018e−01	Rx1	56.7	Pitch_1

of damped frequencies, where one can notice a light upward drift of the frequencies in case of damped modes.

The mode shapes are classified according to the DOF with the dominant share of kinetic energy.

The decay behavior of oscillations associated with damped modes is shown in Fig. 24.

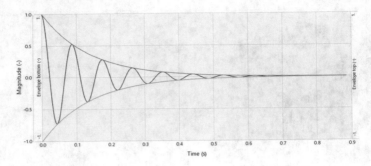

Fig. 24 Decay of oscillations associated with the 1st damped natural mode

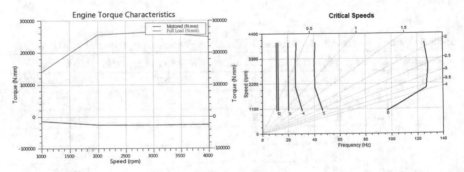

Fig. 25 Engine mean torque (blue curve on the left plot) and critical speeds diagram

As the driveline mean reaction torque changes with the engine speed, the critical speeds diagram shows a spectrum of natural frequencies changing with the engine speed, too (Fig. 25).

Finally, the dynamic forced response is performed under the action of free forces and moments caused by the cranktrain gas and mass excitation (see Figs. 16, 17, and 18), showing the resonances of the system evaluated in terms of motion properties of engine and reactions of mounts (Fig. 26).

Large engine mounting on additional bedplate

For large engines installed in ships as well as in gensets, there can be an additional bedplate placed under the engine/powertrain to better compensate for vibrations caused by the gas loads and by the motion of the powertrain components.

The 12 natural frequencies of a 2-body system modeling a V12 engine-based genset resiliently mounted on a flexibly supported bedplate are listed in Table 8.

Figure 27 shows the 1st natural mode of that "genset-on-bedplate" system at 2.3 Hz.

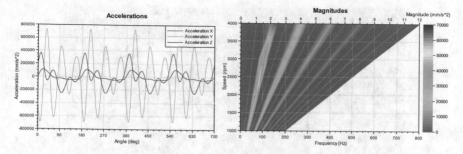

Fig. 26 Accelerations at a point on oilpan surface of an in-line 4-cylinder engine (fringe plot on the right shows the acceleration along x axis)

Table 8 Spectrum of natural frequencies of a large V12 genset engine on bedplate

Speed (rpm)	Mode	Frequency (Hz)	Dominant DOF	Kin. energy (%)	Mode shape
500	1	2.3	Ty2	77.9	Transversal_2
	2	2.9	Tx2	52.6	Longitudinal_2
	3	4.1	Tz2	82.4	Vertical_2
	4	8.7	Rz2	68.0	Yaw_2
	5	8.9	Tz1	37.2	Vertical_1
	6	9.9	Tx2	31.2	Longitudinal_2
	7	10.8	Rx2	47.7	Roll_2
	8	16.5	Rx1	35.1	Roll_1
	9	21.8	Rx1	51.4	Roll_1
	10	21.9	Ry1	56.5	Pitch_1
	11	28.5	Tx1	76.3	Longitudinal_1
	12	40.9	Rz1	53.5	Yaw_1

4.6 Powertrain Torsional Vibration Analysis

By using static analysis, the 3D structured model of crankshaft is automatically transformed into equivalent torsional model (Fig. 28) [5], which can then be included into the torsional vibration system of the whole powertrain (Fig. 29).

As a result of undamped and damped torsional modal analyses we obtain the spectra of torsional frequencies (Table 9) and determine the torsional mode shapes (Fig. 30), as well as the damping factors of different torsional modes. Note a stronger decrease in frequency values of the first three substantially damped modes.

Torsional vibration analysis allows to calculate the vibrations as a result of forced response of the torsional system in frequency domain; the peaks and the resonances of the system under the actual loading conditions in the operation speed range are

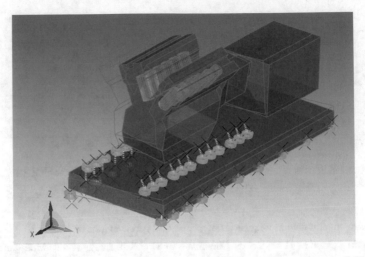

Fig. 27 1st mode of a large V12 engine-based genset model on bedplate with overall 50 mounts

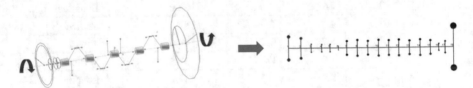

Fig. 28 3D structured model of crankshaft of a 4-cylinder engine (left) and its transformation into equivalent torsional model

determined. The results of the mentioned analysis are used to decrease the vibration levels by selecting a suitable flywheel based on the maximum allowable cyclic speed irregularity (Fig. 31), and a suitable torsional vibration damper based on (a) the maximum allowable total elastic deformation within crankshaft and its orders (Fig. 32) and (b) the maximum allowable deformation of the damper and dissipated power in it, as well as the maximum allowable total torque and total shear stress in the system (Fig. 33).

The frequency-domain simulation also enables to transform the frequency-dependent properties of dampers into equivalent speed-dependent ones for further use in time-domain simulation [6].

4.7 Cranktrain 3D Dynamics in Frequency Domain

To pre-calculate main and axial bearing loads and web section forces and moments and also to roughly estimate the motion of any node of the crankshaft, 3D dynamic

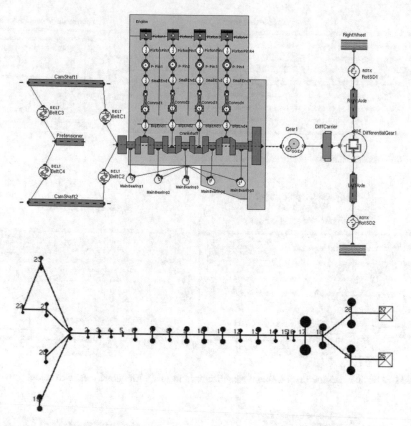

Fig. 29 Powertrain model based on a 4-cylinder engine with double-ring torsional vibration damper: graphical representation (top) and equivalent torsional model (bottom)

Table 9 Spectra of undamped and damped torsional frequencies and the damping factor

Mode No.	Undamped torsional frequency [Hz]	Damped torsional frequency [Hz]	Damping factor [−]	Damped frequency shift [%]
1	6.46	5.98	0.52	−7.4
2	106.29	100.58	0.33	−5.4
3	112.42	105.71	0.35	−6.0
4	338.02	340.57	0.08	0.8
5	477.05	479.86	0.05	0.6
6	559.47	560.22	0.05	0.1
7	730.44	729.65	0.06	−0.1
8	948.54	948.61	0.01	0.0
9	1313.59	1313.68	0.01	0.0

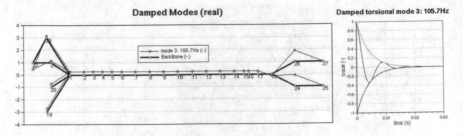

Fig. 30 Torsional damped mode shape and its decay behavior

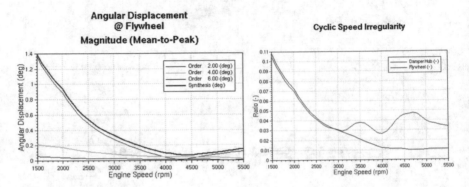

Fig. 31 Absolute angular displacement magnitudes and speed irregularity used to select flywheel

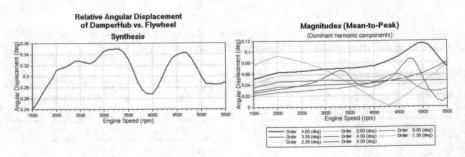

Fig. 32 Relative angular displacement magnitudes used to select torsional vibration damper

forced response is performed in frequency domain, that also takes gyroscopic effects into account [17].

Here the oil film is modeled by linear spring-damper elements distributed over the bearing width (Fig. 34). The main bearing shell is assumed rigid, the main bearing wall can be modeled by springs and hysteretic material damping [3].

The results show radial force, bending moment, and axial thrust load (Fig. 35) versus crank angle for selected bearings. These are summarized then in form of maximum magnitudes (Fig. 36) and harmonic orders (Fig. 37) versus speed.

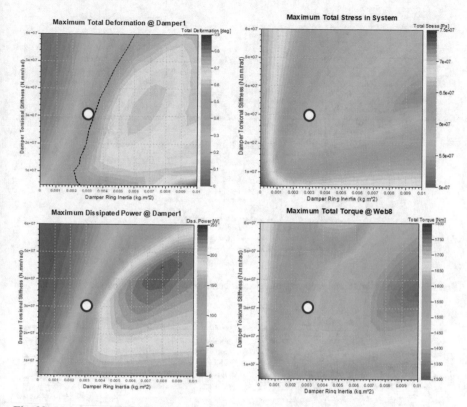

Fig. 33 Main integral dynamic characteristics of torsional vibration damper as a result of simulation with variation of mass and stiffness parameters

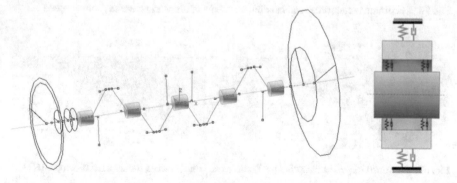

Fig. 34 3D modeling of crankshaft elastically supported by main bearings

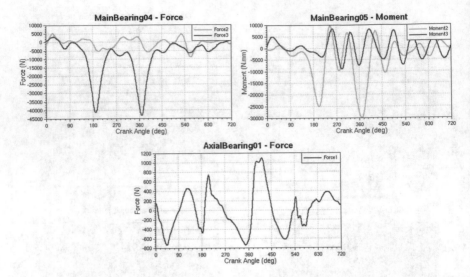

Fig. 35 Dynamic forces and moments of selected bearings of 4-cylinder engine at 3000 rpm

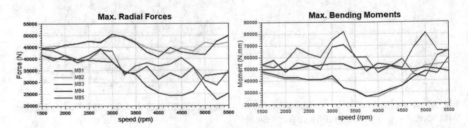

Fig. 36 Maximum radial forces and bending moments in all bearings versus engine speed

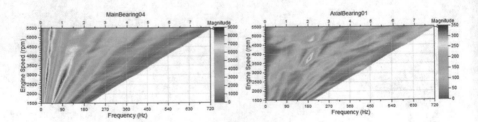

Fig. 37 Campbell diagrams for radial (left) and axial (right) forces for selected bearings, (N)

The solutions obtained by performing this task, can then be used/refined in the further calculations described below.

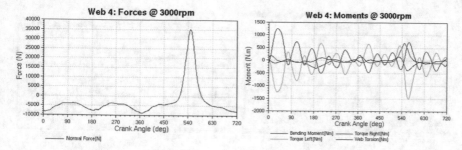

Fig. 38 Web section forces and moments vs. crank angle (as defined in [18]) used for stress evaluation

4.8 Crankshaft Strength and Durability

The strength of crankshafts can be assessed based on durability factors in fillets of main journal and crank pin for each web. To do that, section forces and moments (Fig. 38) as a result of 3D dynamic simulation[6] are used to evaluate combined stresses based on crankshaft material and web/fillet geometry.

Based on "stress-state" diagrams, the durability factors are then evaluated following either von Mises [18] or Gough-Pollard method [19] for each web and engine speed (Fig. 39) and summarized for all the webs in the whole engine speed range (Fig. 40).

While being suitable for rough optimization of durability factors [12], the described approach cannot reliably calculate the absolute levels of stress and strain. To obtain these, fine FE-based modeling of the fillets of separate webs [14] or of the whole crankshaft is needed, followed by the calculation of stresses using FEA software and evaluation of fatigue using specialized tools [4].

4.9 Hydrodynamic Analysis of Slider Bearings

The basic hydrodynamic simulation of separate slider bearings assumes a rigid shell and uses 2D approach, i.e. it does not take the inclination of the journal into account [20]. On input, it uses the load pre-calculated in the 3D dynamic analysis of crankshaft main bearings (see the previous task [17]), or—as in case of the conrod big-end and small-end bearings—by a quasi-static analysis of statically determinate crank throw models.

Such a simplified analysis allows to calculate the time history of the main characteristics relative either to bearing shell or journal/pin, and to summarize the maximum oil film pressure, minimum oil film thickness, friction loss, etc. vs. engine speed (Fig. 41).

---

[6] See the previous paragraph.

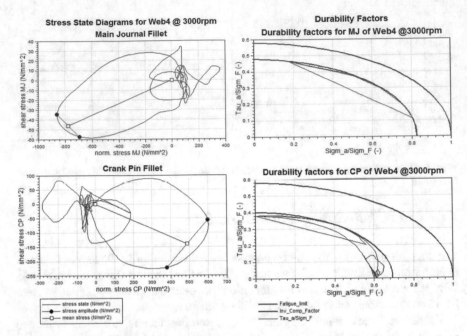

Fig. 39 Stress-state diagrams (left) and durability factor diagrams for fillets at main journal and crank pin of web #4 of in-line 4-cylinder engine at 3000 rpm

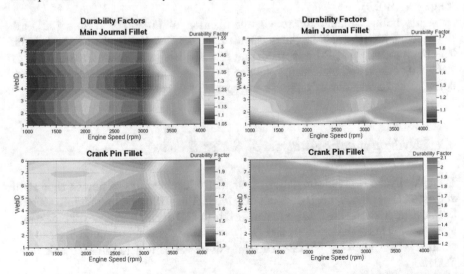

Fig. 40 Durability factors for fillets at main journal and crank pin of all webs of an in-line 4-cylinder engine: based on quasi-static throw-wise approach (left) and on 3D dynamic forced response (right)

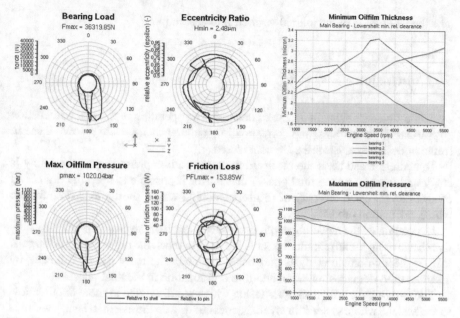

Fig. 41 Rough estimate of main hydrodynamic characteristics of slider bearings versus crank angle (polar plots on the left) or versus engine speed

In the final phase, a high-end 3D time-domain simulation [4] of the powertrain with more advanced elasto-hydrodynamic [21, 22] or thermo-elasto-hydrodynamic [23, 24] modeling of slider bearings with flexible shell is being used which gives a number of additional results (an example is shown in Fig. 42), and also allows to

Fig. 42 Oil film height in the elasto-hydrodynamic analysis of main bearing (example of analysis in final phase with high-end simulation tool [4])

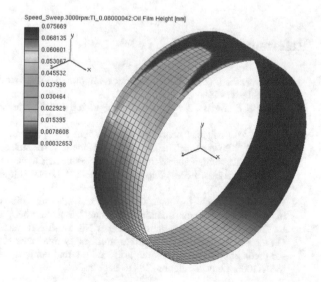

take the inclination of the journal vs shell into account.

5   Conclusion

Typical problems arising in course of designing or operating of internal combustion engine-based powertrains, and corresponding counter-measures to resolve these in frame of design analysis tasks are considered.

To address these tasks the most efficient way, a two-phase project workflow is suggested (early phase + final phase) that makes it possible to keep the high level of accuracy and rich diversity of results combined with a shorter lead time for the whole calculation project.

Depending on the modeling depth necessary in each of the project phases, respective (basic or high-end) simulation tools are used to reach the project specific targets: fast frequency-domain analysis and basic layout in the early phase, followed by accurate time-domain simulation and fine optimization in the final phase.

The early-phase design analysis tasks, which are discussed in more detail, allow to analyze the whole system "in layers" concentrating on the essential issues within a separate task, resulting in early optimal layout of the system and its components. At the same time, it ensures a better understanding of dynamic behavior of the system through its simplification and linearization.

It is also shown, that for fast and smooth transition between the two mentioned project phases, it is advisable to use integrated simulation tools with switchable levels that share common engine configuration and design-related data of engine parts like the crankshaft, with seamless transformation into equivalent models for different levels and tasks.

References

1. Nestorides E. J. (B.I.C.E.R.A. Research Laboratories), A Handbook on Torsional Vibration, Cambridge (1958)
2. Hafner K.E., Maass, H.: Torsionsschwingungen in der Verbrennungskraftmaschine. Springer (1985)
3. EXCITE™ Designer Theory and User Manual, AVL, Release 2021 R1.
4. EXCITE™ Power Unit Theory and User Manual, AVL, Release 2021 R1.
5. Parikyan, T.: Unified approach to generate crankshaft dynamic models for 3D and torsional vibration analyses. ASME Paper ICES2003-591 (2003). https://doi.org/10.1115/ICES2003-0591
6. Parikyan, T., Resch, T., Knaus, O., et al.: Equivalent modeling of torsional vibration dampers in frequency and in time domain for a smooth transition between concept and design phases. In: Torsional Vibration Symposium, Paper No. 3A–2, Salzburg (2017)
7. Offner, G., Priebsch, H.H.: Flexible multi-body dynamics simulation—a powerful method for prediction of structure borne noise of internal combustion engines. In: Proceedings of ISMA2006, Leuven Belgium, 18–20 Sept 2006

8. Offner, G.: Modelling of condensed flexible bodies considering non-linear inertia effects resulting from gross motions. Proc. IMechE Part K J. Multi-body Dyn. **225**(3), 204–219 (2011). https://doi.org/10.1177/1464419311416553

9. Parikyan, T., Resch, T., Priebsch, H.H.: Structured model of crankshaft in the simulation of engine dynamics with AVL EXCITE. ASME Paper 2001-ICE-435, ICE-Vol. 37–3, pp. 105–114 (2001). https://doi.org/10.1115/2001-ICE-435

10. Priebsch, H.-H., Affenzeller, J., Gran, S.: Prediction technique for stress and vibration of nonlinear supported, rotating crankshafts. J. Eng. Gas Turbines Power. **115**(4):711–720 (1993). https://doi.org/10.1115/1.2906764

11. Rasser, M., Resch, T., Priebsch, H. H.: Enhanced Crankshaft Stress Calculation Method and Fatigue Evaluation. CIMAC Congress, Copenhagen (1998) [MTZ (61) 2000, No. 10]

12. Resch, T., Knaus, O., Mlinar, et al.: Application of multi-body dynamics for the crankshaft layout in the concept phase. MTZ Worldw. **65**, 11–14 (2004). https://doi.org/10.1007/BF0322 7711

13. Todorovic G., Parikyan T.: Automated generation of crankshaft dynamic model to reduce engine development time. SAE Paper 2003-01-0926 (2003). https://doi.org/10.4271/2003-01-0926

14. EXCITE™ Shaft Modeler with AutoSHAFT User Manual, AVL, Release 2021 R1

15. Parikyan, T.: Design analysis tasks in simulation of engine and powertrain dynamics: an overview. In: Uhl, T. (eds.) Advances in Mechanism and Machine Science. IFToMM World Congress 2019. Mechanisms and Machine Science, vol. 73. Springer, Cham (2019). https://doi.org/10.1007/978-3-030-20131-9_89

16. Parikyan, T., Naranca, N., Neher, J.: Powertrain resilient mounting design analysis. ASME Paper ICEF2018-9539 (2018). https://doi.org/10.1115/ICEF2018-9539

17. Parikyan, T., Resch, T.: Statically indeterminate main bearing load calculation in frequency domain for usage in early concept phase. ASME Paper ICEF2012-92164 (2012). https://doi.org/10.1115/ICEF2012-92164

18. Calculation of crankshafts for I.C. engines, UR M53 - Rev.4, Aug 2019, IACS

19. Sandberg, E.: Some Classification Aspects on Crankshafts. 24th CIMAC World Congress on Combustion Engine Technology, June, 7–11, 2004, Kyoto, Japan. Proceedings, Paper No. 13 (2004)

20. Butenschön, H.-J.: Das hydrodynamische, zylindrische Gleitlager endlicher Breite unter instationärer Belastung, Diss. Dr.-Ing., Karlsruhe (1976)

21. Ma, M.-T., Offner, G., Loibnegger, B., et al.: A fast approach to model hydrodynamic behaviour of journal bearings for analysis of crankshaft and engine dynamics. Tribol. Ser. **43**, 313–327 (2003). https://doi.org/10.1016/S0167-8922(03)80059-7

22. Allmaier, H., Priestner, C., Six, C., et al.: Predicting friction reliably and accurately in journal bearings—a systematic validation of simulation results with experimental measurements. Tribol. Int. **44**(10), 1151–1160 (2011). https://doi.org/10.1016/j.triboint.2011.05.010

23. Priebsch, H.H., Krasser, J.: Simulation of the oil film behaviour in elastic engine bearings considering pressure and temperature dependent oil viscosity. Tribol. Ser. **32**, 651–659 (1997). https://doi.org/10.1016/S0167-8922(08)70490-5

24. Lorenz, N., Offner, G., Knaus, O.: Fast thermo-elasto-hydrodynamic modeling approach for mixed lubricated journal bearings in internal combustion engines. Proc. IMechE Part J J. Eng. Tribol. **229**(8), 962–976 (2015). https://doi.org/10.1177/1350650115576246

Crankshaft Torsional Vibration Analysis in a Mid-Size Diesel Engine: Simulation and Experimental Validation

Alexandre Schalch Mendes (ID)**, Pablo Siqueira Meirelles** (ID)**, and Douglas Eduardo Zampieri** (ID)

Abstract This chapter presents the theoretical formulation for torsional vibration analysis in internal combustion engines. The methodology can be applied to several crankshaft geometries with different cylinder configurations. The steady-state solution of the equations considers the state transition matrix and the convolution integral. Here, this formulation is applied to the mathematical model of a six-cylinder diesel engine with 7.2 l of volumetric displacement and a viscous damper assembled at the crankshaft front end to reduce the vibration amplitude. With the results of the torsional vibration analyses, it is possible to determine the dynamic torque at each region of the crankshaft. Finally at the end of the chapter, the calculated torsional vibration amplitudes at the crankshaft front end are compared with measured values from an engine to validate the proposed methodology.

Keywords Torsional vibrations · Crankshaft · Internal combustion engines · Viscous damper

1 Introduction

The crankshaft is one of the most important components of an internal combustion engine (ICE). It is subject to many types of periodic excitations that result in considerable stresses at the crankshaft. Some examples of such excitations are the cylinder gas pressure, inertia forces, gyroscopic effect and internal moments due to unbalanced masses. All these loads are responsible for the occurrence of axial, bending and torsional vibrations; the latter of which is the focus of this work.

Nowadays, considering mobility trends worldwide, internal combustion engines are becoming smaller and provide higher power output, which means higher power

A. S. Mendes (✉) · P. S. Meirelles · D. E. Zampieri
University of Campinas (UNICAMP), Campinas, São Paulo, Brazil
e-mail: schalch@m.unicamp.br

P. S. Meirelles
e-mail: pablo@unicamp.br

© The Author(s), under exclusive license to Springer Nature Switzerland AG 2022
T. Parikyan (ed.), *Advances in Engine and Powertrain Research and Technology*,
Mechanisms and Machine Science 114,
https://doi.org/10.1007/978-3-030-91869-9_2

density. From this point of view, engine downsizing is an actual trend. Independently of the size of the engine, the methodology presented in this chapter can be used to obtain accurate results of crankshaft torsional vibration analysis (TVA).

In the past, there were not enough knowledge to clearly understand the root cause of the catastrophic failures occurred in some crankshafts due to torsional stresses and authors, such as [1], started the researches about this phenomenon. Some components presented excessive torsional vibrations at specific engine speeds and [2] was one of the first researchers who studied this subject.

Damping coefficients of ICE were initially estimated by researchers such as [1, 3]. These parameters were obtained empirically and were inaccurate in most cases, generating considerable variations in the dynamic response of the systems. Iwamoto and Wakabayashi [4] proposed hybrid models for damping coefficient estimation. In their research, they proposed analytical relations between the damping and other measurable parameters of the engine to define the torsional damping coefficients with higher accuracy.

In their studies of torsional vibrations, Honda and Saito [5] analysed a six-cylinder diesel engine with a rubber Torsional Vibration Damper (TVD) to reduce the vibration amplitude. They considered the state transition matrix methodology to calculate the responses and observed that, the torsional dynamic stiffness of the damper has a more significant role in the system's characteristics, compared to the internal damping of the engine and the TVD damping itself. As conclusion, the geometry of the rubber and the chemical composition of the elastomer determined the damper dynamic stiffness.

A proposed methodology to predict the torsional vibration amplitude in ICE was developed in [6], which also considers the transient response by the modal superposition method. The study of [7] analyses the coupling between torsional and axial vibrations in the crankshafts. This effect generates vibrations of large amplitude when the axial and torsional natural frequencies are the same, or when the ratio of these frequencies is equal to two.

Excitation torque is generally considered time-invariant for constant engine loads and speeds, and it is also equal in all cylinders. However, this consideration holds true only if the cylinder pressure traces remain constant throughout the engine lifetime. High mileage vehicles have considerable differences in excitation torques. The author of [8] studied these variations through the cylinders due to the wear of piston rings and liner. According to the results, there are situations where the vibration amplitude can diverge significantly from those expected.

Later, Hestermann and Stone [9] concluded in their research that the unexpected torsional vibrations, in multiples of the engine speed, were mainly due to the variable inertia characteristics inherent of the crank mechanism. Reference [10] considered the effect of the non-constant inertia on torsional vibration calculations. The variation of the inertia over the crank throw angular position shall be considered in cases of large displacement engines, where the piston and connecting rod masses are significantly large. According to the vibratory study carried out, torsional vibrations are more critical for the crankshaft compared to the effects of axial and bending vibrations.

In Ref. [11], the authors found accurate indicators of absolute damping coefficient of a single-cylinder engine powered by an electric motor. The system's first two modes of vibration and absolute damping coefficients were determined as a function of the crank angle. Usually, researchers consider the absolute damping as being constant at all engine speeds and in every crank angle.

The effects of variable inertia in ICE were neglected during considerable time and they were not included in the calculations. These secondary effects were analysed by [12], who continued Draminski's previous studies. As a conclusion, he found that these phenomena were responsible for many structural failures of crankshafts.

The TVA results must include, at least, the calculation of the vibration amplitudes at the crankshaft front and rear-ends, dynamic torques between each crank throw, torque at flywheel connection and TVD power dissipation. Of course, depending on the engine design, TVA results at other locations must be evaluated, e.g. timing drive sprocket vibrations, gear trains etc.

Mainly, TVA is performed through the definition of an elastic model which contains the system dynamic characteristics such as: inertia, torsional stiffness and damping. Then, the excitation torque shall be calculated considering the cylinder internal gas pressure and inertial forces of the oscillating and rotating parts. The tangential component of these forces will be responsible for the actuating torque at the crank pin and a Fourier series expansion of this torque is then performed. Calculated harmonics, e.g. the first 24 terms, shall be applied at the corresponding crank throws, considering the engine firing order [13–15].

2 Theoretical Modelling

Crankshafts are a particularly important component of the ICE and they are subjected to high loading generated from the combustion process. Due to the periodic nature of these excitation forces, the dynamic responses of the component, i.e., torsional, axial, and bending vibrations, result in dynamic stresses that must be evaluated accurately to assure the structural integrity of the crankshaft. The focus of the study presented in this chapter is related to the torsional vibration analyses, which is critical for the most part of the existing crankshaft designs and if the calculations are not carried out in a proper way, catastrophic failures can occur inside the engine.

To carry out these analyses, it is necessary to determine first the equivalent mathematical model of the crankshaft and all rotating parts connected to it. It is also called elastic model of the crankshaft and it contains the information of the dynamic characteristics of the system, such as inertia, torsional stiffness and damping coefficients. Figure 1 presents, as an example, the elastic model of a crankshaft.

This study begins by simulating the elastic model without TVD connected to the crankshaft front end, to adjust the engine's internal damping and to check the natural frequencies of the system. In this step, it is recommended to obtain the torsional vibration amplitudes, measured at crankshaft pulley to compare these values with the simulates ones. Thus, the damping coefficients can be adjusted. Some engines do not run without TVD, even in a short period of time, due to the possibility of

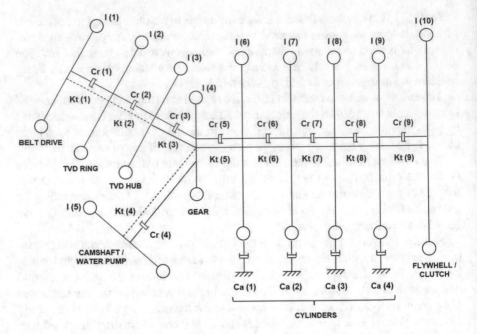

Fig. 1 Example of a complete elastic model of elastic model of an engine rotating parts

structural damage. In these cases, theoretical damping coefficients can be adopted depending on the type of engine configuration.

2.1 Inertia of the System

To obtain the inertia of the system a CAD software is generally used, thus making it quite simple to determine the inertia of each component along its rotating axis. As an example, Fig. 2 shows the model to determine the inertia of a single crank throw of the elastic system. Other components, such as, pulleys, flywheel, clutch, damper ring, etc. have the inertia calculated as the same way, or obtained directly from the manufactures.

In this work we are going to follow as reference for the cylinder numbering, the crankshaft front end as Cylinder 1 and the flywheel side as Cylinder 6.

Some engines have a gear train for power transmission to other components. The inertia of the gear added to the rotating parts of each device shall be also considered in the equivalent model. As the example shown in Fig. 3, the equivalent inertia I_{eq} of a device driven by gear I_2 with a rotational speed n_2 in respect to gear I_1 with a rotational speed n_1 (e.g. the crankshaft gear), can be expressed by Eq. 1.

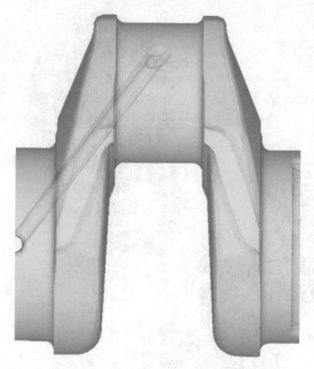

Fig. 2 Example of a model to calculate the inertia of the crank throw

$$I_{eq} = I_2 \left(\frac{n_2}{n_1} \right)^2 \tag{1}$$

The determination of the equivalent inertia shall be done for all components which are driven by the gear train, in respect to the crankshaft gear.

For the inertia of the oscillating masses, composed of complete piston assembly, it is usual to replace by equivalent inertia that must have the same kinetic energy as the piston motion, see [16].

However, this approach does not consider the disturbance due to the variation of the inertia for each crank angle. Here, we will present an approach to consider the inertia oscillation, through a non-constant equivalent inertia that preserves the real kinetic energy at each crankshaft angle. The formulation is developed for a single crank throw, considering that the difference between them is only the relative angular position. The kinetic energy T of a crank throw can be written as follow:

$$T = \frac{1}{2} \left[I_{1_o} \dot{\theta}^2 + m_2 \left(v_{2x}^2 + v_{2y}^2 \right) + I_{2_{CM}} \dot{\beta}^2 + m_3 \dot{X}^2 \right] \tag{2}$$

where θ is the crankshaft angular position, I_{1_o} is the crank throw moment of inertia related to the crankshaft rotation axis, m_2 is the connecting rod mass, v_2 is the velocity

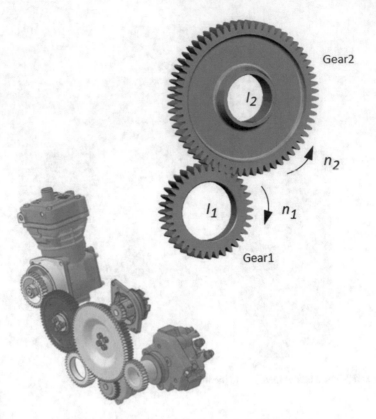

Fig. 3 Example of equivalent inertia calculation (gear 1: crankshaft gear)

of the centre of mass (CM) of the connecting rod , $I_{2_{CM}}$ is the moment of inertia of the connecting rod related to the CM, β is the connecting rod angular position, m_3 is the complete piston mass and X is the piston position referred to the crankshaft centre axis.

Considering the approach presented by Doughty [17] on pages 30 and 254, the kinetic energy can be written depending only on the crankshaft angular speed $\dot{\theta}$ using the velocity coefficients, due to the fact that the crank mechanism is a single Degree of Freedom (DOF). Equation 3 presents this formulation.

$$T = \frac{1}{2}\left[I_{1_o} + m_2\left(\kappa_{2x}^2 + \kappa_{2y}^2\right) + I_{2_{CM}}\kappa_\beta^2 + m_3\kappa_X^2\right]\dot{\theta}^2 \tag{3}$$

In Eq. 3, the expression between square brackets is the equivalent generalised inertia in each cylinder as a function of the crankshaft angle, thus:

$$I_{cyl} = I_{1_o} + m_2\left(\kappa_{2x}^2 + \kappa_{2y}^2\right) + I_{2_{CM}}\kappa_\beta^2 + m_3\kappa_X^2 \tag{4}$$

Following the approach presented in [17], the value of this expression can be easily calculated in a few steps, as long as the geometric and physical proprieties are known, through the next sequence of equations:

$$\beta = \arcsin \left[\frac{r \sin (\theta)}{L} \right] \tag{5}$$

$$X = r \cos (\theta) + L \cos (\beta) \tag{6}$$

$$\kappa_{2x} = -r \sin (\theta) - \kappa_\beta \left[u \sin (\beta) - v \cos (\beta) \right] \tag{7}$$

$$\kappa_{2y} = r \cos (\theta) - \kappa_\beta \left[u \cos (\beta) + v \sin (\beta) \right] \tag{8}$$

$$\kappa_\beta = \frac{r \cos (\theta)}{L \cos (\beta)} \tag{9}$$

$$\kappa_X = X \tan (\beta) \tag{10}$$

The values of u and v are the local coordinates of the connecting rod CM. Complete description of this methodology can be found in [17].

Considering that the torsional vibration amplitudes are small, the inertia of the other cylinders presents the same expression with angular position $\theta_q = \theta_r - \delta_q$, in which δ_q is the difference in angular position between the cylinder q and the reference θ_r, which is a geometric characteristic of each engine type.

Figure 4 shows, as an example, the variation of the con rod and piston inertia in a function of the crank angle. The inertia of the crank throw I_{1o} is not considered in this results.

2.2   Determination of the Torsional Stiffness

The torsional stiffness of all sections of the crankshaft is determined using a finite element model. A constant and known torque is applied at one extremity of the model and the twist angle is calculated considering a counter torque of the same magnitude simulating the reaction torque at the other extremity. The relation considering the value of the applied torque and the difference between the angles of two consecutive sections is equal to the torsional stiffness of this segment, which shall be considered in the equivalent model.

Figure 5 presents, as an example, a finite element model to calculate the torsional stiffness of a crankshaft used in a four-cylinder engine with in-line configuration. Torsional springs in the crank throw intermediate position are used to prevent rigid body motion. The stiffness of these springs shall be as low as possible to avoid interference in crankshaft torsional stiffness determination.

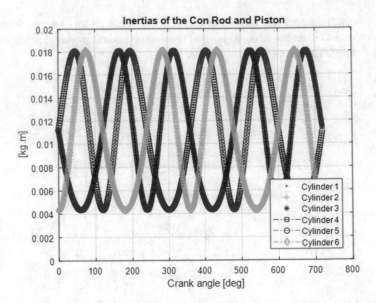

Fig. 4 Variable inertias of the sum of the connecting rod and complete piston

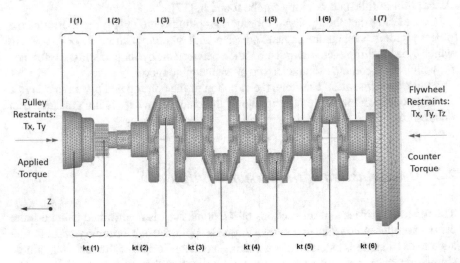

Fig. 5 Example of a finite element model for torsional stiffness calculation

Table 1 Factors for viscous damper torsional stiffness determination

Kinematic viscosity [m²/s]	0.06	0.10	0.15	0.20	0.30	0.50
G_{01} [N/m²]	57.6E−04	20.4E−04	21.0E−04	103E−04	726E−04	2313E−04
B_{01} [K]	3072	3590	3791	3491	3040	2820
a_{01} [−]	2.21	2.25	2.35	2.13	1.93	1.82
a_{11} [K]	384	442	504	454	403	392

Otherwise, the torsional stiffness of the viscous damper k_t, can be determined according to the methodology presented in [18] as a function of the silicone kinematic viscosity.

The dynamic stiffness is then calculated according to the following equations:

$$kt = G_s S \tag{11}$$

$$G_s = G_{01} e^{k_1} f^{k_2} \tag{12}$$

$$k_1 = \frac{B_{01}}{T} \tag{13}$$

$$k_2 = a_{01} - \frac{a_{11}}{T} \tag{14}$$

$$f = n_e n \tag{15}$$

In which, S is the clearance factor obtained from the TVD manufacturer in [m³], T the absolute mean temperature of the silicone film in [K], n is the order number from the Fourier series expansion and n_e the engine speed in [s^{-1}].

The values of the constants can be found in Table 1 according to [18].

According to the previous equations, the torsional stiffness of the TVD is function of the excitation frequency and the silicone temperature. Figure 6 presents an example of the variation of this property. It is possible to observe that lower temperatures and higher excitation frequencies result in higher TVD torsional stiffness.

2.3 Determination of the System Damping

In an ICE there are two types of damping considered in the elastic model. The first one is the absolute damping coefficient ca, which is a function of the friction between the piston, rings and the cylinder liner and it is advisable to obtain this property experimentally. As mentioned previously, absolute damping varies depending on the crank angle. In fact, this variation is small and a mean value of this property can be considered without loss of accuracy.

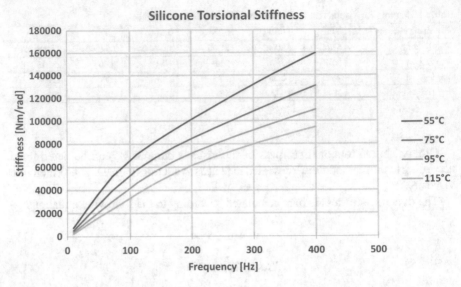

Fig. 6 Example of the variation of the viscous damper torsional stiffness

The second coefficient is the relative damping of the system cr and it can be obtained from different methods. In this work, the definition of this coefficient will be done from the loss angle property, which can be calculated by Eq. 16, considering that ω is the engine's angular velocity.

Figure 7 presents the internal parts of a viscous damper, where it is possible to observe the TVD ring, TVD housing and cover, which are important inertia to be considered in the elastic model.

The average loss factor values for different crankshaft configurations can be found in Refs. [18, 19] and some loss factor values are shown in Table 2. It is important to note that there is a different loss factor, i.e., different damping coefficients for each order of vibration. Generally, we use the mean value in the calculations.

$$\chi = \tan(\delta) = \frac{cr\,\omega}{kt} \tag{16}$$

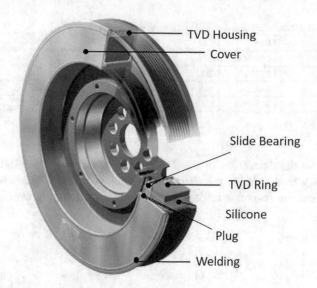

Fig. 7 Viscous damper internal details. *Courtesy* Hasse and Wrede

Table 2 Four-stroke diesel engine average loss factors

Engine configuration	Loss factor d
In-line 4 cylinders (NA)	0.055
In-line 5 cylinders (NA)	0.046
In-line 6 cylinders (NA)	0.056
In-line 8 cylinders (NA)	0.047
In-line 4 cylinders (TC)	0.055
In-line 6 cylinders (TC)	0.035
In-line 8 cylinders (TC)	0.034
V6 cylinders (NA)	0.055
V8 cylinders (NA)	0.047
V10 cylinders (NA)	0.040
V8 cylinders (TC)	0.052
V12 cylinders (TC)	0.042

NA Naturally aspirated, *TC* Turbocharged

At critical crankshaft speeds ω_n, i.e. at resonances, the loss angle is nearly equal to the loss factor and this property is defined according to Eq. 17.

$$d = \frac{cr\ \omega_n}{kt} \qquad (17)$$

Table 3 Factors for determining the TVD damping coefficient

Kinematic viscosity [m²/s]	0.06	0.10	0.15	0.20	0.30	0.50
G_{02} [N/m²]	7.2E−06	0.73	1.01	1.34	7.71	26.4
B_{02} [K]	2974	2332	2353	2356	2050	1760
a_{02} [–]	1.86	1.48	1.50	1.54	1.30	1.11
a_{12} [K]	410	295	321	353	310	277

Similarly to the discussion of the TVD torsional stiffness calculation, there is a methodology to determine the damping coefficient of the TVD. The following equations show the formulas to determine the relative damping coefficient of the viscous damper.

$$cr = \frac{G_v S}{\omega} \tag{18}$$

$$G_v = G_{02} e^{k_3} f^{k_4} \tag{19}$$

$$k_3 = \frac{B_{02}}{T} \tag{20}$$

$$k_4 = a_{02} - \frac{a_{12}}{T} \tag{21}$$

$$\omega = 2\pi n_e n \tag{22}$$

Similarly, S is the clearance factor obtained from the TVD manufacturer in [m³], T the absolute mean temperature of the silicone film in [K], n is the order number from the Fourier series expansion and n_e the engine speed in [s⁻¹].

The constants of the previous equations are obtained experimentally and their values can be found in Table 3 according to [18].

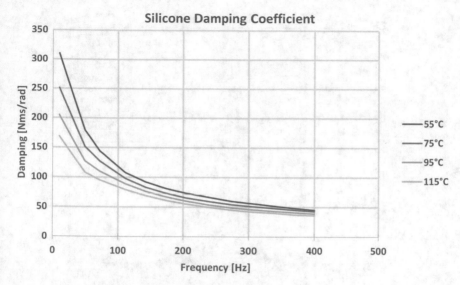

Fig. 8 Example of the variation in silicone damping

Figure 8 shows an example of damping coefficient in respect to the excitation frequency and silicone temperature. It is possible to observe that higher temperatures and excitation frequencies result in lower damping coefficients.

2.4 Crankshaft Excitation Torque

The engine data considered in this study is presented hereunder:

- Four-stroke diesel cycle
- Engine manufacture: MWM Engines and Generators
- Maximum torque: 1100 Nm @ 1200 rpm
- Maximum power: 228 kW @ 2200 rpm
- Maximum engine speed: 2550 rpm
- Firing order: 1-5-3-6-2-4
- Connecting rod length: 207 mm
- Cylinder bore: 105 mm
- Piston stroke: 137 mm
- Oscillating masses: 2.521 kg.

Figure 9 presents the measured cylinder pressure traces for several engine speeds. It is important to highlight that pressure traces were obtained considering engine full load conditions and the values for intermediate engine speeds were determined by interpolating the peak firing pressure (PFP) and the crank angle where PFP values occur (Fig. 10).

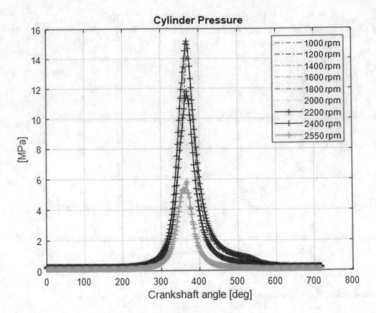

Fig. 9 Measured cylinder pressure traces at several engine speeds

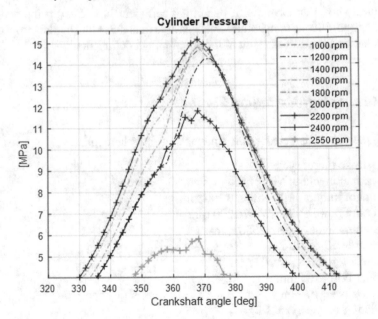

Fig. 10 Detail of cylinder pressure traces at peak firing pressures

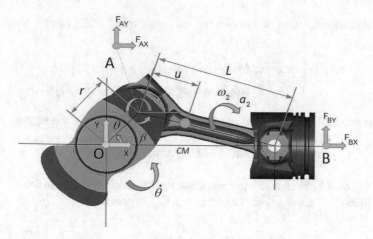

Fig. 11 Crank mechanism dimensions for kinematic and dynamic analyses at global reference system

To determine the excitation torque at the crankshaft, it is necessary to initially calculate the tangential force at the crank pin, considering the kinematics and dynamics of the connecting rod motion. Figure 11 presents the actuating loads and accelerations at the crank mechanism referred to the global coordinate system. The crank angle θ can be assumed as the generalized variable to represent the configuration of the mechanism.

The next equations are used to calculate the positions, velocities, accelerations, and forces to determine the crankshaft excitation torque [20]. The con rod angle β can be determined as a function of the main variable θ according to the equation:

$$\beta = \arcsin\left[\frac{-r \sin(\theta)}{L}\right] \tag{23}$$

The angular velocity ω_2 of the con rod can be obtained by derivation, given:

$$\omega_2 = \frac{-\dot\theta \, \cos(\theta)}{\sqrt{\left(\frac{L}{r}\right)^2 - \sin^2(\theta)}} \tag{24}$$

The angular acceleration a_2 of the connecting rod is obtained by derivation and considering a constant angular velocity for the crankshaft:

$$a_2 = \frac{1}{\cos(\beta)}\left[\dot\theta^2 \left(\frac{r}{L}\right) \sin(\theta) - \omega_2^2 \sin(\beta)\right] \tag{25}$$

Using the angular expressions above, it is possible to obtain the expressions for the position, velocities and accelerations of other points of interest. In this context, the absolute acceleration of the con rod centre of mass (CM) can be obtained as follow:

$$a_{CM} = \left[-r\,\dot{\theta}^2\cos(\theta) - u\,\omega_2^2\cos(\beta) - u\,a_2\sin(\beta)\right] i + \cdots$$
$$\cdots \left[-r\,\dot{\theta}^2\sin(\theta) - u\,\omega_2^2\sin(\beta) + u\,a_2\cos(\beta)\right] j \tag{26}$$

The absolute acceleration of the piston can be determined by the expression:

$$a_B = -r\,\dot{\theta}^2\cos(\theta) - L\,\omega_2^2\cos(\beta) - L\,a_2\sin(\beta) \tag{27}$$

The forces at both ends of the connecting rod are determined from the calculated accelerations and the gas load. The force due to gas pressure is:

$$F_g = \frac{\pi d_p^2}{4} p(\theta) \tag{28}$$

Using these expressions, the forces at the piston end at X and Y directions are calculated, as follows:

$$F_{BX} = -(m_p\,a_B + F_g) - m_p\,g \tag{29}$$

$$F_{BY} = \frac{m_c\,a_{CM}j\,u\,\cos(\beta) - m_c\,a_{CM}i\,u\,\sin(\beta) + I_{2_{CM}}a_2 + F_{BX}\,L\,\sin(\beta)}{L\,\cos(\beta)} \tag{30}$$

The forces at the crank pin at X and Y directions are calculated according to the following equations:

$$F_{AX} = m_c\,a_{CM}i - F_{BX} - m_c\,g \tag{31}$$

$$F_{AY} = m_c\,a_{CM}j - F_{BY} \tag{32}$$

Thus, making a geometric projection in tangential direction of the crank throw, the resultant tangential force at the crank pin is:

$$F_t = F_{AX}\sin(\theta) - F_{AY}\cos(\theta) \tag{33}$$

where

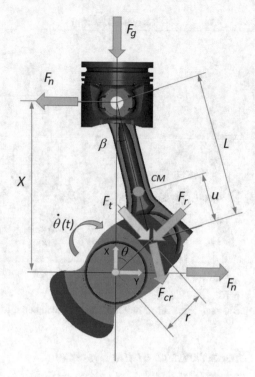

Fig. 12 Crank mechanism dimensions with forces related to crankshaft local system

d_p – piston diameter

$p(\theta)$ – cylinder pressure

g – gravitational acceleration (for engines with vertical cylinder axis)

F_{cr} – con rod force

F_t – tangential force

F_r – radial force

F_n – piston side force

m_c – con rod mass

m_p – complete piston mass

I_{2CM} – moment of inertia of the connecting rod related to the CM.

Figure 12 presents the force decomposition at the crank mechanism highlighting the tangential force. As an example, Fig. 13 shows the torque at each cylinder at rated engine speed.

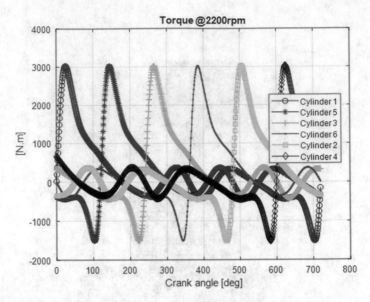

Fig. 13 Actuating torque at each crank throw at 2200 rpm (full load condition)

2.5 Dynamic Characteristics of the System

In mechanical vibrations, the differential equation that represents the dynamic characteristics of the system can be determined according to Eq. 34. Detailed information on this subject is given in Refs. [21–23].

$$[\mathbf{M}]\ddot{\theta}(t) + [\mathbf{C}]\dot{\theta}(t) + [\mathbf{Kt}]\theta(t) = \tau(t) \tag{34}$$

Figure 14 illustrates the crankshaft investigated in this work, containing all components that are considered in the torsional vibration analysis, while Fig. 15 shows the elastic model of the rotating parts.

The inertia matrix of the equivalent model, that has 10 DOF, is written in a diagonal form as follow:

$$[\mathbf{M}] = diag[I(j)], \text{ with } j = 1, \ldots, 10$$

The torsional stiffness matrix (size 10×10) has the following structure:

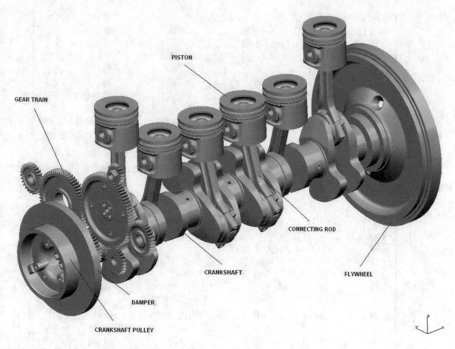

Fig. 14 Illustration of the analysed crankshaft and rotating parts

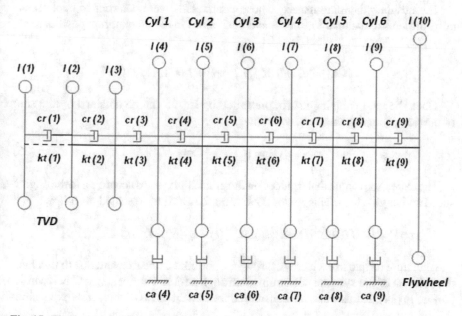

Fig. 15 Elastic model of the crankshaft rotating parts

$$[\mathbf{Kt}] = \begin{bmatrix} kt(1) & -kt(1) & 0 & \ldots & 0 & 0 & 0 \\ -kt(1) & kt(2)+kt(1) & -kt(2) & \ldots & 0 & 0 & 0 \\ 0 & -kt(2) & kt(3)+kt(2) & \ldots & 0 & 0 & 0 \\ 0 & 0 & -kt(3) & \ldots & 0 & 0 & 0 \\ 0 & 0 & 0 & \ldots & 0 & 0 & 0 \\ 0 & 0 & 0 & \ldots & 0 & 0 & 0 \\ 0 & 0 & 0 & \ldots & -kt(7) & 0 & 0 \\ 0 & 0 & 0 & \ldots & kt(8)+kt(7) & -kt(8) & 0 \\ 0 & 0 & 0 & \ldots & -kt(8) & kt(9)+kt(8) & -kt(9) \\ 0 & 0 & 0 & \ldots & 0 & -kt(9) & kt(9) \end{bmatrix}$$

Similarly to the stiffness matrix, the relative damping matrix (size 10×10) can be assembled as follow:

$$[\mathbf{Cr}] = \begin{bmatrix} cr(1) & -cr(1) & 0 & \ldots & 0 & 0 & 0 \\ -cr(1) & cr(2)+cr(1) & -cr(2) & \ldots & 0 & 0 & 0 \\ 0 & -cr(2) & cr(3)+cr(2) & \ldots & 0 & 0 & 0 \\ 0 & 0 & -cr(3) & \ldots & 0 & 0 & 0 \\ 0 & 0 & 0 & \ldots & 0 & 0 & 0 \\ 0 & 0 & 0 & \ldots & 0 & 0 & 0 \\ 0 & 0 & 0 & \ldots & -cr(7) & 0 & 0 \\ 0 & 0 & 0 & \ldots & cr(8)+cr(7) & -cr(8) & 0 \\ 0 & 0 & 0 & \ldots & -cr(8) & cr(9)+cr(8) & -cr(9) \\ 0 & 0 & 0 & \ldots & 0 & -cr(9) & cr(9) \end{bmatrix}$$

The absolute damping matrix, whose coefficients were determined experimentally, presents a diagonal form with average values in the crank throw positions.

$$[\mathbf{Ca}] = diag[ca(j)], \quad with\ j = 1, \ldots, 10$$

Thus, the total damping matrix presented by Eq. 35 can be obtained by the sum of previous damping matrices, as follows:

$$[\mathbf{C}] = [\mathbf{Ca}] + [\mathbf{Cr}] \tag{35}$$

The excitation torque that actuates on the crankshaft varies according to the engine speed and engine load. This vector containing 10 DOF is presented, as follow:

$$\tau(t) = \left\{ 0\ 0\ 0\ M_t^{(1)}(t)\ M_t^{(2)}(t)\ M_t^{(3)}(t)\ M_t^{(4)}(t)\ M_t^{(5)}(t)\ M_t^{(6)}(t)\ 0 \right\}^T$$

Actuating torque in each crank throw is a periodic excitation and due to this fact, the solution of the system is determined through a finite Fourier series which can be revised in [24]. In this work, are considered the first 24 terms of this series expansion.

$$M_t^{(q)}(t) = \frac{A_0^{(q)}}{2} + \sum_{n=1}^{24} \left[C_n^{(q)} e^{in\omega t} + \bar{C}_n^{(q)} e^{-in\omega t} \right] \tag{36}$$

where C_n and $\bar{C}_n$ are the complex coefficient of the Fourier series and its conjugate, with the cylinder number $q = 1, 2, \ldots, 6$.

2.6 System's Steady-State Response

It is possible to express the dynamic behaviour of crankshaft and other parts which includes the system's rotating components, through the first order differential state equations. A detailed explanation of this topic can be found in [21].

$$\dot{x}(t) = \mathbf{A}x(t) + b(t); \quad \dot{x}(t) = \left\{ \begin{array}{c} \theta(t) \\ \dot{\theta}(t) \end{array} \right\} \tag{37}$$

where the state matrix $\mathbf{A}$ and the excitation vector $b(t)$ can be defined as:

$$\mathbf{A} = \begin{bmatrix} 0 & I \\ -\mathbf{M}^{-1}\mathbf{Kt} & -\mathbf{M}^{-1}\mathbf{C} \end{bmatrix} \text{ and } b(t) = \begin{bmatrix} 0 \\ \mathbf{M}^{-1}\tau(t) \end{bmatrix} = \begin{bmatrix} b_1(t) \\ b_2(t) \end{bmatrix}$$

The excitation vector $b(t)$ can be represented as:

$$b_1(t) = \{0\ 0\ 0\ 0\ 0\ 0\ 0\ 0\ 0\ 0\}^T ;$$

$$b_2(t) = \left\{0\ 0\ 0\ \frac{M_t^{(1)}(t)}{I(4)}\ \frac{M_t^{(2)}(t)}{I(5)}\ \frac{M_t^{(3)}(t)}{I(6)}\ \frac{M_t^{(4)}(t)}{I(7)}\ \frac{M_t^{(5)}(t)}{I(8)}\ \frac{M_t^{(6)}(t)}{I(9)}\ 0\right\}^T$$

Considering the expression presented by Eq. 36, the excitation vector can be also written according to Eq. 38:

$$b(t) = \frac{b_0}{2} + \sum_{n=1}^{24} \left[b_n e^{in\omega t} + \bar{b}_n e^{-in\omega t} \right] \tag{38}$$

where

$$b_0 = \left\{ \{0\}_{10x1} \vdots 0\ 0\ 0\ \frac{A_0^{(1)}}{I(4)}\ \frac{A_0^{(2)}}{I(5)}\ \frac{A_0^{(3)}}{I(6)}\ \frac{A_0^{(4)}}{I(7)}\ \frac{A_0^{(5)}}{I(8)}\ \frac{A_0^{(6)}}{I(9)}\ 0 \right\}^T$$

$$b_n = \left\{ \{0\}_{10x1} \vdots 0\ 0\ 0\ \frac{C_n^{(1)}}{I(4)}\ \frac{C_n^{(2)}}{I(5)}\ \frac{C_n^{(3)}}{I(6)}\ \frac{C_n^{(4)}}{I(7)}\ \frac{C_n^{(5)}}{I(8)}\ \frac{C_n^{(6)}}{I(9)}\ 0 \right\}^T$$

$$\bar{b}_n = \left\{ \{0\}_{10x1} \vdots 0\ 0\ 0\ \frac{\bar{C}_n^{(1)}}{I(4)}\ \frac{\bar{C}_n^{(2)}}{I(5)}\ \frac{\bar{C}_n^{(3)}}{I(6)}\ \frac{\bar{C}_n^{(4)}}{I(7)}\ \frac{\bar{C}_n^{(5)}}{I(8)}\ \frac{\bar{C}_n^{(6)}}{I(9)}\ 0 \right\}^T$$

The response of a periodic excited vibratory linear system, represented by its state equation, can be obtained via the fundamental matrix, or transition state matrix, and the convolution integral as depicted in [21, 25].

$$x(t) = \Phi(t)x(0) + \frac{1}{2}\int_0^t \Phi(t-\tau)b_0 d\tau + \sum_{n=1}^{24}\int_0^t \Phi(t-\tau)(b_n e^{in\omega\tau} + \bar{b}_n e^{-in\omega\tau})d\tau$$

$$(39)$$

The transition state matrix is: $\Phi(t) = e^{At}$.

Disregarding the transient response, the constant Fourier terms and solving the harmonic summation, the steady-state response for the asymptotic stable system can be calculated as follows:

$$x_n(t) = \theta_n(t) = g_n e^{in\omega t} + \bar{g}_n e^{-in\omega t} \tag{40}$$

The frequency response vectors are defined as: $g_n = F_n b_n$ and $\bar{g}_n = \bar{F}_n \bar{b}_n$.

The frequency matrices can be written as: $F_n = (in\omega I - A)^{-1}$ and $\bar{F}_n = (-in\omega I - A)^{-1}$.

Thus, the total torsional vibration amplitude of each inertia j of the elastic system, can be calculated according to Eq. 41.

$$\theta_j = \sum_{n=1}^{24} \Theta_{n_j} \cos(n\omega t - \phi_{n_j}) \tag{41}$$

where

$$\Theta_{n_j} = 2\sqrt{Re(g_{n_j})^2 + Im(g_{n_j})^2} = 2\left|g_{n_j}\right| \text{ and } \phi_{n_j} = atan\left[\frac{-Im(g_{n_j})}{Re(g_{n_j})}\right] \tag{42}$$

With: $n = 1, \ldots, 24$ and $j = 1, \ldots, 10$.

The dynamic torque between consecutive inertias can be determined by Eq. 43.

$$\tau_{j-1} = \left|\theta_j - \theta_{j-1}\right| kt_{j-1} \tag{43}$$

It is important to highlight that, the torque due to the constant Fourier term shall be added to elements which represent each cylinder. In fact, this constant torque is the responsible for the engine indicated torque.

Finally, with the calculated angular velocities at each inertia, it is possible also to determine the generated power at the elements. This is especially useful for the evaluation of the TVD power dissipation, what is extremely important for the torsional damper design. Equation 44 computes the generated power at the viscous TVD.

Table 4 Dynamic characteristics of crankshaft elastic model

Element	Inertia [kg m²]	Absolute damping [N m s/rad]	Torsional stiffness [N m/rad]
(1) Damper ring	0.152	0	Calculated (1)/(2)
(2) Pulley, TVD hub	0.097	0	1,106,000 (2)/(3)
(3) Gear train	0.009	0	1,631,000 (3)/(4)
(4) Cylinder 1	0.035	2	1,253,000 (4)/(5)
(5) Cylinder 2	0.021	2	1,253,000 (5)/(6)
(6) Cylinder 3	0.035	2	1,678,000 (6)/(7)
(7) Cylinder 4	0.035	2	1,253,000 (7)/(8)
(8) Cylinder 5	0.021	2	1,253,000 (8)/(9)
(9) Cylinder 6	0.037	2	1,976,000 (9)/(10)
(10) Flywheel, adaptor	2.075	0	–

$$Q_1 = \int_0^t cr_1 \, (\dot{\theta}_1 - \dot{\theta}_2) \tag{44}$$

Based on the viscous TVD power dissipation it is possible to calculate the silicone temperature and detailed information about this procedure can be found in [18]. The next section will present the results of the complete torsional vibration analysis for a crankshaft of an engine chosen to validate the methodology.

3 Simulated and Experimental Results

Table 4 shows the values of the elastic model properties presented in Fig. 15.

Note that torsional stiffnesses are indicated in the table as connections between two consecutive elements. Other variables described in Sect. 2.5 are presented as follows:

- Engine mean loss factor: $d = 0.035$
- Kinematic viscosity of the silicone: $v = 0.15$ m²/s
- Clearance factor: $S = 5.0$ m³
- Reference area of the TVD ring: $Ad = 0.2792$ m²
- TVD maximum temperature: $T_0 = 100\,°C$
- Ambient temperature: $T_{amb} = 50\,°C$
- Constant gear train torque: 86 Nm.

Figures 16, 17, 18, 19, 20, 21, and 22 show the results of the torsional vibration analyses, including the measured amplitudes at engine test bench for comparison. There are other types of graphics to present these results, e.g. Campbell diagrams,

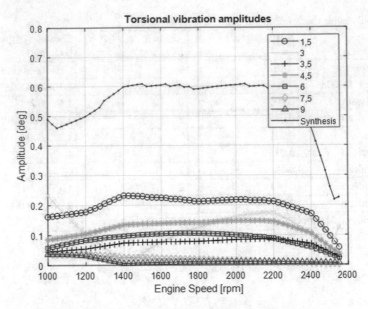

Fig. 16 Calculated torsional vibration amplitudes at crankshaft pulley

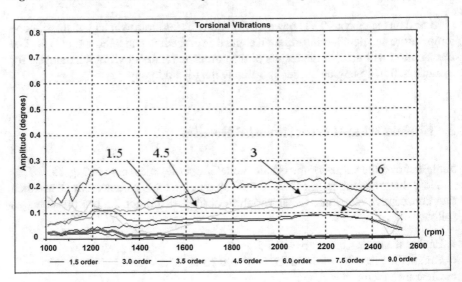

Fig. 17 Measured torsional vibration amplitudes at crankshaft pulley

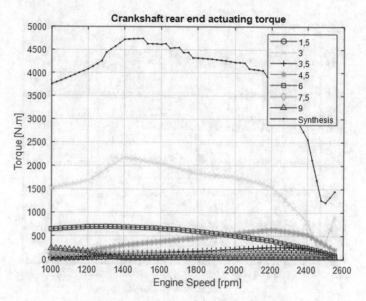

Fig. 18 Dynamic torques at the connection between the flywheel and crankshaft flange

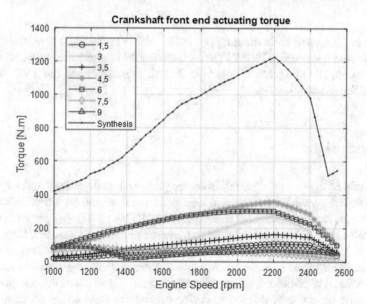

Fig. 19 Dynamic torques at the connection between the pulley and crankshaft flange

but due to its easy and fast interpretation, the selected form is preferred for TVA investigations.

Considering the vibration level and the structural design criteria, the maximum vibration amplitudes at the crankshaft front-end, per main orders, for an in-line six-

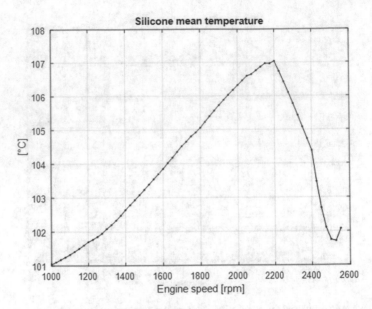

Fig. 20 Average temperature at the silicone oil film

cylinder engine, shall be in the range of 0.20–0.25°. On the TVD side, the silicone temperature cannot exceed 120 °C. Thus, observing the theoretical results and regarding both limits, it is possible to conclude that the crankshaft and the TVD are well designed for this application.

4 Conclusions

An analysis of the results obtained from the proposed methodology and the subsequent comparison with the measured results of a real engine on a test bench leads to the conclusion that theoretical and experimental TVA amplitudes present remarkably similar values. Thus, all adopted hypotheses in this work are, therefore, valid. This study highlights the importance of torsional vibration investigations, considering the increase of the specific power output of the modern engines used in hybrid or conventional applications, as opposed to the mass reduction of the components.

The main objective of this work was to present the state of the art in torsional vibration analysis of crankshafts. Due to this, the design limits of the components are not mentioned here. Here, the reader can find some guidelines for this type of application, but these values shall not be considered as reference for new components design purposes.

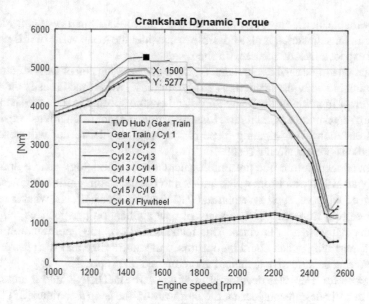

Fig. 21 Dynamic torques between each inertia of the elastic model

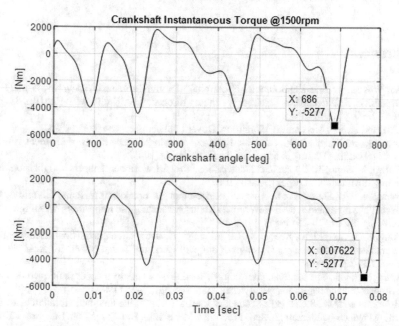

Fig. 22 Instantaneous torque between Cylinder 4 and Cylinder 5 at 1500 rpm

The variable inertia caused by the motion of the oscillating parts was considered in the calculations. In fact, it is also possible to consider the mean value of this property without expect a lack of accuracy in the results.

Excitation torque can be calculated from the series expansion of the resultant force at the crank pin with just few terms [26]. In this case, the torque is easily calculated from a direct function of the crank angle and engine speed. On the other hand, it is possible to determine this excitation load by a more elegant calculations considering Newton equations. Investigating these two approaches, the reader will verify that both methods present similar results.

As mentioned before, the presented calculation methodology can be applied to basically all ICE types. From spark ignition to compression ignition engines, using alternative fuel types, such as: ethanol, CNG, hydrogen or biofuels, what is remarkably interesting from the point of view of new application perspectives, regarding environmentally friendly engines. This technique allows the determination of new design parameters, which could be optimised in shorter time and with fewer tested parts, offering an attractive technical and commercial proposals.

The code for TVA was developed in MATLAB/SIMULINK and it can be also applied in TVD design considering the hardware-in-the-loop technique [27] for cost reduction in durability validation tests.

References

1. Ker Wilson, W.: Practical Solution of Torsional Vibration Problems. Wiley, New York (1963)
2. Draminski, P.: Extended treatment of secondary resonance. Shipbuild. Mar. Eng. Int. **88**, 180–186 (1965)
3. Den Hartog, J.P.: Mechanical Vibrations. Dover Publications, New York (1985)
4. Iwamoto, S., Wakabayashi, K.: A study on the damping characteristics of torsional vibration in diesel engines (Part I). J. Mar. Eng. Soc. **19**, 34–39 (1985)
5. Honda, Y., Saito, T.: Dynamic characteristics of torsional rubber dampers and their optimum tuning. SAE technical paper 870580 (1987)
6. Johnston, P.R., Shusto, L.M.: Analysis of diesel engine crankshaft torsional vibrations. SAE Spec. Pub., presented at SAE Government/Industry Meeting and Exposition, Washington, DC, USA, pp. 18–21, 21–26 (1987)
7. Song, X.G., Song, T.X., Xue, D.X., Li, B.Z.: Progressive torsional-axial continued vibrations in crankshaft systems: a phenomenon of coupled vibration. Trans. ASME Rotat. Mach. Veh. Dyn. 319–323 (1991)
8. Maragonis, I.E.: The torsional vibrations of marine diesel engines under fault operation of its cylinders. Forschung im Ingenieurwesen Eng. Res. **58**, 13–25 (1992)
9. Hestermann, D.C., Stone, B.J.: Secondary inertia effects in the torsional vibration of reciprocating engines—a literature review. Proc. Inst. Mech. Eng. Part C: J. Mech. Eng. Sci. **209**(C1), 11–15 (1994)
10. Brusa, E., Delprete, C., Genta, G.: Torsional vibration of crankshafts: effect of non-constant moments of inertia. J. Sound Vib. **205**(2), 135–150 (1997)
11. Wang, Y., Lim, T.C.: Prediction of torsional damping coefficients in reciprocating engine. J. Sound Vib. **238**(4), 710–719 (2000)
12. Pasricha, M.S.: Effect of the gas forces on parametrically excited torsional vibrations of reciprocating engines. J. Ship Res. **45**(4), 262–268 (2001)

13. Mendes, A.S., Meirelles, P.S., Zampieri, D.E.: Analysis of torsional vibration in internal combustion engines: Modelling and experimental validation. Proc. Inst. Mech. Eng. Part K: J. Multi-body Dyn. **222**(2), 155–178 (2008). https://doi.org/10.1243/14644193JMBD126
14. Meirelles, P.S., Zampieri, D.E., Mendes, A.S.: Experimental validation of a methodology for torsional vibration analysis in internal combustion engines. In: 12th IFToMM World Congress, Besançon, France (2007)
15. Meirelles, P., Zampieri, D.E., Mendes, A.S.: Mathematical model for torsional vibration analysis in internal combustion engines. In: 12th IFToMM World Congress, Besançon, France (2007)
16. Brunetti, F.: Motores de combustão interna (in Portuguese), 1st edn, vol. 2. Blucher, Brazil (2012). ISBN 9-788-52120709-2
17. Doughty, S.: Mechanics of Machines. Wiley, New York (1998). ISBN: 0-471-84276-1
18. Hafner, K.E., Maass, H.: Torsionsschwingungen in der verbrennungskraftmaschine. Springer, Berlin (1985). ISBN: 3-211-81793-X
19. Hafner, K.E., Maass, H.: Theorie der triebwerksschwingungen der verbrennungskraftmaschine. Springer, Berlin (1984). ISBN: 3-211-81792-1
20. Shenoy, P.S.: Dynamic Load Analysis and Optimization of Connecting Rod. M.Sc. Dissertation, University of Toledo, OH, USA (2004)
21. Müller, P.C., Schiehlen, W.O.: Linear Vibrations. Martinus Nijhoff Publishers, Dordrecht, The Netherlands (1985). ISBN: 90-247-2983-1
22. Meirovitch, L.: Principles and Techniques of Vibration. Prentice Hall, Englewood Cliffs, NJ, USA (2000)
23. Inman, D.J.: Engineering Vibration. Prentice Hall, Englewood Cliffs, NJ, USA (2001). ISBN: 0-13-726142-X
24. Arruda, J.R.F., Huallpa, B.N.: Introdução à análise espectral (in Portuguese). Unicamp, Brazil (2002)
25. Mendes, A.S.: Development and Validation of a Methodology for Torsional Vibrations Analysis in Internal Combustion Engines (in Portuguese). M.Sc. Dissertation, Universidade Estadual de Campinas, Campinas, Brazil. http://repositorio.unicamp.br/jspui/handle/REPOSIP/264376 (2005)
26. Taylor, C.F.: The Internal Combustion Engine in Theory and Practice, vol. 2, ch. 8. MIT Press, MA (1985). ISBN 0-262-70027-1
27. Mendes, A.S.: Performance Verification of the Hardware-in-the-Loop Technique Applied to Non-Linear Mechanical Systems (in Portuguese). Ph.D. Thesis, Universidade Estadual de Campinas, Campinas, Brazil. http://repositorio.unicamp.br/jspui/handle/REPOSIP/264387 (2012)

Scuffing Behavior of Piston-Pin/Bore Bearing in Mixed Lubrication

Chao Zhang

Abstract A bench rig is designed, constructed, and used to determine the scuffing mechanism and threshold conditions, and to investigate the effects of surface roughness, clearance, circumferential groove, oil temperature, and oil feeding rate on scuffing in piston-pin/bore contacts. The bench rig simulating conditions include time-dependent dynamic load over a complete engine cycle, an oscillatory rotation, a constant surface velocity corresponding to the maximum surface velocity of the piston pin in real engines, and surface and material properties. Based on the experimental results, it is concluded that the scuffing failure mechanism involves adhesion and fatigue. The deterioration of the surface shear strength by fatigue cracks under a high surface temperature appears to contribute to the final scuffing of the pin-piston contact. The effects of the bore surface roughness and groove on the scuffing behavior are more significant than the clearance. A computer model to study the onset of scuffing conditions at the pin/piston interface is developed. The scuffing factor and the scuffing failure mapping reflect the interaction among the related mechanisms of wear, relevant parameters, and the inaccuracy in modeling.

Keywords Bench tests · Scuffing · Bearings · Piston · Mixed lubrication

1 Introduction

According to the ASTM Terminology standard G40, scuffing is a form of wear occurring in inadequately lubricated tribosystems that is characterized by macroscopically observable changes in texture, with features related to the direction of motion. Scuffing is a complex phenomenon, involving mechanical, thermal and chemical interactions between the contacting bodies, the environment, the lubricant, and other species found at the sliding interface. Previous works [1–13] show that the lubrication breakdown in sliding lubricated contacts is generally recognized as the principal

C. Zhang (✉)

School of Mechatronic Engineering and Automation, Shanghai University, No. 99 Shangda Road, Shanghai 200444, China

e-mail: 13301770659@163.com

© The Author(s), under exclusive license to Springer Nature Switzerland AG 2022

T. Parikyan (ed.), *Advances in Engine and Powertrain Research and Technology*, Mechanisms and Machine Science 114, https://doi.org/10.1007/978-3-030-91869-9_3

cause of scuffing, which is usually treated as a surface-related phenomenon. The mechanisms of lubrication breakdown are due to loss of the successive protective films at the sliding asperity contacts. These include the asperity oil film generated by the micro-elastohydrodynamic action, adsorbed or reactive surface film, oxide film, and finally the thin hard layer of the substrate itself. On the other hand, the surface interactions in mixed and boundary lubrication have severe effects on crack initiation and propagation at significant depths below the surface, which are present before the final adhesive phase of the scuffing process. Scuffing would occur when the surface tangential traction at the interface exceeds the bulk shear strength. Scuffing also is related to the initiation of plastic flow in the contacting asperities, and to crack initiation and propagation on or below the surface. Under dynamic loading, surface and subsurface cracks due to fatigue affect progression of scuffing severely. Although it is generally accepted that in the final stage of scuffing/seizure, all protective films are absent and metal adhesion occurs over large contact areas, the scuffing mechanisms are still not completed understood.

A few scuffing criteria have been developed. The critical surface temperature criterion [14, 15] relates scuffing to a critical temperature condition, but it is not a material property. The critical subsurface stress criterion [16] assumes that ductile rupture occurs when the shear stress at a critical depth under the surface exceeds the temperature-dependent shear strength of the material at this depth. The critical ratio of shear stress/hardness criterion [17] relates scuffing to the gross plastic flow at a subsurface depth. The plastic index criterion [18, 19] relates scuffing to surface roughening as a result of plastic flow and a critical amount of the plastic strain in the sliding track. The damage accumulation time criterion [20] assumes that scuffing occurs when the damage accumulation time necessary for initiation and propagation of cracks is shorter than the time necessary for the damaged material to leave the critical zone of fastest damage accumulation. The critical surface tangential traction [21] assumes that scuffing occurs when the surface tangential traction exceeds the bulk shear strength. The mean value distribution of the wear load (integral value of the asperity contact pressure and sliding speed) is referred to as the scuffing evaluation index [9]. It is clear that any one of the hypotheses or criteria mentioned before may be used only under certain conditions where this wear mechanism dominates the scuffing characteristics.

Being at the heart of engine, the piston assembly, defined as the piston, its rings and the piston pin, is subjected to extremes of thermal and mechanical loading. In the case of piston pin and piston bore bearing, the pin oscillates within small angles, its sliding speed is very slow and squeeze motion is dominant, fully developed hydrodynamic film is usually not present. The bearings run under mixed and boundary lubrication.

In practice, the piston pin is either fixed to or floating between the connecting rod and the piston. In the later designs, the pin and the connecting rod at the small end constitute a wrist bearing and it can effectively avoid scuffing to occur in the piston pin and bore bearing. The possible reason is that when scuffing just begins to happen in any one of the three bearings, higher frictional coefficient will immediately reduce its rotational speed, significantly lowering its bulk and flash temperatures. As a positive feedback process, this bearing will work normally again due to its lower temperature than the critical value. While for the former one, where the piston pin is fixed with the connecting rod, scuffing can occur more readily in the piston pin/bore bearing because of full sliding. However, the fixed mounting is more preferable due to lower cost by eliminating the costly wrist pin bearings. Since the scuffing/seizure mechanisms are not completely understood, predictions of the scuffing/seizure failure are based largely on experience rather than on any analytical model.

Performing tribological tests for scuffing/seizure behavior of piston pin/bore in mixed lubrication using actual engines is neither economical nor efficient. Bench rig simulation either using samples having the same features of the engine components or production parts directly can achieve the goals with less cost. Since scuffing is influenced by numerous factors and can have various manifestations, a successful experimental study of scuffing/seizure must provide both accurate control of these factors and reliable means to identify and reproduce the scuffing process under laboratory conditions. In simulations, one must reproduce the kinematics, load and contact conditions, material properties, and operating conditions as close as possible to those in real automotive engines.

In order to improve understanding of scuffing/seizure mechanisms and its predictive methods, the first goal of the authors' research [22, 23] shown in this chapter is to design and build a bench rig and develop a test procedure to investigate effects of surface roughness, clearance, circumferential groove, and oil feed on scuffing/seizure behavior for four different combinations of speed and load: constant and oscillating journal speed, static and dynamic load. The scuffing condition during a test is identified by transitions in applied load and bulk temperature. Since the production parts, i.e. pistons and piston pins, are used as samples, the simulation of the surface and material properties is automatically satisfied. The second goal is to experimentally investigate the effects of roughness, clearance, circumferential groove, and oil-feed rate on scuffing/seizure failure, and to determine the scuffing/seizure mechanisms and threshold conditions. The third goal is to develop a simple computer model as guidelines to design the piston-pin/piston bore contacts for higher scuffing resistance.

2 Experimental Setup

In this section, the details of the bench test apparatus are presented for the loading system, drive system, lubrication system, instrumentation system, and specimens. A schematic diagram and general views of the bench rig are shown in Figs. 1 and 2a-d, respectively. Figure 3 shows the thermocouple location in the piston. Figure 4 shows the test samples taken from the production piston and piston pin.

As shown in Fig. 1, the dynamic load is applied at the center of the piston pin through a connecting rod, which is connected with a loading beam through two identical needle bearings. The loading beam with three pivot points is used to transfer and amplify 10, 8.8, and 7.2 times, respectively, the compressive load given by the cam-driven leaf spring and Spring 2. Spring 3 exerts a vertical tensile load on the connecting rod to simulate an inertial force applied on a piston pin. The preloads

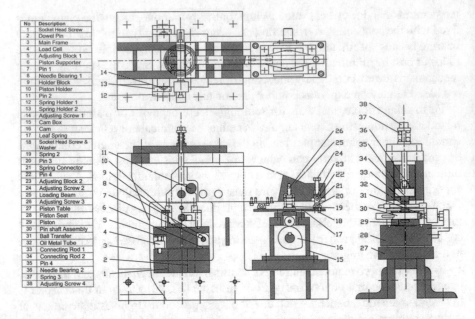

No	Description
1	Socket Head Screw
2	Dowel Pin
3	Main Frame
4	Load Cell
5	Adjusting Block 1
6	Piston Supporter
7	Pin 1
8	Needle Bearing 1
9	Holder Block
10	Piston Holder
11	Pin 2
12	Spring Holder 1
13	Spring Holder 2
14	Adjusting Screw 1
15	Cam Box
16	Cam
17	Leaf Spring
18	Socket Head Screw & Washer
19	Spring 2
20	Pin 3
21	Spring Connector
22	Pin 4
23	Adjusting Block 2
24	Adjusting Screw 2
25	Loading Beam
26	Adjusting Screw 3
27	Piston Table
28	Piston Seat
29	Piston
30	Pin shaft Assembly
31	Ball Transfer
32	Oil Metal Tube
33	Connecting Rod 1
34	Connecting Rod 2
35	Pin 4
36	Needle Bearing 2
37	Spring 3
38	Adjusting Screw 4

Fig. 1 Schematic diagram of part of the bench rig

of these three springs can be regulated by adjusting Screws 24, 26 and 38. In the present study, the load applied by the cam-driven springs is amplified approximately ten times. The cam is designed to give a load variation over a cycle approximately equal to that in actual four-stroke engines. A cycle length for the load variation in the piston is two times the cycle length for the load variation in the cam. The loading beam and the main frame are very stiff in order to make the loading curve applied on the piston very similar to that given by the cam-driven springs. Two sets of a piston holder and two holder blocks set the piston always in contact with the piston table. Two springs held by the spring supporters set the piston table in contact with the load cell. The force of Spring 1 is adjusted by the Adjusting Screw 1 and its total force is bigger than the maximum tensile load used. The natural frequency of the loading system is more than three times the maximum oscillating frequency.

As shown in Fig. 2a–c, both the camshaft and the pin shaft are driven by an 18.7 kW AC motor, which has a speed range from 600 to 6000 rpm and is controlled by a variable frequency drive motor speed controller. Two sets of timing belts and an oscillatory assembly for oscillation case are used to transfer power. The pulley ratio is.

2.813:1 from the camshaft to the motor shaft, 1.333:1 from the camshaft to the pin shaft under constant pin speed, and 2:1 for the camshaft to the oscillator drive shaft under oscillation. A flexible coupling is used to connect the camshaft with the driven pulley shaft to accommodate misalignment between the two shafts for the dynamic loading. For static loading, these two shafts are disconnected by separating components of the coupling. Two universal joint couplings are used to connect the

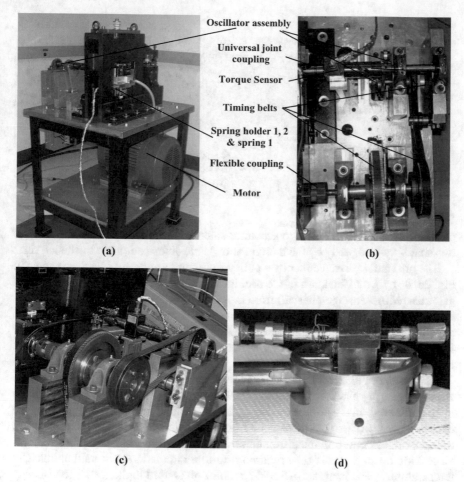

(a)

Oscillator assembly

Universal joint coupling

Torque Sensor

Timing belts

Spring holder 1, 2 & spring 1

Flexible coupling

Motor

(b)

(c)

(d)

Fig. 2 General view of the bench rig

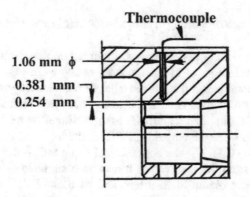

Thermocouple

1.06 mm φ

0.381 mm

0.254 mm

Fig. 3 Thermocouple location in the piston

Fig. 4 Specimens

pin shaft with the pulley shaft to accommodate variable misalignment between the two shafts, caused partly by the thermal and elastic deformations of the bench rig.

The pin shaft is driven either by a pulley shaft with a constant speed, as shown in Fig. 2c, or by a rocking shaft of the oscillatory assembly, as shown in Fig. 2a, b. A ball transfer prevents the pin shaft from moving out axially. The oscillator assembly is designed based on the principle of a crank-rocker mechanism and it can generate a motion closely approximating the oscillation of the piston pin in actual engines. A second set of identical crank and rocker mechanism is also mounted below to balance the inertial forces.

As shown in Fig. 2, a filtered closed-loop lubrication system provides oil for the wrist pin bearing and the piston-pin/bore bearings. The area of the upstream hole in the oil metal tube can be adjusted by a line bonded on the tube to control the distribution of oil feed into these three bearings. A hole is drilled in the piston skin to make oil leak out from the piston inner volume. A groove in the piston seat leads all leaking oil to a plastic tube returning oil to a circulation tank held at constant temperature. A commercial SAE 5W-30 motor oil is used in the test. The viscosity of the oil is 64.2 cSt at 40 °C and 11.0 cSt at 100 °C, the density is 0.875 g/cm^3 at 25 °C.

The cam and the two supporting bearings in the cam box are lubricated and cooled by oil stored in the box.

A load cell with ∼ 22.4 kN capacity is mounted on the piston table and its top surface contacts the bottom of the piston seat, as shown in Fig. 1, to measure the applied load. The piston seat and the piston table are connected by two needle bearings and the two centers of the load cell and bearings are symmetrical about the piston. Hence, the ratio of the applied load to the load measured by the load cell is almost 2:1. The relation between the output signal from the load cell and the applied load is $F = 2616.5(X - 0.4)$, F—the applied load (N) and X—the signal (millivolt). The output signal from the load cell is transmitted simultaneously to a three-pen strip-chart recorder, a digital multi-meter, and an oscilloscope with a differential amplifier. Then the filtered and amplified signal is transferred to a computer-based

Table 1 Composition of tested piston alloys (wt%)

Silicon	14.5–17.0	Nickel	0.4–0.8	Magnesium	0.6–1.0
Iron	0.9 max	Zinc	1.0 max	Titanium	0.25 max
Copper	1.5–2.5	Manganese	0.2–0.5	Total other elements	0.5 max

data acquisition system. Although for the dynamic loading case, the three pens strip-chart recorder and the digital multi-meter can't respond to the dynamic load signal, they can still be used to monitor the scuffing/seizure process.

As shown in Fig. 2b,c, a torque sensor with ~ 56.5 N m capacity is also mounted between two universal joint couplings for measuring the total friction torque of the wrist pin bearing and the piston-pin/bore bearings.

A small hole drilled vertically from the top surface shown in Fig. 3 is very close to the bore inner surface and the potential scuffing zone. An ANSI Type K (chromel–alumel) thermocouple is inserted into the hole and pressed on the subsurface to measure bulk temperature. A Three-pen strip-chart recorder and a digital multi-meter are used to monitor and record the signal from the thermocouple.

Piston pins are made of SAE 1016 steel with a diameter of 21.9964 mm and root-mean-square (RMS) of surface roughness (σ or Sigma) in the range of $\sigma = 0.17$–$0.19 \, \mu m$. Two identical grooves machined symmetrically at one end of the pin enable the pin shaft to be driven by two keys. Pistons are made of high silicon aluminum alloy material with the composition shown in Table 1. Piston bores machined by boring have different diameter, roughness, and oil groove geometry. The roughness of unworn smooth pistons is 0.27–0.7 μm, which becomes 0.17–0.27 μm after running in. Figure 4 shows the test samples taken from the production pistons and piston pin. Pistons with a flat top surface and without any ring groove are used in this study to increase piston stiffness.

3 Test Procedure

The ambient temperature for the test rig is kept at 20 °C room temperature. Heat transfer from the test rig to the static ambient air is due to heat convection. The heat generated in the bearing caused mainly by asperity contact can be more than that dissipated by conduction and convection, resulting in increases in the temperature as well as thermal expansion in the test assembly with time. As a response to the thermal expansion, through the loading arm, the cam-driven springs are compressed a distance ten times the vertical thermal expansion at the load cell center. In return, the applied load and the heat generated increase. However, it does not influence reliability and comparability of the results because the scuffing failure is strongly dependent on the flash temperatures and local contact pressures. The effect of temperature equilibrium delays the time when scuffing occurs.

3.1 Journal Oscillation and Dynamic Load

After a new piston and a piston pin are installed, running in is performed for 20 min with an initial static load, 2224 N, and at 1000 rpm oscillator drive shaft speed. After running in, the oscillator drive shaft speed is kept at 1500 rpm and the static loads of two cam-driven springs are increased stepwise until scuffing occurs. Figure 5 shows the pin oscillating surface linear velocity variation corresponding to 1500 rpm of the drive shaft speed. Figure 6 shows a typical test chart during a test. Both the measured bulk temperature and the applied load increase with time during running in due to the thermal expansion of the test piston assembly. When scuffing occurs, the measured bulk temperature first increases significantly and then decreases pronouncedly and at that time the scuffing manifestations could be observed from the surfaces of the pin and the bore. From Fig. 6, it can be observed that the three-pen strip chart recorder can record evolution of the maximum applied loads versus time. If the load is further increased, the bulk temperature stops to decrease and begins to increase again. In order to compare the effects of the different factors on scuffing failure, when the scuffing happens, the corresponding applied load is referred to as the scuffing load as shown in Fig. 7 and recorded by the computer. The test rig stops only after the applied load is released in order to avoid additional scuffing and severe wear occurring when the rig stops immediately without releasing the applied load.

Figures 8 and 9 show the oscillating surface velocity of the piston pin and the load acting on the piston-pin/bore bearing of Ford 4.6 L V8 2 V four-stroke engine calculated from the cylinder pressure and engine parameters provided by Zollner Co.

The diagrams mentioned in this section show that the bench tests can simulate the kinetic and dynamic characteristics of the pin/bore bearings in an engine.

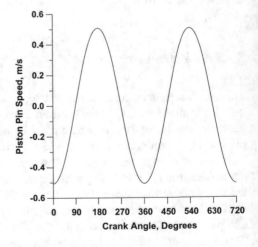

Fig. 5 Pin speed for 1500 rpm of the oscillator drive shaft sped

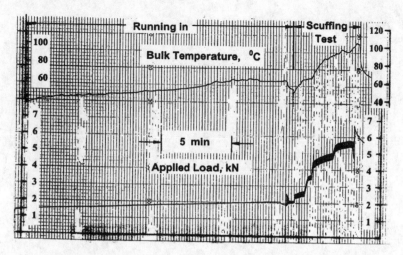

Fig. 6 Scuffing test chart for piston pin/bore bearing (# XB-11)

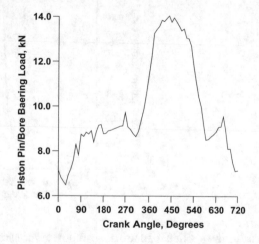

Fig. 7 Applied load corresponding Scuffing (# XB-11)

3.2 Constant Pin Speed and Dynamic Load

Because the tangential surface velocity of the pin for the constant pin speed cases is higher than that in the oscillating case, the heat caused by the asperity contact is higher in the first case. Therefore, the static applied load for running in is lower in the constant speed case than that in the oscillation case to prevent thermohydrodynamic seizure from occurring. Running in is performed at the constant pin speed 660 rpm and under an initial static applied load of 1957 N for 20 min. After running in, the constant pin speed is kept at 1000 rpm and the initial loading of two cam-driven springs is fixed until seizure occurs. Because seizure evolves very quickly, stopping

Fig. 8 Pin speed in an engine

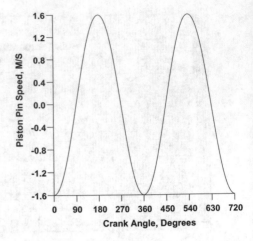

Fig. 9 Bearing load for Fig. 8

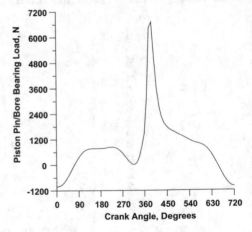

the test rig and releasing the applied load are done almost at the same time according to the bulk temperature recorded by a three-pen strip chart recorder. Since at the very beginning seizure is very sensitive to the applied load, as soon as seizure trends occur, releasing the applied load can prevent it from occurring. Figures 10 and 11 show a typical test chart and the applied load corresponding to seizure.

4 Test Results and Discussions

Figure 12a–c show pictures of the piston bore and pin surfaces taken after scuffing and seizure failures. As shown in Fig. 12a, b, on the bore surfaces dark layers, which are deposits of carbonized oil, are formed. The light-colored portions of the surface

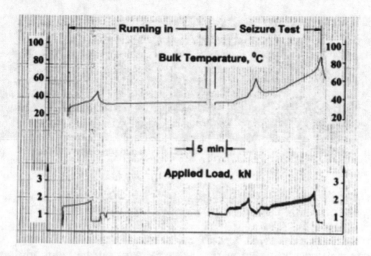

Fig. 10 Seizure test chart for piston pin/bore bearing (# XB-17)

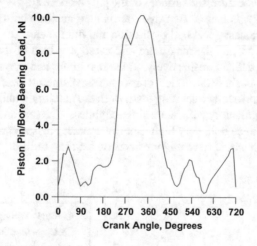

Fig. 11 Applied load corresponding seizure (# XB-17)

are deep valleys, which is characteristic of sections of the surface removed by fatigue and scuffing. The pin surfaces are covered with the layers of the piston material in the contact zones, while on the worn bore surfaces, severe tearing scars are commonly seen and some material is transferred from the piston bore to the pin surface due to solid-phase welding. Mild and severe wear, initial and final stages of scuffing can be found on the bore and pin surfaces. When the Al-Si material strength in a surface layer is stronger than the adhesive strength between Al-Si and steel, wear occurs on the sliding contact surfaces. On the other hand. if the reverse is true, then scuffing occurs.

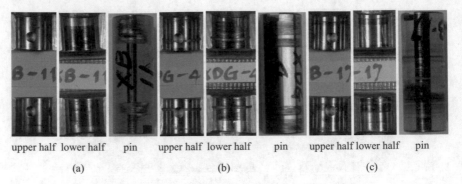

upper half lower half   pin   upper half lower half   pin   upper half lower half   pin

(a)   (b)   (c)

Fig. 12 Pictures of piston and pin surfaces after scuffing and seizure: **a** # XB-11, without groove (scuffing); **b** #XDG-4, with groove (scuffing); **c** #XB-17, without groove (seizure)

After seizure occurs, severe tearing scars are seen on the worn bore surfaces and a color turning blue is observed on the pin surface at the connecting rod bearing. Because just before seizure, the surfaces of the pin and bore are very smooth, seizure, in the present study, is caused by rapid thermal expansion due to the small radial clearance and more heat generated by friction and fluid shear.

When scuffing occurs, the maximum and average values of the applied load in one load cycle are defined, respectively, as the maximum and average scuffing load.

Figures 13 and 14 show the effects of roughness, diametral clearance, and groove on the maximum and average scuffing load. Both the maximum and average scuffing loads decrease significantly with increasing roughness. The reason is that the surface roughness can produce both very high asperity contact pressure and very high flash temperature, which would have major effects on scuffing failure.

Fig. 13 Effects of roughness and groove on scuffing resistance

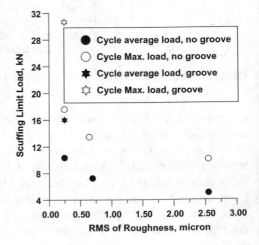

Fig.14 Effects of clearance and groove on scuffing resistance

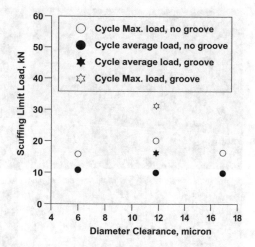

The average scuffing load decreases slightly with increasing clearance, while the maximum scuffing load first increases and then decreases slightly with increasing clearance.

A circumferential 360° groove may increase the scuffing resistance significantly due to its better lubricating and cooling effects. Under the test conditions, only a very small part of feeding oil can flow into the bearings without the circumferential groove due to the very small radial clearance, while almost all the feeding oil can flow into the bearings with the circumferential groove. The higher the bulk temperature, the lower the scuffing resistance. Because the feed temperature is very low compared with the flash temperature, feed temperature has a very weak effect on the scuffing resistance for the case without the oil groove due to the very small oil flow, while it has a significant effect for the case with the oil groove.

Figure 15a–d show the scanning electron microscope (SEM) micrographs of the piston bore surfaces in mild and severe wear, as well as a scuffing area, for the originally smooth and rough bore surfaces. The comparison between Fig. 15a and c shows that, in the area where a mild wear occurs, the originally rough surface yields more material flow and tearing, as well as more visible agglomerate debris. Figures 15b and d show that on the severely worn surfaces, there are a few shallow abrasive grooves. In the scuffing regimes, a more heavily deformed stratified patch is observed in Fig. 15b than in Fig. 15d. This observation suggests that roughness shortens the time for the transition from severe wear to scuffing. Obviously, roughness has a considerable effect on the wear process and scuffing.

Figures 16 and 17 show the optical and SEM micrographs of the cross sections of the tested piston bores along the sliding direction. In the cross sections shown in Fig.16a–c, the brightest, darker, and darkest parts are the aluminum substrates, the primary silicon particles, and cracks, respectively. An examination of Fig. 16a, b suggests that under the scuffed area, a transformed layer and a plastically deformed region may appear. Because of the cyclic loading and the oscillating motion of the

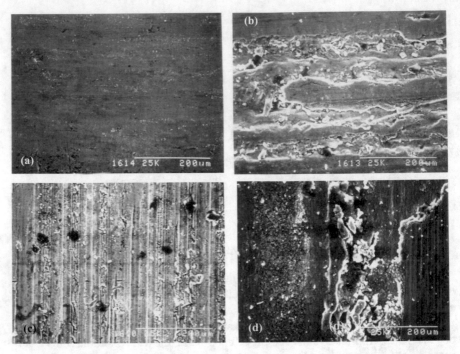

Fig. 15 SEM micrograph of a piston bore surface after scuffing: **a** smooth original surface (#XB-1 1), mild wear regime; **b** smooth original surface (#XB-11), severe wear and scuffing regimes; **c** rough original surface (#XER-4), mild wear regime; **d** rough original surface (#XER-4), severe wear and scuffing regimes

Fig. 16 Optical micrograph of a cross section of a piston bore after scuffing: **a** shear and powder bands; **b** stratified bands containing extensively fragmented silicon particles; **c** transfer layer

pin, the piston bores undergo cyclic stress variations. As a result, the primary larger silicon particles are fragmented into smaller particles, as shown in Fig. 17a, some of which becomes powder bands parallel to the surface as a result of the plastic flow in the sliding direction, as shown in Fig. 16a, b. Voids and fatigue cracks might have nucleated at the interface between the fragmented silicon particles and the Al matrix and propagated along the intersection of the shear and powder bands. They are also observed in the stratified bands containing fragmented silicon particles and in the transfer layer as shown in Figs. 16a, b and 17a, d. Some surface-originated cracks stem from surface asperities as shown in Fig. 17d. Cracks weaken the surface

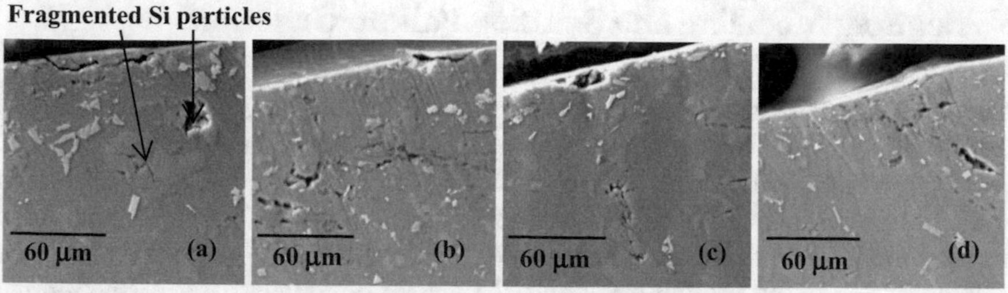

Fragmented Si particles

Fig. 17 SEM micrograph of cross sections of a piston bore after scuffing: **a** fragmentation of primary silicon particles; **b** crack nucleation and propagation along the Si particle-Al matrix interfaces and plastic flow; **c** connection of cracks among neighbor fragmented Si particles; **d** crack propagation into transfer layer

and substrate and release debris from the base alloy subsurface into the interface, as shown in Fig. 16c, which are embedded again into the surface. It is obvious that the fatigue cracks strongly affect the wear and scuffing resistance of the material. In other words, the deterioration of the material strength by fatigue cracks contributes to the final scuffing.

The loss of all of the successive protective layers of the sliding asperity finally results in scuffing. These protective films include the lubricant film generated by the microelastohydrodynamic action, an adsorbed or reactive surface film, and an oxide layer. Asperity sliding friction may also play an important role in the propagation of surface scuffing by producing frictional heating. Scuffing may initiate on a single asperity when its flash temperature exceeds a critical temperature. Microscuffing occurs if the single-asperity scuffing does not propagate much beyond its own initial contact area. When the thermal influence of the scuffed asperity raises its neighboring asperity temperatures to a critical value, those asperities would be affected and more asperities may scuff. The thermal influence of newly scuffed asperities also affects their neighboring asperities, and then scuffing may propagate to the entire contact area, resulting in scuffing failure, i.e., macroscuffing.

On the other hand, the deterioration of the material strength by the flash temperature and the occurrences of material transfer, plastic deformation, fatigue cracks, and accumulated damage in the surface and substrate contribute to the final scuffing, although work hardening may increase the material strength [24]. Such material deterioration is especially important for the piston-pin/bore bearing because the contact is under the oscillating motion of the connecting rod, and the contacting materials are subjected to cyclic stresses. Scuffing would occur when the shear stresses in a certain substrate volume exceed the combined bulk shear strength there. Therefore, scuffing is governed by a group of mechanisms in the complex interaction among the phenomena mentioned above.

5 Scuffing Modeling and Scuffing Failure Criterion

The preceding discussion shows that shear stresses in a certain substrate volume and the combined bulk shear strength are two major parameters to predict scuffing failure. However, it is difficult to calculate the real stress in the worn Al-Si material, which is altered by wear and other deterioration effects mentioned before. A simplified scuffing model for industrial applications is proposed here. This model involves the journal motion in the bearing clearance, the degree of contact between the journal and the bushing, the asperity contact pressure, the maximum flash temperature, and the surface tangential traction.

A journal center orbit describes the journal motion in the bearing clearance volume. It can be used to determine the degree of contact between the journal and the bushing. A number of numerical methods are available for the center-orbit determination [25–29]. Booker's mobility method [25] is a simple and convenient method. Goenka [26] presents a set of curve-fitting formulas for the mobility vectors of finite journal bearings. The mobility concept [25] and Goenka's [26] curve fit method for bearing analysis which is programmed by Shen [28] are used here. The assumptions are that the lubricant is isothermal, incompressible, and Newtonian; the surfaces of journal and bearing are rigid, and the misalignment is neglected.

The contact model developed by Polonsky and Keer [30] is used to determine the relationship between the average contact pressure and the average gap. The local asperity temperature, i.e. the maximum asperity temperature composed of bulk temperature and asperity flash temperature, has been recognized as a major contributing factor to wear deterioration due to its influences on mechanical properties of materials, surface chemical composition, and lubricant-surface interaction. Classical theories of flash temperature focus on the average surface temperature using the average friction [31]. Although micro flash temperature and its distribution can be calculated at high computing cost [32], the simple equations for estimating the average value of the flash temperature by Kuhlmann-Wilsdorf [31] are used here in the way suggested by Wang and Cheng [33, 34]. The maximum oscillating speed and the maximum asperity contact pressure corresponding to the minimum nominal film thickness are utilized here to calculate the maximum heat source in the Kuhlmann-Wilsdorf's equations.

The temperature-dependent shear strength of Al-Si alloy is shown by the solid dots in Fig. 18. Although a high sliding velocity induces a high local shear strain rate, raising the hardness and hence the shear strength, working hardening is not considered here. The material strength at the critical spots is determined using the calculated maximum asperity temperature.

Based on the suggested scuffing mechanism, an improved criterion for scuffing failure is proposed as follows, indicating that scuffing failure would occur when the maximum surface tangential traction is larger than the modified shear strength:

$$\tau_{sm} \geq f S_m \tag{1}$$

Fig. 18 Max. tangential traction and bulk shear strength versus max. flash temperature, surface roughness, and frictional coefficient

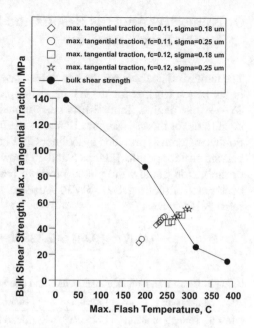

here τ_{sm} is the maximum surface tangential traction determined by the product of the frictional coefficient and the maximum calculated average contact pressure, S_m is the shear strength of the piston material corresponding to the maximum asperity temperature without considering working hardening, f is a scuffing factor designated as follows:

$$f = \tau_{sm}^* / S_m^* \tag{2}$$

where τ_{sm}^* and S_m^* are the critical values of τ_{sm} and S_m, which correspond to the maximum asperity temperature when scuffing starts without considering working hardening. The scuffing factor reflects the interaction among the related mechanisms of wear and the inaccuracy in modeling. With Eq. (2), a set of f can be determined using τ_{sm}^* and S_m^* corresponding to the individually measured initial scuffing loads, the realistic piston and piston pin geometries and their material properties, and the actual operating conditions.

A set of the contour plots of the scuffing factors in the scuffing failure mapping show scuffing failure regions, the variation ranges of the parameters of the frictional coefficients, material properties, geometrical characteristics, roughness, lubricant, and operating conditions, etc.

6   Numerical Analysis Results and Discussions for the Bench Test Case

The numerical analysis for the bench test case, which uses the operating conditions in the bench tests, is for the XB10 piston-pin/bore bearing case in Ford 4.6L V8 2V 4-stroke engine. Table 2 lists bearing and lubricant parameters, and operating conditions for this case. Figure 18 shows the bulk shear strength versus temperature for this material as provided by Zollner Co. Figure 19 shows the measured running in bearing surface profile. Figures 5 and 20 show the oscillating surface linear velocity of the piston pin and the piston pin/bore bearing load measured just prior to the occurrence of scuffing. SAE-5W30 is used as lubricant and its viscosity-temperature relation is taken as:

$$lg(lg(v + 0.6)) = 7.82841 + (-3.03329 \times lgT) \tag{3}$$

Table 2 Bearing and lubricant parameters, and operating conditions for the bench test case

Bearing diameter	22 mm	Bearing length	16.175 mm
Bearing Young's modulus	71 GPa	Bearing Poisson's ratio	0.333
Pin Young's modulus	206 GPa	Pin Poisson's ratio	0.292
Typical asperity width	10 μm	Dry friction coefficient	0.15
Crankshaft speed	1500 rpm	RMS asperity height	0.18 μm
Lubricant	SAE-5W30	Bulk and operating oil temperature	110 °C
Bearing thermal expansion coefficient			10.8 mm/(mm·°C)
Pin thermal expansion coefficient			23.94 mm/(mm·°C)

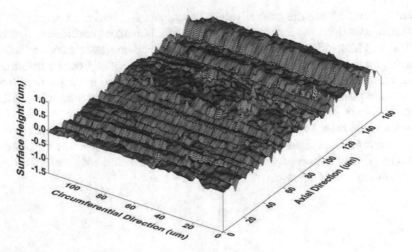

Fig. 19 Measured piston bore surface after running in

Fig. 20 Bearing load (bench test case)

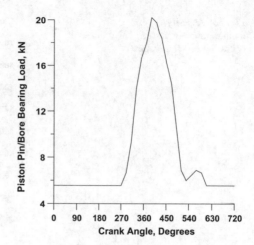

where v is viscosity in cSt, T is temperature in deg K.

Figures 21 and 22 show effects of operating oil temperature on minimum nominal film thickness, average and maximum asperity temperature rise, respectively. Figures 23, 24, and 25 show effects of nominal radial clearance on evolution of minimum nominal film thickness, eccentricities, and maximum nominal film pressure versus crank angle, respectively. Table 3 summarizes the numerical analysis results for the bench test case.

Figure 23 presents a typical variation of the minimum nominal film thickness in one load cycle. The valley value of the minimum film thickness in a cycle corresponds to the peak value of the applied load. Comparison between Figs. 20 and 23 reveals that in one load cycle, mixed lubrication occurs at the peak load. The existence of the mixed lubrication may be explained by the minimum film thickness in Fig. 23, which is about the same in magnitude as the roughness of the bearing surface. Such an

Fig. 21 Min. nominal film thickness versus operating oil temperature

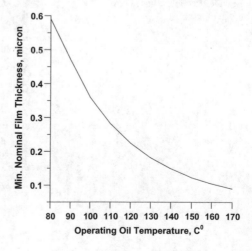

Fig. 22 Asperity
temperature rise versus
operating oil temperature

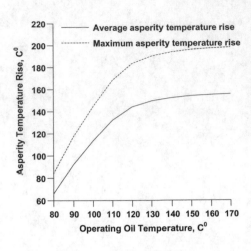

Fig. 23 Min. nominal film
thickness in one cycle of load

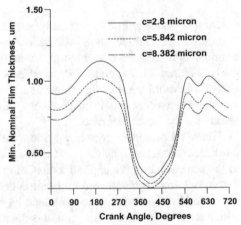

Fig. 24 Eccentricities in one
cycle of load

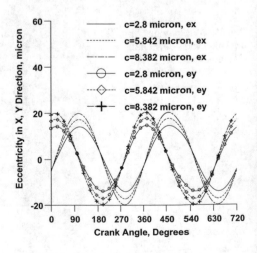

Fig. 25 Max. film pressure in one cycle of load

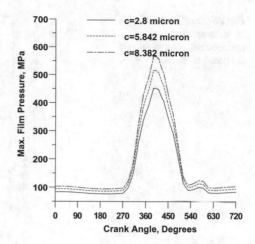

Table 3 Numerical analysis results for the bench test case (C—nominal bearing radial clearance; h_{min} - minimum nominal film thickness; I1—Average temperature rise; I2—Max. temperature rise; I3—Average flash temperature; I4—Max. flash temperature; I5—Max. tangential traction)

C, μm	h_{min}, μm	I1, °C	I2, °C	I3, °C	I4, °C	I5, MPa
2.8	0.328	124.9	159.0	234.9	269.0	52.47
5.842	0.283	132.2	168.3	242.2	278.3	60.50
8.382	0.253	135.6	172.7	245.6	282.8	62.88

argument agrees well with the pictures of the scuffed bore surfaces given in Fig. 12, where obvious wear and scuffing marks exist only near the center of the lower half of the bearing surface.

Figures 23, 24 and 25 and Table 3 show that with an increase in the nominal radial clearance, the minimum film thickness decreases, and the maximum film pressure and maximum flash temperature all increase.

Figure 18 and Table 3 show that for the piston XB10 with its actual nominal radial clearance 5.842 μm, the calculated maximum tangential traction (60.5 MPa) at scuffing is just slightly larger than the bulk shear strength at the calculated maximum flash temperature.

Figures 18 and 26 show the calculated maximum surface tangential traction, measured bulk shear strength, and the scuffing factor at the nominal bearing clearance versus the maximum asperity temperature for different coefficients of friction and roughnesses. These results are calculated based on the individually measured scuffing load, actual bearing geometric parameters, and the operating conditions of the bench tests. From these figures, one can see that the maximum surface tangential traction, the maximum asperity temperature, and the scuffing factor increase significantly with the increases in roughness and the coefficient of friction. The scuffing factor varies from about 0.3 to about 1.7 with the increase in the asperity temperature. The variation of the scuffing factor suggests that the interaction among the factors that

Fig. 26 Scuffing factor
versus max. flash
temperature, roughness, and
frictional coefficient

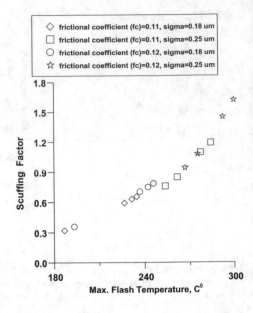

influence wear and scuffing, as well as modeling errors, is different for different test
cases. Figure 26 also shows that the higher the maximum asperity temperature, the
larger the scuffing factor. It reveals that the direct comparison between the maximum
surface tangential traction and asperity temperature-dependent shear strength without
considering the scuffing uncertainty is insufficient to predict scuffing failure.

Figure 27 presents the scuffing failure mapping for the bench test case. It shows that
for the same bore surface roughness and the given nominal bearing radial clearance,
the value of the scuffing factor increases with the increase in the frictional coefficient

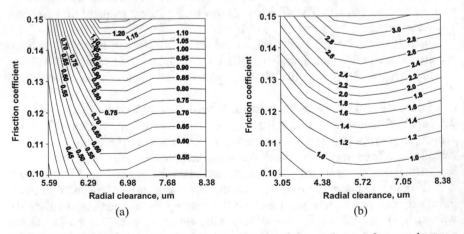

Fig. 27 Scuffing failure maps (bench test case)—scuffing factor vs bore surface roughness σ,
nominal bearing radial clearance C and frictional coefficient μ: **a** σ = 0.18 μm; **b** σ = 0.25 μm

to avoid scuffing, indicating that the upper region of the contour plot of the scuffing factor is unsafe, i.e. scuffing failure, while the lower region of that contour plot, the variation ranges of the parameters of the frictional coefficients, material properties, geometrical characteristics, roughness, lubricant, and operating conditions, etc., is safe, i.e. unscuffing failure. Comparison between Fig. 27a and b reveals that for the same nominal bearing radial clearance and frictional coefficient, increasing the bore surface roughness results in the increase in the value of the scuffing factor. The reason is that bigger frictional coefficient or bore surface roughness means tighter contact, bigger surface tangential traction, higher flash temperature, and smaller shear strength of the material, resulting in bigger value of the ratio of the maximum surface tangential traction to the shear strength of the material, i.e. the scuffing factor.

The data plotted in Figs. 25, 26 and 27 are the results averaged from the bench test under the same bearing geometry and operating conditions. The asperity frictional coefficient depends on the material pair and operating conditions, which is considered to be 0.1–0.2 for the current scuffing modeling based on knowledge about the frictional coefficients of the Al-Si alloy and steel in mixed lubrication and initial scuffing [16, 35].

7 Numerical Analysis Results and Discussions for the Engine Case

The numerical example chosen here for the engine case, which corresponds to actual engine operating conditions, is also for XB10 piston-pin/bore bearing case in Ford 4.6L V8 2V 4-stroke engine. Table 4 lists parameters for engine, bearing, lubricant, and operating conditions. The bearing surface profile in Fig. 28 amplified from that shown in Fig. 19, the oscillating surface linear velocity of the piston pin shown in Fig. 8, and the piston pin/bore bearing load shown in Fig. 9 are used here.

Table 4 Engine, bearing, lubricant parameters, and operating conditions for the engine case

Bearing diameter	22 mm	Bearing length	16.175 mm
Bearing Young's modulus	71 GPa	Bearing Poisson's ratio	0.333
Pin Young's modulus	206 GPa	Pin Poisson's ratio	0.292
Typical asperity width	10 μm	Dry friction coefficient	0.15
Crankshaft speed	4600 rpm	RMS asperity height	0.18 μm
Crankshaft radius	45 mm	Connecting rod length	150.7 mm
Lubricant	SAE-5W30	Bulk and operating oil temperature	110 °C
Bearing thermal expansion coefficient			10.8 mm/(mm °C)
Pin thermal expansion coefficient			23.94 mm/(mm °C)

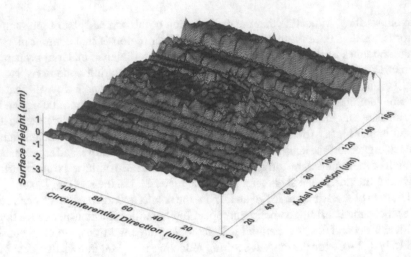

Fig. 28 Piston bore surface profile used for $\sigma = 0.4\ \mu$m

Figures 29, 30 and 31 show effects of nominal radial clearance and roughness on the minimum nominal film thickness, maximum film pressure, and eccentricity respectively. Table 5 summarizes the numerical analysis results.

Figures 29, 30 and 31 and Table 5 show that with an increase in the nominal radial clearance, the minimum film thickness decreases, and the maximum film pressure increase significantly, while the asperity temperature rise and maximum tangential traction only change mildly. With an increase in roughness, the minimum film thickness increases due to the contribution of asperity contact load and the maximum film pressure decreases quite significantly. However, the asperity temperature rise and maximum tangential traction only change mildly.

Figure 32 shows the scuffing failure mapping for the engine case. Figure 32a, d

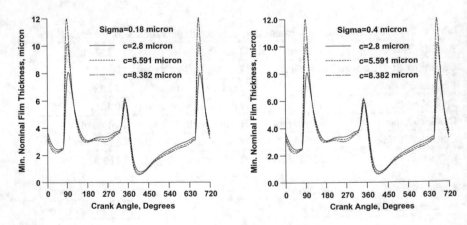

Fig. 29 Effects of bore roughness and nominal radial clearance on min. nominal film thickness

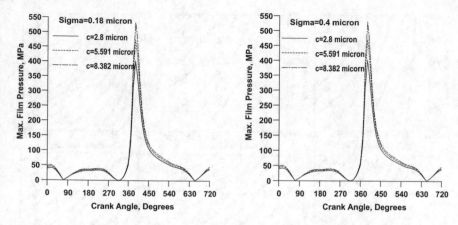

Fig. 30 Effects of bore roughness and nominal radial clearance on max. film pressure

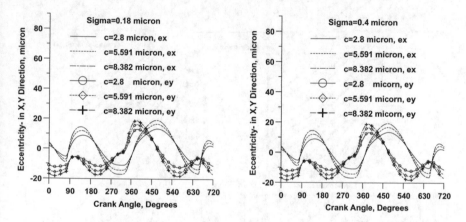

Fig. 31 Effects of bore roughness and nominal radial clearance on eccentricity (engine case)

Table 5 Numerical results for the engine case (C—nominal bearing radial clearance; σ—RMS asperity height; h_{min}—minimum nominal film thickness; I1—Average temperature rise; I2—Max. temperature rise; I3—Average flash temperature; I4—Max. flash temperature; I5—Max. tangential traction)

C, μm	σ, μm	h_{min}, μm	I1, °C	I2, °C	I3, °C	I4, °C	I5, MPa
2.8	0.18	0.773	0	0	0	0	0
5.591	0.18	0.646	128.5	163.7	238.5	273.7	30.2
8.382	0.18	0.567	174.9	222.6	284.9	332.6	35
2.8	0.4	0.790	547.4	696.9	657.4	806.9	65
5.591	0.4	0.668	573.1	729.7	683.1	839.7	71.8
8.382	0.4	0.593	583.1	742.4	693.1	852.4	72.9

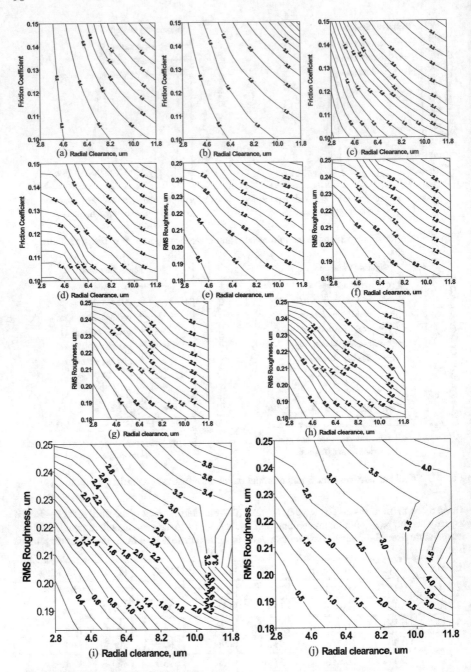

Fig.32 Scuffing failure maps (engine case)—scuffing factor vs bore surface roughness σ, nominal bearing radial clearance C, and frictional coefficient μ: **a** σ = 0.18 μm; **b** σ = 0.20 μm; **c** σ = 0.22 μm; **d** σ = 0.25 μm; **e** μ = 0.1; **f** μ = 0.11; **g** μ = 0.12; **h** μ = 0.13; **i** μ = 0.14; **j** μ = 0.15

show the same features and trends as in Fig. 27a, b. Figure 32e, f show that for the same frictional coefficient and the given nominal bearing radial clearance, increasing the bore surface roughness increases the value of the scuffing factor to avoid scuffing. Comparison between Fig. 32e and f reveals that for the same nominal bearing radial clearance and bore surface roughness, increasing frictional coefficient results in the increase in the value of the scuffing factor to avoid scuffing. Figure 32g, h show that when $C > 10$ μm and $\mu > 0.14$, the value of the scuffing factor does not always increase monotonically with the increase in the bore surface roughness to avoid scuffing. It is mainly due to the comprehensive effects of the bearing radial clearance and the bore surface roughness on a journal center orbit. Therefore, in order to build the more accurate and effective scuffing failure mapping, more related important parameters of material property, geometry, lubricant, and operating conditions etc. are needed and their number and values of each parameter needs to be optimized, and more accurate numerical models and better numerical tools should be used.

8 Some Typical Measures to Mitigate Scuffing Failure

The first mitigation measure is the floating pin design, in which there are two types of bearing systems: one is the piston wrist pin and the piston pin bores; the other is the piston wrist pin and the small bearing of the connecting rod. When scuffing initiation occurs on either bearing, the higher friction force will stop the pin's rotational motion relative to that bearing; however, the other bearing can still operate effectively. On the scuffed bearing, due to no sliding motion, the working condition becomes less severe and scuffing may not propagate due to a halt of the sliding motion. The piston wrist pin will fail only if both bearings scuff. Therefore, the floating pin has a superior scuffing resistance over fixed pins [7, 8].

The second mitigation measure is an innovative electro-erosion machining process for the critical high-precision interface between the piston and the pin of heavy-duty diesel engines [36]. Wide variations in thermal and mechanical load during the engine operating cycle necessitate complex geometry for the pin bore to maintain satisfactory contact conditions. Therefore, precise control of the dimensions, shape, and texture of the bore is critical. Very precise profiling of the pin bore to distribute the load over a sufficient contact area can prevent the piston and pin from running 'steel-on-steel,' which risks scuffing and joint seizure. Traditional machining methods are unable to maintain the required tolerances beyond a very short batch size, typically 25 parts, because of rapid tool wear. This introduces a heavy overhead burden due to frequent tool replacement, machine resetting, and the sacrifice of the first part in every batch during setup. The lack of robustness in the process dictates 100% inspection and the scrapping of non-compliant parts. The geometry of the pin bore causes further problems: the profile required is not cylindrical but trumpet-shaped towards the inner and outer end of each boss. Also, the shape is not circumferentially continuous throughout its length leading to an interrupted cut that can cause further tool vibration. These limitations mean that the typical geometric surface tolerance is 10 to 12 microns for such a difficult form, or 6–8 μm under the very best conditions. However, this application requires a form tolerance of just a few microns, to ensure that every piston functions reliably and avoids the early-life failure known as 'infant mortality'. High-precision electro-erosion machining can entirely eliminate conventional metal cutting from the profiling operation. It removes material electrochemically: the workpiece becomes the anode of a DC circuit, the shaped tool the cathode, and a low-cost salt-water electrolyte completes the circuit and conveys the eroded material away from the workpiece. No significant heat is introduced during the process, so dimensional stability and workpiece material properties remain unaffected. In fact, fully hardened material presents no additional challenge, in contrast with conventional machining. Cycle time per piece for this application is only half that of conventional machining, because although the mounting and alignment take a little longer, the metal removal time is much shorter. In-process monitoring ensures that tolerances are held throughout the production run, largely eliminating the inspection burden associated with traditional methods. Because no tools contact the workpiece, there is no deflection, no chatter, no tool wear, and no production disruption to change tools. Furthermore, the capital investment required for a piston pin bore line can be one-third to one-half as much as a conventional high precision boring facility.

The last mitigation measure is three different pin's surface treatments: CrN and diamond-like carbon (DLC) coatings applied to standard pins directly over the carburized surface using physical vapor deposition (PVD) by closed field unbalanced magnetron sputtering, and laser surface texturing (LST) [6, 37]. The standard pins were carburized steel with a minimum case depth of 0.4 mm and a minimum surface hardness of Rockwell 15 N 89. The CrN coating was 1.4 μm thick, had a modulus of 235 GPa and a hardness of 16 GPa. The DLC coating contained 11 at.% chromium, was 2.2 μm thick, had a modulus of 70 GPa and a hardness of 8 GPa. The LST texture consisted of shallow spherical shape dimples having diameter of about 100 μm, depth of 3 μm and area density of about 10%. The LST pins were lapped by a special soft pad lapping (SPL) technology to remove bulges that are formed around the dimples during the laser texturing. The treated pins had a measured surface roughness of Ra 0.02 μm while the standard pins had Ra of 0.04 μm. All the three treated pins gave lower friction coefficient compared to the reference standard pin. DLC coated pin gave lower friction coefficient than CrN coated pin and the LST performed best. In case of with SN 90 base oil, scuffing load with the LST pin had 38% improvement over the reference standard pin and CrN coating, and 20% improvement over the DLC coating.

9 Conclusions

A bench rig is designed and built which can simulate the kinetic and dynamic characteristics of the pin/bore bearings in an engine, providing reliable measurements of scuffing/seizure load, bulk temperature, and frictional torque. By using this rig, effects of surface roughness, clearance, circumferential groove, and oil-feed on scuffing/seizure resistance can be studied for four different combinations of constant or oscillating pin speed under static or dynamic loads. Based on the experimental observations, the following conclusions can be drawn:

For the case of constant pin speed and dynamic load, seizure occurs due to rapid thermal expansion of the pin. For the case of oscillating pin speed and dynamic load, seizure usually does not occur and predominate mode of failure is scuffing.

Scuffing resistance is reduced significantly by an increase in piston bore roughness but changes slightly with an increase in piston pin/bore clearance. The addition of a circumferential groove is found to have a marked improvement in scuffing resistance due to its better lubricating and cooling effects.

For the scuffed area, the subsurface is divided into three characteristic regions: transformed layer, plastically and elastically deformed regions. The large silicon particles are fragmented into smaller particles, some of which near surface becomes power bands. Voids and fatigue cracks nucleate and propagate next to silicon particles and along the intersection of shear bands and in the radial direction during the scuffing process. The surface-originated cracks stem from direct contact of surface asperities and have significant effects on bearing fatigue. The deterioration of the surface shear strength by fatigue cracks under a high surface temperature appears to contribute to the final scuffing of the pin-piston contact.

A scuffing criterion showing that the scuffing occurs when the surface tangential traction exceeds the effective bulk shear strength of a surface layer, a scuffing factor reflecting the influence of factors that affect wear and scuffing and the inaccuracy in modeling, and a scuffing failure mapping, which indicates the variation ranges of the parameters of material property, geometry, lubricant, and operating conditions etc. for scuffing failure, are developed, validated, and appear to be acceptable for both of scuffing failure diagnosis and parameter design of friction pair, lubricant, and operation conditions.

The scuffing factor increases as the maximum asperity temperature, frictional coefficient, and roughness increase. The effect of the radial clearance on the scuffing factor is not linear, and the trends of variations correspond to those of the change of the minimum film thickness. Optimization of the parameters of material property, geometry, lubricant, and operating conditions and application of more accurate numerical models are needed to build more accurate and effective scuffing failure mapping.

The floating pin design, an innovative electro-erosion machining process for the critical high-precision interface between the piston and the pin, and three different pin's surface treatments: CrN and DLC coatings, and LST are typical measures to mitigate scuffing failure.

References

1. Castro, J., Seabra, J.: Scuffing and lubricant film breakdown in FZG gears Part 2. New PV scuffing criteria, lubricant and temperature dependent. Wear **215**, 114–122 (1998)
2. Hershberger, J., Ajayi, O.O., Zhang, J., Yoon, H., Fenske, G.R.: Evidence of scuffing initiation by adiabatic shear instability. Wear **258**, 1471–1478 (2005)
3. Lee, Y.-Z., Kim, B.-J.: The influence of the boundary lubricating conditions of three different fluids on the plastic fatigue related mechanisms of wear and scuffing. Wear **232**, 116–121 (1999)
4. Sheiretov, T., Yoon, H., Cusano, C.: Scuffing under dry sliding conditions—Part I: experimental studies. STLE Tribol. Trans. **41**(3), 435–446 (1998)
5. He, X.: Experimental and analytical investigation of the seizure process in Al-Si alloy/steel tribocontacts, Ph.D. Dissertation, University of Northwestern University (1998)
6. Etsion, I., Halperin, G., Becker, E.: The effect of various surface treatments on piston pin scuffing resistance. Wear **261**, 785–791 (2006)
7. Abed, G., Zou, Q., Barber, G., Zhou, B., Wang, Y., Liu, Y., Shi, F.: Study of the motion of floating piston pin against pin bore. SAE International, 01–1215 (2013)
8. Zhang, R., Zou, Q., Barber, G., Zhou, B., Wang, Y.: Scuffing test rig for piston wrist pin and pin bore. SAE International, 01–0680 (2015)
9. Kobayashi, T.: Prediction of piston skirt scuffing via 3D piston motion simulation. SAE International, 01–1044 (2016)
10. Obert, P., Müller, T., Füßer, H.-J., Bartel, D.: The influence of oil supply and cylinder liner temperature on friction, wear and scuffing behavior of piston ring cylinder liner contacts—a new model test. Tribol. Int. **94**, 306–314 (2016)
11. Wan, S., Li, D., Zhang, G., Tieu, A.K., Zhang, B.: Comparison of the scuffing behaviour and wear resistance of candidate engineered coatings for automotive piston rings. Tribol. Int. **106**, 10–22 (2017)
12. Tas, M.O., Banerji, A., Lou, M., Lukitsch, M.J., Alpas, A.T.: Roles of mirror-like surface finish and DLC coated piston rings on increasing scuffing resistance of cast iron cylinder liners. Wear **171**, 1558–1569 (2017)
13. Sangharatna, M.R., Chelladurai, H.A.: Diagnosis of liner scuffing fault of a diesel engine via vibration and acoustic emission analysis. Tribol. Online **15**(1), 9–17 (2020)
14. Blok, H.: Theoretical study of temperature rise at surfaces of actual contact under oiliness lubricating conditions. In: Proceedings of the General Discussion on Lubrication and Lubricants in London, 13–15 Oct. The Institute of Mechanical Engineers, pp. 222–235 (1937)
15. Cuotiongco, E.C., Chung, Y.W.: Prediction of scuffing failure based on competitive kinetics of oxide formation and removal: application to lubricated sliding of AISI 52100 steel on steel. STLE Tribol. Trans. **37**, 622–628 (1994)
16. Somi Reddy, A., Pramila Bai, B.N., Murthy, K.S.S., Biswas, S.K.: Wear and seizure of binary Al-Si alloys. Wear **171**, 115–127 (1994)
17. Biswaas, S. K.: Sliding wear of materials. In: Proceeding of the First Word Tribology Conference, pp. 159–175 (1997)
18. Greenwood, J.A., Williamson, B.P.: Contact of nominally flat surfaces. Proc. R. Soc. Lond. Ser. A. **295**, 300–319 (1966)
19. Hirst, W., Hollander, A.E.: Surface finish and damage in sliding. Proc. R. Soc. Lond. Ser. A. **337**, 379–394 (1974)
20. Sheiretov, T., Yoon, H., Cusano, C.: Scuffing under dry sliding conditions—Part II: theoretical studies. STLE Tribol. Trans. **41**(3), 447–458 (1998)
21. Yoon, H., Sheiretov, T., Cusano, C.: Scuffing of area contacts under starved lubrication conditions. STLE Tribol. Trans. **44**, 435–446 (2001)
22. Zhang, C., Cheng, H.S., Qiu, L., Knipstein, K.W., Jay, B.: Scuffing behavior of piston-pin/bore bearing in mixed lubrication—Part I: experimental studies. STLE Tribol. Trans. **46**(2), 193–199 (2003)

23. Zhang C., Cheng, H.S., Wang, Q.J.: Scuffing behavior of piston-pin/bore bearing in mixed lubrication—part II: scuffing mechanis,m and failure criterion. STLE Tribol. Trans. **47**, 149–156 (2004)
24. Lim, S.C., Asheby, M.F.: Wear-mechanism maps. Acta Metall. **35**(1), 1–24 (1987)
25. Booker, J.F.: Dynamically loaded bearings: mobility method of solution. ASME J. Basic Eng. 537–546 (1965)
26. Goenka, P.K.: Analytical curve fits for solution parameters of dynamically loaded journal bearings. ASME J. Tribol. **106**, 421–428 (1984)
27. Zhang, C., Zhang, H., Qiu, Z.: Fast analysis of crankshaft bearings with a database including shear thinning and viscoelastic effects. STLE Tribol. Trans. **42**(4), 922–928 (1999)
28. Shen, M.C.: A computer analysis of lubrication of dynamically loaded journal bearings including effects of asperity contacts. Master thesis, Northwestern University (1986)
29. Paranjpe, R.S., Han, T.Y.: A transient thermohydrodynamic analysis including mass conserving cavitation for dynamically loaded journal bearings. ASME Journal of Tribology **117**, 369–378 (1995)
30. Polonsky, I.A., Keer, L.M.: A fast and accurate method for numerical analysis of elastic layered contacts. ASME J. Tribol. **122**, 30–35 (2000)
31. Kuhlmann-Wilsdorf, D.: Temperatures at interfacial contact spot: dependence on velocity and on role reversal of two materials in sliding contact. ASME J. Tribol. **109**, 321–329 (1987)
32. Qiu, L., Cheng, H.S.: Temperature rise simulation of three-dimensional rough surfaces in mixed lubricated contact. ASME J. Tribol. **120**, 310–318 (1998)
33. Wang, Q., Cheng, H.S.: A mixed lubrication model for journal bearings with a soft coating, Part I: contact and lubrication analysis. STLE Tribol. Trans. **38**, 654–662 (1995)
34. Wang, Q., Cheng, H.S.: A mixed lubrication model for journal bearings with a soft coating, part II: flash temperature analysis and its application to tin coated Al-Si bearings. STLE Tribol. Trans. **38**, 517–524 (1995)
35. Wang, Y., Zhang, C., Wang, Q., Lin, C.: TEHD analyses of journal bearing under severe operating conditions. Tribol. Int. **35**, 395–407 (2001)
36. William J. Z.: Electrochemical process machines complex steel pistons. Adv. Mater. Process. **26** (2010)
37. Etsion, I.: Improving tribological performance of mechanical components by laser surface texturing. Tribol. Lett. **17**, 733–737 (2004)

A Review on the Rotor Dynamics of Automotive Turbochargers

Thales Freitas Peixoto ⓘ and Katia Lucchesi Cavalca ⓘ

Abstract Since the Industrial Revolution (1760–1820), 2040 gigatonnes of carbon dioxide (CO_2) were emitted to the atmosphere and the half of these emissions occurred between 1970 and 2010. Worldwide, greenhouse gas (GHG) emissions increased over 1970–2010, with larger absolute increases between 2000 and 2010 and 14% of this total amount was released due to transportation. Furthermore, 65% of GHG emissions are related to CO_2 and, in order to mitigate its effect on global warming, stricter legislations lowered the CO_2 emission limits for automobiles. To achieve these stricter emission limits, the strategy of engine downsizing with the use of a turbocharger has been adopted by the automotive industry. The turbocharger is characterized as a low weight and high-speed rotor, achieving rotational speeds as high as 350 krpm. To withstand these extremely high rotating speeds, the shaft is commonly supported by floating ring bearings, leading to complex rotor dynamic phenomena. This work summarizes the rotor dynamics of automotive turbochargers, presenting its expected, general behavior, along with a literature review on the most important topics regarding turbocharger dynamic analyses. The nonlinear effect of the floating ring bearing on the shaft lateral vibrations is discussed, along with the effect of different bearing systems on the turbocharger response, including the thrust bearing effect on the rotor axial and lateral dynamics. Current researches on turbocharger modelling and investigation are also presented. Most works rely on developing high fidelity models with low computational costs, including several different effects, such as temperature variations and mass-conserving cavitation algorithms in the lubricated bearings, investigations on newer geometries of the floating ring and thrust bearings and optimal solutions to reduce friction losses.

Keywords Automotive turbocharger · Floating ring bearing · Thrust bearing · Rotor dynamics · Hydrodynamic lubrication · Oil whirl/whip

T. F. Peixoto (✉) · K. L. Cavalca
School of Mechanical Engineering, University of Campinas, Campinas, Brazil
e-mail: thalesfp@fem.unicamp.br

© The Author(s), under exclusive license to Springer Nature Switzerland AG 2022
T. Parikyan (ed.), *Advances in Engine and Powertrain Research and Technology*,
Mechanisms and Machine Science 114,
https://doi.org/10.1007/978-3-030-91869-9_4

97

1 Introduction

The use of turbochargers on downsized engines has become the new standard on automotive engines. The current practice of engine downsizing focuses on higher performance, using smaller engines to provide the same power of larger ones, ultimately reducing fuel consumption, as requested by consumers, and to abide to emission regulations, as required by government laws [1]. This ever-lowering demand for reductions on emissions and fuel consumption is achieved with smaller engine displacements or fewer numbers of cylinders, reducing the vehicle weight and resulting in lower friction between pistons and cylinders, which decreases the driving losses [2]. Less fuel consumption translates to less greenhouse gas emissions, such as carbon dioxide (CO_2) and nitrogen oxides (NO_x). Nonetheless, it also translates to smaller engine power. A boosting device increases the engine specific power, and the turbocharger can use part of the energy contained in the engine hot exhaust gases, increasing the engine efficiency [3].

The average energy flow in a typical passenger car is illustrated in Fig. 1a. Only 10–15% of the energy from the fuel is used for driving power, while the rest of the available energy can be either required for thermal cooling (27–33%), lost due to friction or load changes (10–14%) or used to power control devices, electric motors and entertainment devices (2–4%). In addition, a great part of the total energy (30–35%) is lost in the thermal exhaust gases. The hot exhaust gases temperature may reach 820–850 °C for diesel and 950–1050 °C for gasoline, so its enthalpy energy is expressively high, but completely lost to the environment [3]. The turbocharger (TC) uses part of this energy to boost the smaller engine and increase its specific power.

The exhaust-gas turbocharger (TC), schematically illustrated in Fig. 2, comprises a core unit CHRA (center housing and rotating assembly), turbine, compressor and actuator. The compressor wheel (CW) and turbine wheel (TW) are mounted on the rotating shaft, which is supported by a pair of journal bearings, to sustain the lateral vibrations inherently to every rotating system, and a double-acting thrust bearing, to sustain an axial force imbalance from the gas flows in the CW and

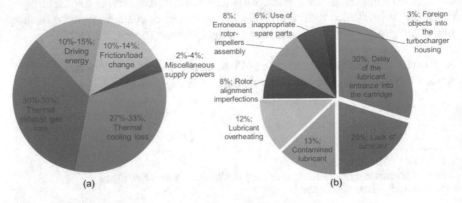

Fig. 1 a Energy flow of passenger car [3] and b turbochargers failure causes [4]

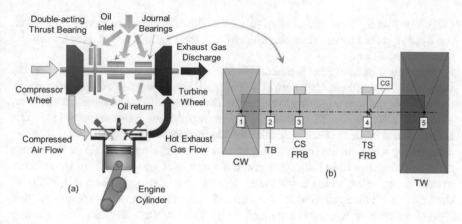

Fig. 2 **a** Scheme of turbocharger supported by thrust and journal bearings charging an ICE and **b** Finite element model of the turbocharger rotor system

TW [5]. The TC rotor is commonly supported by hydrodynamic bearing systems, which exhibits a highly nonlinear behavior. A special attention is required to analyze these components, as the TC lubricating system is responsible for about 75% of turbocharger failures (Fig. 1b) and there is still a significant potential to increase its efficiency in the context of mechanical losses [2].

The increase of the turbocharged engine specific power is achieved with an expansion of the hot exhaust gases in the TW, generating rotational kinetic energy and driving the rotor, compressing the intake air in the CW. The atmospheric air entering the centrifugal compressor is first compressed, increasing its pressure and density, flows through an intercooler, reducing its temperature and further increasing its density, and enters the cylinder at the intake port, wherein the intake valve injects fuel. After combustion of the air and fuel mixture, the hot burnt gases leave the cylinder by the exhaust valve, being directed at high speeds by vanes into the turbine wheel blades and produce work, wherein the enthalpy energy contained in the hot gases is delivered. The turbine drives the compressor [6]. The use of turbochargers improves the engine specific power and reduces emissions. Turbocharged downsized engines can reduce the cylinder volume by 25%, resulting in nearly 10% of fuel consumption [3]. The gains with respect to engine power can be even more expressive, as the use of a turbocharger on a commercial 3.9 L diesel engine increased the non-turbo engine power by 85% [7].

As illustrated on Fig. 2, the standard automotive TC design comprises a compressor and turbine wheel overhung. Other configurations are possible, but they are not common [8]. A finite element (FE) model of a typical TC rotor system, based on [5], is shown in Fig. 2b. The FE model of the rotating shaft is illustrated on Fig. 2b. Nodes 1 and 5 represent the impellers modeled as rigid discs, wherein node 1 represents the CW and node 5, the TW. Nodes 3 and 4 represent the lubricated floating ring bearings (FRB) supporting the rotating shaft, at compressor side (CS) and turbine side (TS), respectively. Node 2 has the double-acting thrust bearing

(TB). The rotating shaft is modelled using the Timoshenko beam theory [9, 10] and the hydrodynamic forces from the fluid film in the bearings are obtained from the Reynolds equation.

The typical turbocharger is assembled from three parts: the rotating shaft and the compressor and turbine impellers. The FE model must capture the essential rotor-dynamic phenomena of such components. The compressor impeller is made of an aluminum alloy, while the shaft and turbine impeller are made of steel [11]. The turbine wheel is much more important for the transient response of the TC than the compressor wheel or the shaft, as it accounts for approximately 70% of the inertia of the rotating assembly [12]. As a result, the center of gravity (CG) of the rotating system is very close to the turbine side floating ring bearing, as seen in Fig. 2b. In the TC rotor FE discretization, the turbine and compressor impellers are modelled as rigid disks, as their diameters are much larger than the shaft diameter. The bending stiffness of both wheels is, consequently, considerably higher than that of the shaft. Therefore, rotor bending occurs mainly on the shaft, while parameters of both wheels affect mostly the inertia of the rotary system, so the impellers are approximated as rigid disks [11]. Table 1 summarizes the important information on the turbocharger FE discretization, along with the hydrodynamic bearings main parameters and the physical properties of the lubricating oil.

As previously mentioned, the TC rotor is commonly supported by hydrodynamic bearings, of several types: single oil film bearings, two oil film, fully floating or semi-floating bush bearings, and ball bearings supported by squeeze film dampers [13]. For automotive applications, due to manufacturing costs and performance, the most common configuration is the fully floating ring bearing (FRB), as the rotational speed may reach up to 350 krpm [3], as it reduces the parasitic power loss. At high rotor speeds, the semi-floating ring bearings (SFRB) may provide a better stability and vibration attenuation than the FRB, but it cannot support any static load without ring motion [11] and its power loss increase ranges from 15 to 37% [13]. Due to insufficient damping characteristics, ball bearings are not widely used in TCs [11]. However, they are becoming more common nowadays, as better throttle responses may be achieved at low rotational speeds, due to smaller friction forces [14]. Nonetheless, at high rotational speeds, not only FRBs have less friction, but the acoustic emission of ball bearings is quite intense, negatively affecting its application in automobiles.

Figure 3 illustrates both types of commonly used bearings in turbochargers for automotive applications. Figure 3a represents the (fully) floating ring bearings supporting the rotating shaft and Fig. 3b represents the semi-floating ring bearing. The SFRB is usually composed of a single body, and a locking pin prevents the rotation of the floating ring. As a result, the outer film acts purely as a squeeze film damper, in contrast with the free-to-rotate fully floating ring, where the outer film will have both important mechanisms of oil pressure increase mostly found in a radial bearing (the squeeze and wedge effects). Figure 3c illustrates a cross section view of the FRB, along with its main variables. It is clear the floating ring divides the radial clearance in two, acting as a mass pedestal between both oil films. The SFRB has

Table 1 Turbocharger rotor-bearing system details

Shaft parameters		
Density (kg m^{-3})	7860	
Young's modulus (GPa)	200	
Poisson's ratio	0.30	
Diameter (mm)	11.0	
Length (mm)	16.5–16.5–39.0–33.0	
Rigid discs parameters	Compressor Wheel	Turbine Wheel
Mass (kg)	0.118	0.326
Polar MoI (10^{-6} kg m^2)	44.0	81.0
Transverse MoI (10^{-6} kg m^2)	32.7	77.0
Floating ring bearing parameters	Inner Film	Outer Film
Radial clearance (μm)	34.0	44.0
Bore radius (mm)	5.534	8.000
Length (mm)	6.5	9.0
Ring mass (g)	7.2	
Ring's polar MoI (10^{-7} kg m^2)	3.381	
Thrust bearing parameters		
Inner radius (mm)	5.5	
Outer radius (mm)	10.0	
Shoulder height (μm)	20.0	
Pad angular extent (°)	100	
Ramp angular extent (°)	75	
Number of pads	3	
Lubricating oil parameters		
Type	Synthetic oil 0W20	
Density (kg m^{-3})	786.93	
Specific heat capacity (J kg^{-1} °C^{-1})	2149.2	
Thermal conductivity (W m^{-1} °C^{-1})	0.125	
Viscosity-temperature relation	$\mu = \mu_0 \exp \frac{\lambda}{T-T_0}$	
Reference viscosity μ_0 (10^{-3} Pa.s)	1.1119	
Viscosity coefficient λ (°C)	91.835	
Reference temperature T_0 (°C)	39.390	

the same cross-section geometry, with the exception that the locking pin imposes a null rotation to the floating ring, i.e., $\Omega_r \equiv 0$.

It is worth noticing that the rotor dynamics of turbochargers significantly changes depending on its size and operation condition [14]. While turbochargers for industrial, locomotive and marine applications operate below 20 krpm, with substantial mass and inertia effects [15], automotive turbochargers are characterized as a low-weight,

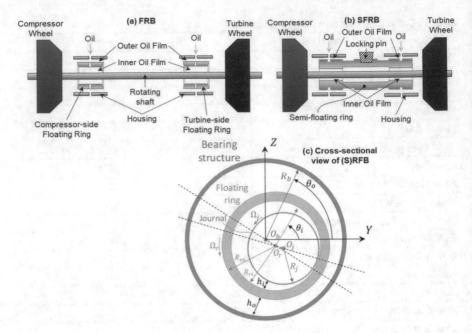

Fig. 3 Turbocharger rotor supported by **a** FRB and **b** SFRB and **c** cross-sectional view of the floating ring bearing and its main variables

high-speed rotating system. In large, low speed TCs, the rotor dynamics is quasilinear, whereas it is highly nonlinear for small, high speed TCs. It is also a multi-physics problem, wherein aerodynamic forces act on the rotor wheels, fluid-structure interaction occurs in the lubricated bearings and the rotating shaft oscillations are within the field of solid mechanics. The high nonlinearities of the turbocharger dynamics make its modelling complex and a theoretical analysis of such system demands assumption of several hypothesis, altogether with experimental validation of the proposed models.

2   Linear Analysis of Turbochargers

The first studies on rotors supported by FRB models characterized the bearings in the rotating system and its dynamic response in terms of the bearings equivalent dynamic coefficients [16]. Further studies revealed the FRB superior stability to suppress vibration over conventional cylindrical bearings was due to the large damping effect of the outside film [17]. In the predicted unstable regions given by linear analysis, rotor motion was verified to operate within limit cycles [18, 19]. These first results showed the necessity to consider nonlinear bearing models in TC theoretical analyses, as the linear predictions are not accurate and may not even predict all unstable modes of the TC-FRB rotating system. Nonetheless, linear analysis is useful to investigate the overall TC behavior and helps in explaining nonlinear results in a linear way [20]. Also, modal analysis is considerably cheaper and faster than nonlinear simulations.

Several linearization procedures of the FRB may be employed to investigate the linear TC behavior. Overall, the linearization process consists in finding the static equilibrium position of the journals within the bearings, for a given rotor speed and static load. After this position is known, the bearing hydrodynamic forces are expanded in a Taylor series and the linear terms in this expansion are identified as the bearing equivalent stiffness and damping coefficients [21]. A review of linearization techniques commonly employed in TC-FRB models is given in [22]. The simplest models neglect the floating ring mass in the equations of motion of the system. Given that the inner film clearance is smaller than the outer clearance, the inner film is stiffer than the outer film and the FRB dynamic coefficients are approximated by the outer film equivalent coefficients (Fig. 4a). In this configuration, only the direct coefficients may be considered (as in [23, 24]), or both direct and cross-couple terms are considered [25, 26]. If the ring mass is neglected [19], as in Fig. 3b, both oil films are connected in series [25] or the total impedance of the inner and outer film may act as a single journal bearing [26], in which case the bearing is characterized by the equivalent stiffness coefficient as:

$$\frac{1}{k_{mn}} = \frac{1}{k_{mn}^i} + \frac{1}{k_{mn}^o} \tag{1}$$

wherein the subscript may refer to $m, n = y, z$ and the superscript i, o refers, respectively, to the inner and outer films. An analogous expression is obtained for the equivalent damping coefficient c_{mn}. The bearing stiffness and damping matrices are:

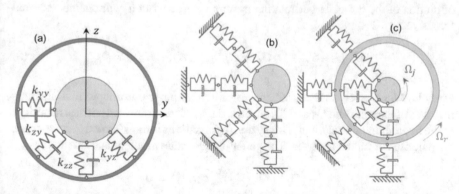

Fig. 4 Different models for the FRB dynamic coefficients: **a** journal supported on orthotropic bearings; **b** both oil films linearization, neglecting the ring mass; **c** oil films linearization, considering the ring mass

$$K_b = \begin{bmatrix} k_{yy} & k_{yz} \\ k_{zy} & k_{zz} \end{bmatrix}, \; C_b = \begin{bmatrix} c_{yy} & c_{yz} \\ c_{zy} & c_{zz} \end{bmatrix} \tag{2}$$

From a physical point of view, a better linearization of the FRB considers the floating ring mass in finding the static equilibrium position of the system [22, 27]. The static equilibrium position may be sought assuming either the ring speed is known or it is also an independent degree of freedom. Analytical expressions may be derived based on short bearing theory with the half-Sommerfeld condition. An overly simplified equation assuming nearly centered operation gives the ring speed ratio [28] as:

$$\overline{\Omega} = \frac{\Omega_r}{\Omega_j} = \frac{1}{1 + \frac{\mu_o}{\mu_i} \frac{L_o}{L_i} \frac{c_i}{c_o} \left(\frac{D_o}{D_i} \right)^3} \tag{3}$$

A slightly better approximation considers the inner and outer relative eccentricities [3]:

$$\overline{\Omega} = \frac{\Omega_r}{\Omega_j} = \frac{1}{1 + \frac{\mu_o}{\mu_i} \frac{L_o}{L_i} \frac{c_i \sqrt{1-\varepsilon_i^2}}{c_o \sqrt{1-\varepsilon_o^2}} \left(\frac{D_o}{D_i} \right)^3} \tag{4}$$

Finally, the best approach is to add the ring speed as an independent parameter and look for the static equilibrium position of the system altogether with the ring speed [22]. All the linearization processes can be used to do the modal analysis of the turbocharger. The thrust bearing may also be considered in the linearized equations of motions of the rotating system, as in [5]. Irrespective of the linearization process to approximate the dynamic characteristics of the fluid film bearings, the equations of motion of the turbocharger rotating system may be written, without any external forces, as:

$$M_r \ddot{q} + (C_r + \Omega_j G_r + C_b) \dot{q} + (K_r + K_b) q = 0 \tag{5}$$

wherein M_r, C_r, G_r and K_r are the inertia, damping, gyroscopic and stiffness matrices of the rotor and C_b and K_b are the equivalent damping and stiffness matrices of the hydrodynamic bearings. This system of equations is transformed to state-space for purposes of modal analysis. The general eigenvalue problem is written:

$$[A(\Omega_j) - \lambda_k(\Omega_j) I] u_k(\Omega_j) = 0 \tag{6}$$

wherein $A(\Omega_j)$ is the system matrix and I is the identity matrix. The eigenvalues are $\lambda_k(\Omega_j)$ and the associated eigenvectors are $u_k(\Omega_j)$, both are dependent on the rotational speed Ω_j. The eigenvalues are complex numbers, generically represented as:

$$\lambda_k(\Omega_j) = a_k(\Omega_j) + ib_k(\Omega_j) \tag{7}$$

The k-th eigenfrequency of the system is given by the imaginary part b_k, while the real part (called growth/decay rate) characterizes the stability of the system. Defining the modal damping factor as:

$$\xi_k(\Omega_j) = -\frac{a_k(\Omega_j)}{|b_k(\Omega_j)|} \tag{8}$$

the stability of the system is evaluated according to the following rules [22]:

- $\forall k$: $\xi_k(\Omega_j) \geq 0$, the system is stable;
- $\exists k$: $\xi_k(\Omega_j) < 0$, the system is unstable.

Moreover, each eigenvalue $\lambda_k(\Omega_j)$ has a corresponding complex eigenvector $\mathbf{u_k}(\Omega_j)$. The coordinates corresponding to the lateral modal displacements may be expressed as:

$$v_j = v_{j,r} + iv_{j,i}, \ w_j = w_{j,r} + iw_{j,i} \tag{9}$$

Here, v_j and w_j represent the modal displacements of node j in the horizontal and vertical directions, respectively. The precession direction is determined after a rather lengthy algebraic manipulation. The sign of the indicator:

$$\Psi_j = v_{j,i}w_{j,r} - v_{j,r}w_{j,i} \tag{10}$$

defines the precession of the motion of node j, according to [22]:

- $\forall j$: $\Psi_j > 0$, the shaft whirls with forward (FW) precession;
- $\forall j$: $\Psi_j < 0$, the shaft whirls with backward (BW) precession;
- $\exists i, j$: $\Psi_i > 0$ and $\Psi_j < 0$, the shaft whirls with a combined precession.

These results may be summarized in a Campbell diagram. A holistic Campbell diagram, as presented in [29], may be useful in investigating the modal properties of the turbocharger, as it depicts the natural frequencies, modal damping and precession of mode shapes altogether in a single figure.

The turbocharger model proposed in [5] is the same as the one described in [30], with the inclusion of a thrust bearing to account for axial displacements of the rotor. The most precise linearization of the fluid film forces estimates the inner and outer film equivalent coefficients, solving altogether for the ring speed for each rotational speed, and also approximates the thrust bearing dynamic characteristics by its equivalent coefficients in the axial and rotational degrees of freedom, as described in [5, 31]. Figure 5 presents the holistic Campbell diagram of the turbocharger, altogether with selected mode shapes and the predicted unstable modes. In this diagram, the

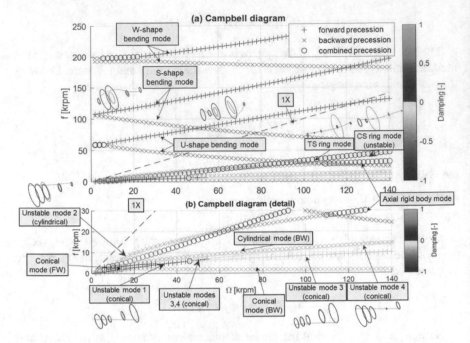

Fig. 5 Campbell diagram of turbocharger supported by both FRB and thrust bearings

blue-green markers represent the (in a linear sense) stable modes, while red-yellow markers indicate unstable ones.

The Campbell diagram predicts that up to three unstable modes may exist for rotational speeds above 45 krpm. More than one critical speed may be excited during operation and excitation of the unstable modes creates distinct audible noises. Turbocharger failure is very often caused by these unstable sub synchronous modes. The general results may be summarized as follows:

- The threshold speed of instability of the Unstable mode 1 is around 30 krpm. This unstable mode is a conical mode, whose amplitudes of the compressor and turbine wheel (CW and TW, respectively) are nearly the same. The predicted amplitude orbits of both journals and rings are approximately the same. At a rotating speed around 55 krpm, this unstable mode bifurcates onto two different unstable modes. While unstable mode 3 predicts a higher eccentricity for the turbine side floating ring bearing (which for the rest of the paper will be referred simply as TS FRB), unstable mode 4 predicts an extremely low amplitude for the TS FRB journal and ring. The difference between unstable modes 1 and 3 is that the amplitudes of the CW are smaller for unstable mode 3 than unstable mode 1. The most dangerous Unstable mode 4 predicts much higher amplitudes for the CW and CS FRB, leading to wear on the ring inner surface and contact of the compressor impeller on the volute surface. The unstable modes 1, 3 and 4 are illustrated on Fig. 6, along with the stable FW and BW conical modes.

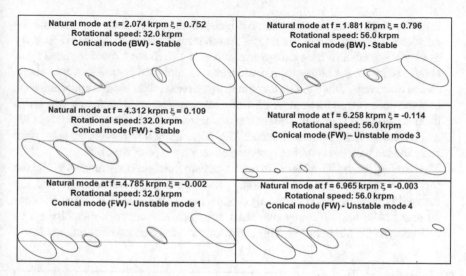

Fig. 6 Conical modes: stable BW and FW conical modes and Unstable modes 1, 3 and 4

- Unstable mode 2 is a cylindrical mode, whose orbits of the CW and journal and ring at CS FRB (compressor side floating ring bearing) are greater than the orbits of the nodes on turbine side. This unstable mode exists for every rotational speed. At high rotational speeds, the orbits of both rings are much larger than the shaft orbits, so the unstable cylindrical mode of the rotating system reduces to the unstable CS ring mode. The unstable shaft mode is still cylindrical, but the ring orbit is much higher than the shaft orbits. This is prejudicial to the turbocharger system and may lead to outer ring and bearing wear. Unstable mode 2 and the ring modes are illustrated on Fig. 7.
- The frequency of the axial rigid body mode increases with the rotating speed, as the oil film in the thrust bearings becomes stiffer for higher rotational speeds (smaller film thickness). The predicted effect of the thrust bearing (TB) on the lateral modes

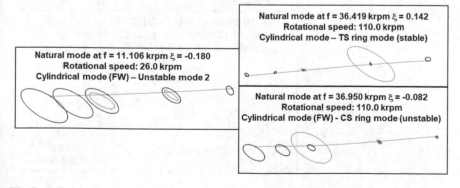

Fig. 7 Cylindrical modes: Unstable mode 2 and TS (stable) and CS (unstable) ring modes

is almost negligible, because, for an aligned thrust collar, the TB cross-coupled coefficients are negligibly small [29]. Another observation is that the rigid body axial mode predicts a backward precession of the shaft for a rotating speed up to about 110 krpm. For higher rotational speeds, a combined precession mode shape can be observed. Although this combined precession mode seem counterintuitive, as some nodes are in forward whirl, while other nodes are in backward whirl, this phenomenon can appear in rotor systems supported on anisotropic bearings [32] and has been reported in the literature [22, 30]. Figure 8 illustrates the axial rigid body modes with the combined precession mode shape of the shaft.

- The bending modes have higher frequencies than the rigid body modes. While the forward frequencies of S and W bending modes are higher than the shaft speed, the forward U bending mode frequency crosses the shaft speed (1X line on Fig. 5a) around 124 krpm, indicating that shaft bending may occur during high-speed operation of the turbocharger. Figure 9 illustrates the first two bending modes.

Fig. 8 Axial rigid body modes: backward and combined precession modes at different rotational speeds

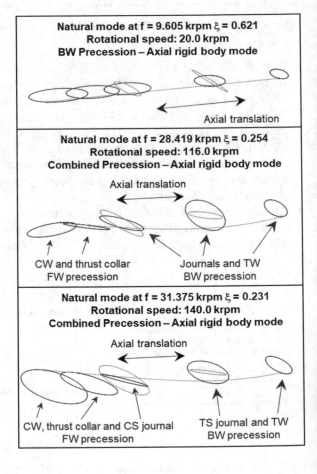

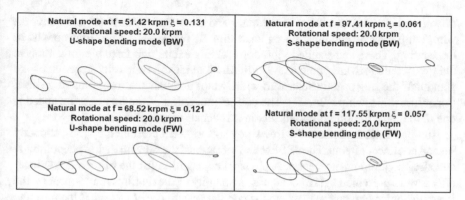

Fig. 9 Bending modes (stable): U-shape BW and FW and S-shape BW and FW

The linear predictions are useful to gain a deep insight into the overall turbocharger behavior and may be helpful to explain nonlinear simulated and experimental results [30], even though the high amplitude limit cycle motion cannot be linearly approached [20] and some frequency components of nonlinear unstable bearings cannot be associated with any unstable mode in the Campbell diagram [22]. Nonetheless, linear results may aid in the explanation of nonlinear phenomena in a linear way.

3 Turbocharger Nonlinear Response

The typical automotive turbocharger behavior is often investigated in waterfall plots, to reveal the main frequency components of the turbocharger nonlinear oscillations as a function of the shaft rotational speed. The transient response may be studied in run-up or run-down conditions, either empirically, via extensive and costly experiments, or numerically, with affordable computational simulations. Usually, as seen on the previous section, the turbocharger operating speed range is higher than the first two rigid body modes and may be above the first bending mode [13]. Furthermore, because the rotating shaft is supported in fully floating ring bearings, the phenomenon of oil whirl and oil whip (self-excited vibrations due oil film instabilities, generated by the rotor motion in the journal bearing [33]) may appear in the inner and outer films, leading to high amplitude sub synchronous oscillations, whose amplitudes are, in most applications, greater than the synchronous oscillations [24]. In properly balanced rotors, the amplitudes of sub synchronous vibrations can be as high as three times the synchronous components [14]. These oscillations are also the root cause of constant tone noise at the automobile interior, being also the main reason for acoustic problems on vehicle applications [1].

In the fully floating ring bearing configuration, the floating bush is free to rotate and its rotational speed is in the range of 10–50% of the shaft speed, due to shear driven torque (lubricant drag) in the inner and outer films. These explains why oil whirl/whip may be encountered on each film, as two spinning oil films coexist in the bearing. These self-excited vibrations mainly excite two unstable rotor modes, the gyroscopic forward conical and cylindrical modes [34], and its frequencies are about half the ring speed (outer film whirl/whip) and half the rotor plus ring speed (inner film whirl/whip) [26, 27]. The correct prediction of the floating ring speed is one of the most important parameters in turbocharger analyses [35].

Run-up measurements of the frequency spectra were presented by [36], showing several nonlinear effects. The self-excited vibrations, the oil whirl/whip phenomenon and also the jump phenomenon. A physical explanation for the observed nonlinear effects were presented in [23]. Roughly, instability of the oil films may appear on the inner and outer film and may excite the conical or cylindrical modes. If the unstable inner film excites the forward conical mode, we refer to the *sub1* frequency. If it excites the forward cylindrical mode, then we denote it the *sub2* frequency. If it is the unstable outer film that excites the conical mode, we refer to the *sub3* frequency [36]. Theoretically, a fourth component may exist. The *sub4* frequency appears due to the unstable outer film exciting the cylindrical mode, even though no measurements and simulations were, to the best of the authors knowledge, found. Finally, synchronization of both the inner and outer oil film whirl/whip may excite a rotor natural frequency. The resulting motion is denoted total instability (TI) [23] or critical limit cycle (CLC) oscillations [30] and a long time operation in this condition leads to damage and failure of the turbocharger. Which of the mentioned sub synchronous frequencies appears in the response spectrum depends heavily on the rotor/bearing system (rotor inertia, impeller mass, shaft stiffness, bearing parameters, unbalance levels, lubricating oil properties etc.).

Several distinct frequency spectra of turbocharger run-ups are encountered on [22, 23], along with an explanation of the different jump phenomena encountered in this type of system. In [27, 36], it is shown that although two unstable modes may exist together during the turbocharger run up, the inner film instability might disappear rather suddenly. This almost instantaneous change of motion may be explained by bifurcation theory and it is referred to as *jump* (phenomenon), due to its peculiar appearance in frequency spectrum [22], as illustrated on Figs. 13, 14 and 15 for the response, respectively, of a perfectly aligned TC rotor, with small and high level of unbalance.

3.1 Effect of Self-excited Vibrations

In order to investigate the strong effect of fluid-induced instability in the floating ring bearings, the turbocharger rotor-bearing system described in [5] will be considered for the simulations. The foregoing simulations consider the full, nonlinear hydrodynamic forces and details of its modelling and implementation are described in [5, 31]. However, to emphasize the jump phenomena and the different sub synchronous frequencies that may appear in the turbocharger run up frequency spectra, the inner and outer film radial clearances will be considered equal to 34 and 44 μm, respectively. In this section, the effect of oil whirl/whip in the FRB inner and outer films is investigated assuming a perfectly aligned turbocharger, i.e., no unbalance is prescribed to the compressor and turbine wheels. The run-up numerical simulations assume the turbocharger speed linearly increasing from 0 to 180 krpm in 10 s.

Figure 10 presents the turbocharger dynamical response in the absence of any residual unbalanced mass during run up. Figure 10a presents the (dimensionless) vertical displacement of the compressor and turbine wheels. These oscillations indicate that in the very beginning of the transient run, shaft motion is stable, around the static equilibrium position. Around 0.3 s, the first instability occurs, as the oscillation amplitudes become much greater. Figure 10a, d present the eccentricities of the inner and outer films of both FRBs. In the very beginning of the numerical results, the shaft speed is almost negligible and the turbocharger center of mass is located almost on the TS FRB position. After the numerical solution starts, the abrupt growth of the eccentricities of the TS FRB indicates this bearing is supporting practically the entirety of the static load in the rotating system. After the first instability begins,

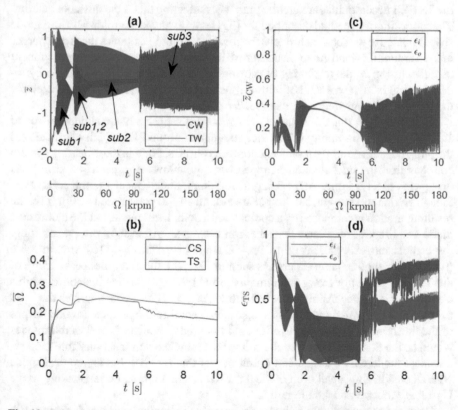

Fig. 10 Perfectly balanced turbocharger run up: **a** dimensionless vertical displacement of CW and TW, **b** ring speed ratio of both floating rings at CS and TS and **c, d** eccentricities of inner and outer films of CS FRB and TS FRB

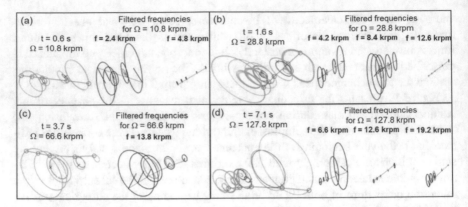

Fig. 11 Turbocharger resulting motion and filtered mode shapes at selected time intervals: **a** 0.6 s (shaft speed: 10.8 krpm), **b** 1.6 s (shaft speed: 28.8 krpm), **c** 3.7 s (shaft speed: 66.6 krpm) and **d** 7.1 s (shaft speed: 127.8 krpm)

the TS FRB eccentricities are greater than 0.8, indicating this is the unstable bearing. The mode filtering method described by [37] let us evaluate the rotor whirling mode shapes related to the sub synchronous frequencies. Figure 11a shows the turbocharger motion around 0.6 s and the corresponding filtered mode shapes. The main component of turbocharger in the beginning of motion is a conical mode. The unstable cylindrical mode is also present, but with a much smaller influence. This characterizes the sub synchronous frequency as the *sub1* component.

As seen on Fig. 10a, the unstable *sub1* motion disappears rather suddenly around 1.2 s and another instability takes place, beginning around 1.3 s. This characterizes the jump phenomenon commonly encountered on TCs. For this rotor-bearing system configuration, the *sub1* component appears at a quite low rotor speed (4.5 krpm) and disappears at 21.6 krpm. At 24.3 krpm, the dominant frequency is the *sub2*, even though *sub1* is still present, but disappears around 40.5 krpm. As seen on Fig. 11b, the resulting motion is composed by a conical and a cylindrical mode, but the cylindrical mode amplitudes are greater than the conical mode. After *sub1* disappears, only the conical mode is present up to 95.4 krpm, as seen in Fig. 11c. Around 5.3 s (95.4 krpm), another jump occurs. As seen in Fig. 10d, the outer film eccentricity of the TS FRB suddenly grows and remains around 0.75, indicating instability of this oil film. As it excites mainly a conical mode (Fig. 11d), this characterizes the *sub3* frequency. The filtered mode shapes also correspond to the predicted mode shapes of the linear analysis, even though it could not predict which of the unstable modes would be the dominant mode. In the *sub3* frequency, the dominant unstable mode is a conical mode with a large whirl amplitude of the TS FRB, but the conical mode with a considerably small orbit of the TS FRB is also present, corresponding to the Unstable Modes 3 and 4, in Figure 5.

The jump phenomenon can also be observed in the floating ring speeds. Figure 10b shows the ring speed ratio (the ratio of the floating ring rotating speed by the shaft rotating speed). It is possible to observe jumps on the ring speed in the transition to the *sub2* (1.3 s) and *sub3* (5.3 s) frequencies. We also notice the ring speed remains well below 50% of the shaft speed, reaching values as high as 30% of the shaft speed and as low as 17% of the shaft speed.

The results in time domain of the shaft vertical displacements shown in Fig. 10a may also be investigated in frequency domain, with the aid of the short-time Fourier transform (STFT) of the signal. The displacement signal in time domain is divided in equal length overlapped segments, multiplied by a Hann window function and, then, the short-time Fourier transform is computed [38]. This results in the waterfall diagrams shown in Fig. 12. In these plots, the dashed red line denotes the shaft rotational speed (Ω_j), half the shaft speed ($0.5\Omega_j$) and half of both the ring speeds ($0.5\Omega_r$). The frequency f axis is shown only up to 40 krpm, for better visualization of the curves. It is evident from this figure that the sub synchronous oscillations are closely related to the ring speed. Nonetheless, it is easier to evaluate this waterfall diagram, the frequency spectra and the jump phenomena in a top view of Fig. 12, as well as shown in Fig. 13.

It should be noted that the presented results for turbocharger run up are illustrative and do not embrace all possible bifurcation sequences that may appear. In addition, it should be emphasized the floating ring bearing is heavily subjected to oil whirl/whip, as the shaft oscillations are due to the self-excited vibrations of the inner and outer films, as no unbalance was considered in these simulations [39]. In addition, it should

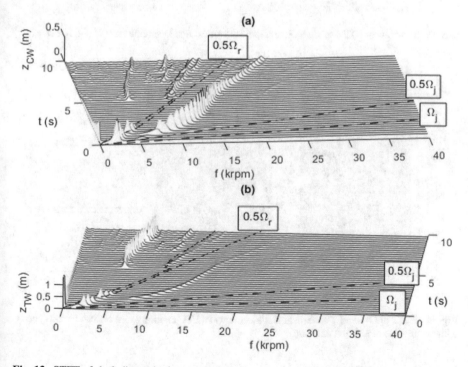

Fig. 12 STFT of shaft dimensionless vertical displacements: **a** CW and **b** TW

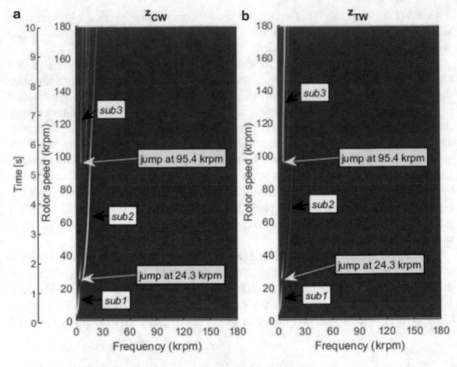

Fig. 13 Top-view of STFT of shaft dimensionless vertical displacements: **a** CW and **b** TW

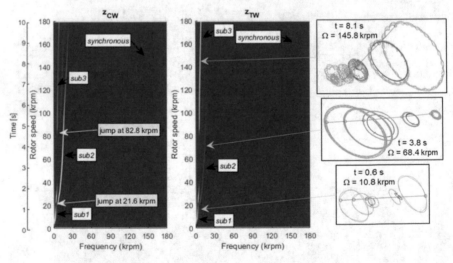

Fig. 14 STFT of CW and TW vertical displacements and TC resulting motion at selected speeds, with small value of residual unbalance

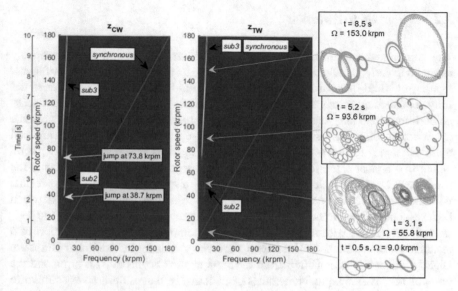

Fig. 15 STFT of CW and TW vertical displacements and TC resulting motion at selected speeds, with a high unbalance level

be noticed that the fluid-induced instability excites, generically, two unstable modes: a conical mode and a cylindrical mode [40].

Another important consideration is that the waterfall diagram may be obtained via run-up or run-down simulations [23, 29, 30] or in a series of steady-state responses [5, 27]. With respect to the bearing parameters, the geometric configuration of the bearing is more important than the oil properties [41]. Almost 40% of the turbocharger global response is controlled by the design of the outer clearances and, if the bearing width is also considered, these parameters are able to control about 55% of the system response [42]. The outer film clearance has a great influence on the turbocharger response, the jumps and frequency components [30]. Also, stability of the outer film is mainly determined by the outer clearance, while the inner film stability is more sensitive to the inner-to-outer clearances ratio [43]. The turbocharger vibrations may also be reduced controlling the manufacturing tolerance of the FRB clearances [44]. However, the *sub2* and *sub3* frequencies are not totally avoidable and, even though the *sub2* could be listed under the comfort-issue problems, being a root cause for turbocharger noise, the *sub3* with extended amplitudes may lead to rotor destruction [45].

3.2 Effect of Unbalance

The compressor and turbine impellers, modeled as rigid disks, were previously assumed in perfectly aligned conditions. This can be useful for the purpose of investigating the fluid-induced instabilities from the fluid film bearings, but it does not fully represent an actual turbocharger rotor model. The effect of a residual mass on both impellers may lead to different turbocharger responses. Figure 14 presents the waterfall diagrams of the turbocharger run up with a small level of induced mass unbalance for the CW and TW, respectively, of 0.59 g.mm and 1.63 g.mm, which is a typical value for automotive turbochargers. Figure 9 also presents the resulting TC motion at select time instants and rotor speeds. Comparing these results with those obtained in Sect. 3.1, practically no differences can be noticed between both responses. The turbocharger motion is predominantly determined by the self-induced vibrations, with a small synchronous component. With the addition of a residual unbalance on the TC impellers, the onset speeds for the nonlinear jumps are slightly reduced. The transition from the *sub1* to *sub2* begins at 21.6 krpm and the onset of *sub3* is 82.8 krpm. Nonetheless, qualitatively, the results are very similar to the results without unbalance.

If, however, a high level of unbalance is assumed, and the CW and TW unbalances are set to 3.3 g.mm and 9.1 g.mm, respectively, the run-up results are quite different, as shown in Fig. 15. The high unbalance suppresses the first instability (*sub1*) on the very beginning of the transient, wherein the shaft performs only unbalanced induced vibrations. Additionally, the onset speed for beginning of the *sub2* instability increases to 38.7 krpm. Despite this apparently beneficial effect of a high unbalance on the overhung impellers, the onset speed for the *sub3* decreases further to 73.8 krpm and the amplitude oscillations steadily grows with the rotor speed, which may lead to TC destruction.

These results show that excitation of the rotor unstable modes is inherently caused by the unstable FRBs. Hence, improvements on rotor unbalance cannot be effective in reducing the sub synchronous whirl motion in a typical TC. However, the actual turbocharger response and the bifurcation sequences observed during run up and run down conditions depend on the impellers residual unbalance levels [46]. Nonetheless, the pursue to reduce the amplitude oscillations in TCs is still a current research open issue.

Experimental investigations showed the superior load capacity of the FRB compared with custom designed fixed geometry bearings [47], and the same authors proposed a pioneer method to suppress the TC unstable vibrations, inducing either the compressor or turbine unbalance at a certain level [48]. Another possibility for satisfactory vibration performance was investigated with specially designed tilting pad bearing, possible only when the (very small) bearing clearance and proper oil flow were achieved [7].

3.3 Influence of Radial Bearing System

The proposed tilting pad geometry [7] is not suitable to TC applications, due to the high cost and tight tolerances necessarily employed in order to reach safe operation condition. This fact, however, does not limit the bearing applications to supporting the TC rotating shaft. Even though the FRB is the most common bearing used in automotive turbochargers, for reasons of low cost and practically infinite lifetime, other bearings have been proposed to overcome the problem of high amplitude oscillations due to fluid-induced instabilities. While the FRBs were already in use in aircraft engines from 1920 to 1930 [28], a much newer bearing appearing in the past 20 years is the semi-floating (non-rotating) ring [39]. In the semi-floating ring bearing (SFRB), the ring rotation is somewhat prevented, usually with an anti-rotation pin locking the bearing. Because the ring does not rotate, only the inner film presents whirl/whip phenomena. In addition, the SFRB requires a lower oil pressure supply [4].

Figure 16 presents the run up results of the same TC considered in the previous sections, but supported on SFRBs. The ring rotation is prevented and, clearly, it is identically null throughout the whole simulation (Fig. 16b). Because the outer film acts purely as a squeeze damper, it is only effective in reducing vibrations if there is dynamic excitation, either from the inner film self-excited vibration or due to rotor unbalance. The unstable inner film may excite the conical (*sub1*) or cylindrical (*sub2*) modes (Fig. 16a) and the film eccentricities are smaller (Fig. 16c, d). At high rotor

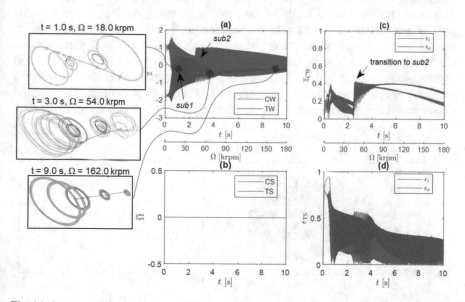

Fig. 16 Run up of TC supported on SFRB: **a** vertical displacement of CW and TW, **b** floating ring speeds and **c, d** eccentricities of inner and outer films of CS SFRB and TS SFRB

speeds, however, the SFRB may present higher power losses in comparison with the traditional FRB [13].

Other types of bearings are also possible to support the TC and have been investigated by several authors. Ball bearings are significantly more rigid and stable [49], but their application is limited, as it possess almost no damping characteristics. Rolling bearings generate more noise and the maximum achievable shaft speed is smaller than traditionally lubricated fluid film bearings. A ring with a multi-lobed clearance [50–53] or an elliptical ring [54] was also investigated and verified to suppress shaft whirl and increase the onset shaft speed of ring whirl. The effect of the axial and circumferential grooves on the floating ring have also been investigated [55–58]. Other designs for lubrication groove may also reduce the sub synchronous vibrations and power losses in the bearings [59].

3.4 Impact of Thrust Bearing on Turbocharger Response

The thrust bearing (TB) was empirically verified to account for a great part of the TC bearing system friction losses [60, 61]. The actual TB operation depends on the axial load it must support, wherein the resulting axial load acting on the TC rotor results from the air flow in both impellers. The resulting thrust load can be estimated by the methods proposed in [3, 62]. Numerical results have shown the thrust bearing effect on the turbocharger lateral oscillations [5, 63–65]. The TB mainly affects the conical mode of the rotating system [66]. Figure 17a presents the CW vertical displacement with and without the TB support. We notice the TB influence on the lateral motion of the system. For this particular TC-TB configuration, the TB completely suppress fluid-induced vibrations up to 4.8 s. Around 6.2 s, sub synchronous oscillations dominate, but the amplitudes are slightly smaller with the inclusion of the TB. Figure 17b presents the CW axial displacement. The external axial thrust is directed from the turbine to the compressor and it is assumed to increase

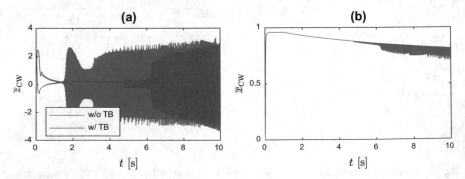

Fig. 17 **a** Comparison of vertical displacement of CW during run up of TC, with and without a thrust bearing supporting the rotating shaft and **b** Axial displacement of CW

from 0 to 100 N in the first second, remaining constant throughout the rest of run up simulation. We see the fluid film in the axial bearing is squeezed in order to sustain the external axial load and, once the external thrust remains constant and the shaft speed increases, the axial position diminishes, increasing the minimum film thickness on the thrust bearing located at compressor side (CS TB). Because the external thrust is directed from the TW to the CW, it is the CS TB that sustains the external thrust.

The correct modelling of the thrust bearing is crucial to turbocharger design. In turbochargers supported by FRB, the thrust bearing affects mainly the first and third sub synchronous oscillations (*sub1* and *sub3* components), since they are related to the unstable conical modes [63]. The thrust bearing may also reduce the ring speeds during transient operation of the turbocharger [5]. For the design process, an investigation on two major parameters (the thrust bearing location along the shaft and the shaft diameter) showed that the thrust bearing position may have either positive, neutral or negative influence on shaft motion and that certain combinations of shaft diameter and thrust bearing position have a negative influence on the thrust bearing itself, reducing the thrust bearing load capacity [64]. Another critical point on thrust bearing design is the reduction on friction losses. It is mandatory to evaluate friction losses based on time transient simulations, as stationary designed friction loss thrust bearings are likely to result in transient properties not corresponding to the original design goal [65]. Savings in friction losses of about 20% were achieved by means of Genetic Algorithm optimization of the thrust bearing geometry, while maintaining existing bearing manufacturing technology [67]. Unfortunately, the lubricant flow, especially in the anti-thrust side, may not be accurately predicted by solutions of standard forms of the Reynolds equation for the thrust bearing, due to the greater thickness of the bearing gap, which may require the use of CFD models to accurately model thick lubricating gaps [68]. It has been shown that the accurate estimate of the load carrying capacity of a typical turbocharger thrust bearing requires the inclusion of angular misalignment effects on the film gap description and the effect of temperature variations due to fluid shear in the lubricated bearing. Also, linearization of the thrust bearing fluid film forces is not possible to accurately describe the supported load [31].

3.5 Temperature Variations in Turbochargers

Inclusion of temperature variations in the TC dynamic model may have a significant change on the turbocharger response. The TC oscillation amplitudes are generally greater for THD models, given the lower load carrying capacity of the bearings accounting for temperature variations of the oil film. Thermal effects on the lubricated bearings are important to check oil degradation, oil coking and burning. Temperature variations also change the oil viscosity and operating bearing clearances, affecting the turbocharger behavior [69]. THD models of FRBs have been presented in the literature, showing temperature variations should be accounted for in a reliable TC model [5, 27, 31, 69–73]. Inclusion of temperature variations in the thrust bearing model must account for conduction and convection in all three spatial directions. Neglecting some terms in the Energy equation to ease computations is only suitable for low rotational speeds (up to about 20 krpm) and, therefore, simpler models have a limited range of applicability [74].

4 Current Trends in Rotor Dynamics of Turbocharger

The high computational cost of hydrodynamic bearing models including important effects such as estimation of hydrodynamic forces of finite length bearings, temperature variations within the oil films, turbulence effects, inclusion of feeding grooves, newer bearing geometries, among others, drives part of the current research in developing reliable and faster computational models. Analytical solutions [75, 76], approximate solutions with correcting polynomials [77], weak Galerkin methods [78] and lookup table methods [68, 79–82] have been proposed. Analytical and approximate solutions are suitable for cylindrical bearings and have been employed to estimate the radial bearing loads, assuming a constant temperature of the oil film. Galerkin methods can only be used for isothermal analysis. The use of lookup table methods seems more robust and is capable to include several different effects on the hydrodynamic bearing modelling. One such work compared the use of lookup table methods against a direct solution of the Reynolds equation. Not only the lookup table method could include temperature variations in the thrust bearing models, it also greatly reduced the computational cost of the time transient analysis. The simulations were about 1,600 times faster with the lookup table method instead of solving the pressure and energy equations at every time step of the time transient analysis [79]. Faster simulations with semi implicit approaches for time integration of the resulting nonlinear equations of motion have also been investigated [83]. To reduce friction losses, newer floating ring bearing geometries have been proposed. Changing the layout of lubrication grooves can reduce the friction losses on floating ring bearings, without modifications on the current manufacturing technology to produce such bearings [59]. As the thrust bearing power losses are considerable in a typical TC, optimum solutions for the thrust bearing geometry have also been sought and could also be proposed, under the same condition of using the current equipment to manufacture thrust bearings [67, 84]. Internally skewed bearings were also shown to suppress the oil whip instability and could replace floating ring bearings in turbochargers. An optimum configuration for the bearing angle was designed, achieving a better stability response for turbochargers [85]. The influence of temperature and injection pressure of the lubricating oil in the hydrodynamic bearings supporting turbochargers have been experimentally investigated. It was found that the oil working temperature should be the temperature indicated by the manufacturer for longer service life. On the other hand, pressure variation due to engine power demands did not considerably change the results. Low oil temperatures (under 70 °C) and pressure of oil injection (under 300 kPa) can be further investigated [86]. A new mass-conserving cavitation algorithm has also been proposed, eliminating the numerical problems of Elrod's algorithm to model cavitation, while being computationally fast (only about 20% more effort than Gümbel's model). The proposed algorithm was validated against literature and measurements, being very efficient and accurate [87]. Thermo-elastic-hydrodynamic models of the bearings have also been considered [88]. Engine excitation may also influence the turbocharger response [20, 89, 90]. Experimental results have verified the shaft axial displacement does not influence the floating ring speeds and the frequency of the sub synchronous oscillations are close to the floating ring at turbine side [91]. Further empirical investigations revealed pressure fluctuations on the compressor outlet air flow may induce synchronous axial oscillations [92]. Measurement of the floating ring speed is a difficult task, and a method for ring speed measurement using small sized eddy current sensors have been proposed, showing the advantages of eddy current sensors against other speed sensor concepts, as it does not disturb the bearing behavior [93]. Overall, the turbocharger rotor dynamic analysis is quite complex, due to its high nonlinear nature and the presented works can promisingly model with sufficient accuracy the time transient response of this dynamic system.

References

1. Lokesh, C., Praveen Kumar, S., Prasanth, V.R., Subramani, D.A.: Investigation of subsynchronous noise & vibration on turbocharger fully floating hydro-dynamic bearings—test & prediction. In: Uhl, T. (ed.) Advances in Mechanism and Machine Science, pp. 3489–3498. Springer International Publishing, Cham (2019)

2. Perge, J., Stadermann, M., Pischinger, S., et al.: Influence of non-linear rotor dynamics on the bearing friction of automotive turbochargers. Lubricants **5**, 29 (2017). https://doi.org/10.3390/lubricants5030029

3. Nguyen-Schäfer, H.: Rotordynamics of Automotive Turbochargers. Springer International Publishing, Cham (2015)

4. Polichronis, D., Evaggelos, R., Alcibiades, G., et al.: Turbocharger lubrication—lubricant behavior and factors that cause turbocharger failure. IJAET **2**, 40–54 (2013)

5. Peixoto, T.F., Cavalca, K.L.: Thrust bearing coupling effects on the lateral dynamics of turbochargers. Tribol. Int. **145**:106166 (2020).https://doi.org/10.1016/j.triboint.2020.106166

6. Heywood, J.: Internal Combustion Engine Fundamentals, 2nd edn. McGraw-Hill Education, New York (2018)

7. Kirk, R.G.: Experimental evaluation of hydrodynamic bearings for a high speed turbocharger. J. Eng. Gas Turbines Power **136** (2014).https://doi.org/10.1115/1.4026535

8. Sjöberg, E. Friction characterization of turbocharger bearings. Master Thesis, KTH Royal Institute of Technology, School of Industrial Engineering and Management (ITM) (2013)

9. Nelson, H.D.: A finite rotating shaft element using Timoshenko beam theory. J. Mech. Des. **102**, 793–803 (1980). https://doi.org/10.1115/1.3254824

10. Nelson, H.D., McVaugh, J.M.: The dynamics of rotor-bearing systems using finite elements. J. Eng. Ind. **98**, 593 (1976). https://doi.org/10.1115/1.3438942

11. Liang, F., Zhou, M., Xu, Q.: Effects of semi-floating ring bearing outer clearance on the subsynchronous oscillation of turbocharger rotor. Chin. J. Mech. Eng. **29**, 901–910 (2016). https://doi.org/10.3901/CJME.2016.0421.057

12. Subramani, D.A., Dhinagaran, R., Prasanth, V.R.: Introduction to turbocharging—a perspective on air management system. In: Lakshminarayanan, P.A., Agarwal, A.K. (eds.) Design and

Development of Heavy Duty Diesel Engines: A Handbook, pp. 85–193. Springer, Singapore (2020)

13. Cao, J., Dousti, S., Allaire, P., Dimond, T.: Nonlinear transient modeling and design of turbocharger rotor/semi-floating bush bearing system. Lubricants **5**, 16 (2017). https://doi.org/10.3390/lubricants5020016

14. Sass, P.: Literature research on experimental investigations of automotive turbocharger rotordynamics. Acta. Tech. Jaur **12**:268–293 (2019). https://doi.org/10.14513/actatechjaur.v12.n4.498

15. Chen, W.J.: Rotordynamics and bearing design of turbochargers. Mech. Syst. Signal Process. **29**, 77–89 (2012). https://doi.org/10.1016/j.ymssp.2011.07.025

16. Orcutt, F.K., Ng, C.W.: Steady-State and dynamic properties of the floating-ring journal bearing. J. Lubrication Tech. **90**, 243–253 (1968). https://doi.org/10.1115/1.3601543

17. Tanaka, M., Hori, Y.: Stability Characteristics of floating bush bearings. J. Lubr. Technol. **94**, 248–256 (1972). https://doi.org/10.1115/1.3451700

18. Li, C.-H., Rohde, S.M.: On the steady state and dynamic performance characteristics of floating ring bearings. J. Lubrication Tech. **103**, 389–397 (1981). https://doi.org/10.1115/1.3251687

19. Li, C.-H.: Dynamics of rotor bearing systems supported by floating ring bearings. J. Lubr. Technol. **104**, 469–476 (1982). https://doi.org/10.1115/1.3253258

20. Tian, L., Wang, W.J., Peng, Z.J.: Dynamic behaviours of a full floating ring bearing supported turbocharger rotor with engine excitation. J. Sound Vib. **330**, 4851–4874 (2011). https://doi.org/10.1016/j.jsv.2011.04.031

21. Lund, J.W.: Review of the concept of dynamic coefficients for fluid film journal bearings. J. Tribol. **109**, 37 (1987). https://doi.org/10.1115/1.3261324

22. Dyk, Š., Smolík, L., Rendl, J.: Predictive capability of various linearization approaches for floating-ring bearings in nonlinear dynamics of turbochargers. Mech. Mach. Theory **149**:103843 (2020). https://doi.org/10.1016/j.mechmachtheory.2020.103843

23. Schweizer, B.: Total instability of turbocharger rotors—physical explanation of the dynamic failure of rotors with full-floating ring bearings. J. Sound Vib. **328**, 156–190 (2009). https://doi.org/10.1016/j.jsv.2009.03.028

24. Schweizer, B.: Oil whirl, oil whip and whirl/whip synchronization occurring in rotor systems with full-floating ring bearings. Nonlinear Dyn **57**, 509–532 (2009). https://doi.org/10.1007/s11071-009-9466-3

25. Alsaeed, A.A.: Dynamic stability evaluation of an automotive turbocharger rotor-bearing system. Master Thesis, Virginia Tech (2005)

26. Holt, C., San Andrés, L., Sahay, S., et al.: Test response and nonlinear analysis of a turbocharger supported on floating ring bearings. J. Vib. Acoust. **127**, 107–115 (2005). https://doi.org/10.1115/1.1857922

27. San Andrés, L., Kerth, J.: Thermal effects on the performance of floating ring bearings for turbochargers. Proc. Inst. Mech. Eng. Part J: J. Eng. Tribol. **218**, 437–450 (2004). https://doi.org/10.1243/1350650042128067

28. Shaw, M.C., Nussdorfer, T.J. Jr: An analysis of the full-floating journal bearing (1947)

29. Woschke, E., Daniel, C., Nitzschke, S.: Excitation mechanisms of non-linear rotor systems with floating ring bearings—simulation and validation. Int. J. Mech. Sci. **134**, 15–27 (2017). https://doi.org/10.1016/j.ijmecsci.2017.09.038

30. Tian, L., Wang, W.J., Peng, Z.J.: Effects of bearing outer clearance on the dynamic behaviours of the full floating ring bearing supported turbocharger rotor. Mech. Syst. Signal Process. **31**, 155–175 (2012). https://doi.org/10.1016/j.ymssp.2012.03.017

31. Peixoto, T.F., Cavalca, K.L.: Investigation on the angular displacements influence and nonlinear effects on thrust bearing dynamics. Tribol. Int. **131**, 554–566 (2019). https://doi.org/10.1016/j.triboint.2018.11.019

32. Friswell, M.I., Penny, J.E.T., Garvey, S.D., Lees, A.W.: Dynamics of Rotating Ma-chines. Cambridge University Press, Cambridge (2010)

33. Kamesh, P.: Oil-whirl instability in an automotive turbocharger. PhD thesis, University of Southampton, Insitute of Sound and Vibration Research (2011)

34. Kirk, R.G., Alsaeed, A.A., Gunter, E.J.: Stability analysis of a high-speed automotive turbocharger. Tribol. Trans. **50**, 427–434 (2007). https://doi.org/10.1080/10402000701476908
35. San Andrés, L., Rivadeneira, J.C., Gjika, K., et al.: Rotordynamics of small turbochargers supported on floating ring bearings—highlights in bearing analysis and experimental validation. J. Tribol. **129**, 391–397 (2007). https://doi.org/10.1115/1.2464134
36. Schweizer, B., Sievert, M.: Nonlinear oscillations of automotive turbocharger turbines. J. Sound Vib. **321**, 955–975 (2009). https://doi.org/10.1016/j.jsv.2008.10.013
37. Vistamehr, A.: Analysis of automotive turbocharger nonlinear response including bifurcations. Master Thesis, Texas A&M University (2010)
38. Smolik, L.: Autofft—FFT analyzer. In: GitHub (2021). https://github.com/CarlistRieekan/autoffft/releases/tag/1.2.5.1). Accessed 21 Jan 2021
39. Bonello, P.: Transient modal analysis of the non-linear dynamics of a turbocharger on floating ring bearings. Proc. Inst. Mech. Eng. Part J: J. Eng. Tribol. **223**, 79–93 (2009). https://doi.org/10.1243/13506501JET436
40. Inagaki, M., Kawamoto, A., Abekura, T., et al.: Coupling analysis of dynamics and oil film lubrication on rotor—floating bush bearing system. JSDD **5**, 461–473 (2011). https://doi.org/10.1299/jsdd.5.461
41. Koutsovasilis, P., Schweizer, B.: Parameter variation and data mining of oil-film bearings: a stochastic study on the Reynolds's equation of lubrication. Arch. Appl. Mech. **84**, 671–692 (2014). https://doi.org/10.1007/s00419-014-0824-3
42. Koutsovasilis, P., Driot, N., Lu, D., Schweizer, B.: Quantification of sub-synchronous vibrations for turbocharger rotors with full-floating ring bearings. Arch Appl Mech **85**, 481–502 (2015). https://doi.org/10.1007/s00419-014-0924-0
43. Smolík, L., Hajžman, M., Byrtus, M.: Investigation of bearing clearance effects in dynamics of turbochargers. Int. J. Mech. Sci. **127**, 62–72 (2017). https://doi.org/10.1016/j.ijmecsci.2016.07.013
44. Wang, L., Bin, G., Li, X., Zhang, X.: Effects of floating ring bearing manufacturing tolerance clearances on the dynamic characteristics for turbocharger. Chin. J. Mech. Eng. **28**, 530–540 (2015). https://doi.org/10.3901/CJME.2015.0319.034
45. Koutsovasilis, P., Driot, N.: Turbocharger rotors with oil-film bearings: sensitivity and optimization analysis in virtual prototyping. In: 11th International Conference on Vibrations in Rotating Machines—SIRM 2015. Magdeburg, Germany (2015)
46. Tian, L., Wang, W.J., Peng, Z.J.: Nonlinear effects of unbalance in the rotor-floating ring bearing system of turbochargers. Mech. Syst. Signal Process. **34**, 298–320 (2013). https://doi.org/10.1016/j.ymssp.2012.07.017
47. Kirk, R.G., Alsaeed, A., Mondschein, B.: Turbocharger vibration shows nonlinear jump. J. Vib. Control **18**, 1454–1461 (2012). https://doi.org/10.1177/1077546311420460
48. Kirk, G.R., Alsaeed, A.A.: Induced Unbalance as a Method for Improving the dynamic stability of high-speed turbochargers. Int. J. Rotating Mach. **2011**, 1–9 (2011). https://doi.org/10.1155/2011/952869
49. Brouwer, M.D., Sadeghi, F., Lancaster, C., et al.: Whirl and friction characteristics of high speed floating ring and ball bearing turbochargers. J. Tribol. **135**:041102 (2013). https://doi.org/10.1115/1.4024780
50. Eling, R., van Ostayen, R., Rixen, D.: Multilobe floating ring bearings for automotive turbochargers. In: Pennacchi, P. (ed.) Proceedings of the 9th IFToMM International Conference on Rotor Dynamics, pp. 821–833. Springer International Publishing, Cham, (2015)
51. Bernhauser, L., Heinisch, M., Schörgenhumer, M., Nader, M.: The effect of non-circular bearing shapes in hydrodynamic journal bearings on the vibration behavior of turbocharger structures. Lubricants **5**:6 (2017). https://doi.org/10.3390/lubricants5010006
52. Strzelecki, S.: Floating ring journal bearings of multilobe design in turbo-chargers. IOP Conf. Ser. Mater. Sci. Eng. **421**:042077 (2018) https://doi.org/10.1088/1757-899X/421/4/042077
53. Zhang, C., Wang, Y., Men, R., et al.: The effect of three-lobed bearing shapes in floating-ring bearings on the nonlinear oscillations of high-speed rotors. Proc. Inst. Mech. Eng. Part J. J. Eng. Tribol. 135065011987380 (2019).https://doi.org/10.1177/1350650119873802

54. Zhang, C., Wang, Y., Men, R., et al.: Dynamic behaviors of a high-speed turbocharger rotor on elliptical floating-ring bearings. Proc. Inst. Mech. Eng. Part J. J. Eng. Tribol. 135065011984974 (2019).https://doi.org/10.1177/1350650119849743

55. Zhang, C., Men, R., He, H., Chen, W.: Effects of circumferential and axial grooves on the nonlinear oscillations of the full floating ring bearing supported turbocharger rotor. Proc. Inst. Mech. Eng. Part J. J. Eng. Tribol. 233:741–757 (2019).https://doi.org/10.1177/135065011880 0581

56. Zhang, C., Men, R., Wang, Y., et al.: Experimental and numerical investigation on thermodynamic performance of full-floating ring bearings with circumferential oil groove. Proc. Inst. Mech. Eng. Part J J. Eng. Tribol. 233, 1182–1196 (2019). https://doi.org/10.1177/135065011 9826421

57. Nowald, G., Boyaci, A., Schmoll, R., et al.: Influence of axial grooves in full-floating-ring bearings on the nonlinear oscillations of turbocharger rotors. In: SIRM 2015—11th International Conference on Vibrations in Rotating Machines, p. 9. Magdeburg, Germany (2015)

58. Nowald, G., Boyaci, A., Schmoll, R., et al.: Influence of circumferential grooves on the nonlinear oscillations of turbocharger rotors in floating ring bearings. In: Proceedings of the 14th IFToMM World Congress (2015). https://doi.org/10.6567/IFToMM.14TH.WC.OS14.006

59. Smolík, L., Dyk, Š.: Towards efficient and vibration-reducing full-floating ring bearings in turbochargers. Int. J. Mech. Sci. 175:105516 (2020). https://doi.org/10.1016/j.ijmecsci.2020. 105516

60. Deligant, M., Podevin, P., Descombes, G.: Experimental identification of turbocharger mechanical friction losses. Energy 39, 388–394 (2012). https://doi.org/10.1016/j.energy.2011. 12.049

61. Hoepke, B., Uhlmann, T., Pischinger, S., et al.: Analysis of thrust bearing impact on friction losses in automotive turbochargers. J. Eng. Gas Turb. Power 137:082507 (2015). https://doi. org/10.1115/1.4029481

62. Luddecke, B., Nitschke, P., Dietrich, M., et al.: Unsteady thrust force loading of a turbocharger rotor during engine operation. 10 (2015)

63. Chatzisavvas, I., Boyaci, A., Koutsovasilis, P., Schweizer, B.: Influence of hydrodynamic thrust bearings on the nonlinear oscillations of high-speed rotors. J. Sound Vib. 380, 224–241 (2016). https://doi.org/10.1016/j.jsv.2016.05.026

64. Koutsovasilis, P.: Automotive turbocharger rotordynamics: interaction of thrust and radial bearings in shaft motion simulation. J. Sound Vib. 455, 413–429 (2019). https://doi.org/10.1016/j. jsv.2019.05.016

65. Koutsovasilis P (2020) Impact of thrust bearing pad design and allocation on automotive turbocharger rotordynamics. J. Sound Vib. 115546 https://doi.org/10.1016/j.jsv.2020.115546

66. Vetter, D., Hagemann, T., Schwarze, H.: Potentials and limitations of an extended approximation method for nonlinear dynamic journal and thrust bearing forces. In: Volume 7B: Structures and Dynamics. American Society of Mechanical Engineers, Oslo, Norway, p V07BT34A022 (2018)

67. Novotný, P., Hrabovský, J., Klíma, J., Hort, V.: Improving the thrust bearing performance of turbocharger rotors using optimization methods and virtual prototypes. In: 12th International Conference on Vibrations in Rotating Machinery, 1st ed. CRC Press, pp 431–442 (2020)

68. Novotný, P., Hrabovský, J.: Efficient computational modelling of low loaded bearings of turbocharger rotors. Int. J. Mech. Sci. 174:105505 (2020). https://doi.org/10.1016/j.ijmecsci. 2020.105505

69. San Andrés, L., Yu, F., Gjika, K.: On the influence of lubricant supply conditions and bearing configuration to the performance of (semi) floating ring bearing systems for turbochargers. J. Eng. Gas Turb. Power 140:032503 (2018). https://doi.org/10.1115/1.4037920

70. Kim, S., Palazzolo, A.B.: Effects of thermo hydrodynamic (THD) floating ring bearing model on rotordynamic bifurcation. Int. J. Non-Linear Mech. 95, 30–41 (2017). https://doi.org/10. 1016/j.ijnonlinmec.2017.05.003

71. Li, Y., Liang, F., Zhou, Y., et al.: Numerical and experimental investigation on thermohydrodynamic performance of turbocharger rotor-bearing system. Appl. Therm. Eng. 121, 27–38 (2017). https://doi.org/10.1016/j.applthermaleng.2017.04.041

72. Liang, F., Li, Y., Zhou, M., et al.: Integrated three-dimensional thermohydrodynamic analysis of turbocharger rotor and semifloating ring bearings. J. Eng. Gas Turb. Power **139**:082501 (2017). https://doi.org/10.1115/1.4035735

73. San Andrés, L., Barbarie, V., Bhattacharya, A., Gjika, K.: On the effect of thermal energy transport to the performance of (semi) floating ring bearing systems for automotive turbochargers. In: Volume 5: Manufacturing Materials and Metallurgy; Marine; Microturbines and Small Turbo-machinery; Supercritical CO_2 Power Cycles, pp. 561–570. American Society of Mechanical Engineers, Copenhagen, Denmark (2012)

74. Peixoto, T.F., Daniel, G.B., Cavalca, K.L.: Thermo-Hydrodynamic model influence on first order coefficients in turbocharger thrust bearings. In: Cavalca, K.L., Weber, H.I. (eds.) Proceedings of the 10th International Conference on Rotor Dynamics—IFToMM pp. 16–31. Springer International Publishing (2019)

75. Chasalevris, A.: Finite length floating ring bearings: operational characteristics using analytical methods. Tribol. Int. **94**, 571–590 (2016). https://doi.org/10.1016/j.triboint.2015.10.016

76. Chasalevris, A.: An investigation on the dynamics of high-speed systems using nonlinear analytical floating ring bearing models. Int. J. Rotating Mach. **2016**, 1–22 (2016). https://doi. org/10.1155/2016/7817134

77. Dyk, Š, Smolík, L., Hajžman, M.: Effect of various analytical descriptions of hydrodynamic forces on dynamics of turbochargers supported by floating ring bearings. Tribol. Int. **126**, 65–79 (2018). https://doi.org/10.1016/j.triboint.2018.04.033

78. Chatzisavvas, I.: Efficient thermohydrodynamic radial and thrust bearing modeling for transient rotor simulations. Ph.D. Thesis, Technische Universi-tät (2018)

79. Peixoto, T., Cavalca, K.: Influence of thrust bearings in lateral vibrations of turbochargers under axial harmonic excitation. In: 12th International Conference on Vibrations in Rotating Machinery, 1st ed. CRC Press, pp. 39–51 (2020)

80. Novotný, P., Škara, P., Hliník, J.: The effective computational model of the hydrodynamics journal floating ring bearing for simulations of long transient regimes of turbocharger rotor dynamics. Int. J. Mech. Sci. **148**, 611–619 (2018). https://doi.org/10.1016/j.ijmecsci.2018. 09.025

81. Novotný, P., Hrabovský, J., Juračka, J., et al.: Effective thrust bearing model for simulations of transient rotor dynamics. Int. J. Mech. Sci. **157–158**, 374–383 (2019). https://doi.org/10.1016/ j.ijmecsci.2019.04.057

82. Chasalevris, L.: Evaluation of transient response of turbochargers and turbines using database method for the nonlinear forces of journal bearings. Lubricants **7**, 78 (2019). https://doi.org/ 10.3390/lubricants7090078

83. Busch, M., Esmaeili, L., Koutsovasilis, P., et al.: Advanced rotordynamic simulation of turbochargers using coupled multibody and finite element models. In: 10th International Conference on Turbochargers and Turbocharging, pp. 159–171. Elsevier (2012)

84. Novotný, P., Jonák, M., Vacula, J.: Evolutionary Optimisation of the thrust bearing considering multiple operating conditions in turbomachinery. Int. J. Mech. Sci. **195**:106240 (2021).https:// doi.org/10.1016/j.ijmecsci.2020.106240

85. Ibrahim. M.S., Dimitri, A.S., Bayoumi, H.N., El-Shafei, A.: Stable turbocharger bearings. In: 12th International Conference on Vibrations in Rotating Machinery, 1st ed. CRC Press, pp. 588–597 (2020)

86. Sandoval, O.R., Machado, L.H., Caetano, B., et al.: Influence of Temperature and Injection Pressure of Lubrication in the Vibration of an Automotive Turbo-charger. In: Cavalca, K.L., Weber, H.I. (eds.) Proceedings of the 10th International Conference on Rotor Dynamics – IFToMM, pp. 388–397. Springer International Publishing, Cham (2019)

87. Nitzschke, S., Woschke, E., Daniel, C.: Application of regularised cavitation algorithm for transient analysis of rotors supported in floating ring bearings. In: Cavalca, K.L., Weber, H.I. (eds.) Proceedings of the 10th International Conference on Rotor Dynamics—IFToMM, pp 371–387. Springer International Publishing, Cham (2019)

88. Gu, C., Yuan, Z., Yang, Z., et al.: Dynamic characteristics of high-speed gasoline engine turbocharger based on thermo-elasto-hydrodynamic lubrication bearing model and flexible

multibody dynamics method. Sci. Progress **103**:003685041989771 (2020). https://doi.org/10.1177/0036850419897712

89. San Andrés, L., Maruyama, A., Gjika, K., Xia, S.: Turbocharger nonlinear response with engine-induced excitations: predictions and test data. In: GT2009, Vol. 6, Structures and Dynamics, Parts A and B, pp. 637–647 (2009)

90. Chiavola, O., Palmieri, F., Recco, E.: Vibration analysis to estimate turbo-charger speed fluctuation in diesel engines. Energy Procedia **148**, 876–883 (2018). https://doi.org/10.1016/j.egypro.2018.08.107

91. Leichtfuss, S., Durbiano, L., Kreschel, M., et al.: Advanced measurements and model update of an automotive turbocharger with full-floating journal bearings. In: Santos, I. (ed.) Proceedings of the 13th International Conference Dynamics of Rotating Machinery—SIRM, Copenhagen. Copenhagen, Denmark (2019)

92. Peixoto, T.F., Nordmann, R., Cavalca, K.L.: Dynamic analysis of turbochargers with thermo-hydrodynamic lubrication bearings: abstract. J. Sound Vib. 116140 (2021).https://doi.org/10.1016/j.jsv.2021.116140

93. Daniel, C., Woschke, E., Nitzschke, S.: Simulation and measurement of ring speed of full floating ring bearing in an automotive turbocharger. In: Santos, I. (ed) Proceedings of the 13th International Conference Dynamics of Rotating Machinery—SIRM, Copenhagen. Copenhagen, Denmark (2019)

94. IPCC (2014) Climate change 2014: synthesis report. Contribution of Working Groups I, II and III to the Fifth Assessment Report of the Intergovernmental Panel on Climate Change. IPCC, Geneva, Switzerland

Combustion/CFD/Emissions/Fuels

Development of Gasoline Low Temperature Combustion Engine System Compatible with SULEV30 Emissions

Hanho Yun and Madhusudan Raghavan

Abstract With evolving global CO_2 emissions legislation and the increased adoption of electrified powertrains such as hybrid electrics, there is a growing need to extract more fuel efficiency from the Gasoline SI engine. By applying gasoline low temperature combustion over typical driving cycle conditions and traditional stoichiometric spark-ignition combustion over the high speed and high load operating conditions, reduced peak pressures and low friction losses can be maintained across the entire operating ranges, consistent with standard SI engine vehicle practices and expectations. This research is focused on combining three enabling technologies for synergistic integration: (a) downsized boosting to address parasitic losses, (b) lean, low temperature combustion to address heat and work extraction losses, and (c) physics-based cylinder-pressure driven controls to streamline the calibration and implementation process. In summary, this research shows that an advanced low temperature combustion system integrated with modern downsize boosted engine technology can deliver a significant fuel consumption benefit (~20%) over conventional natural-aspirated, homogeneous stoichiometric spark-ignition engines. Furthermore, this research proves that this can be done with commercially available fuels, without impacting the customer experience consistent with the most stringent emissions requirements (SULEV30).

Keywords Negative valve overlap · Positive valve overlap · Gasoline low temperature combustion · Passive ammonia SCR system (PASS) · Hybrid-suitable engine

Abbreviations

SI   Spark-ignition
EGR   Exhaust Gas Recirculation

H. Yun (✉) · M. Raghavan
Propulsion Systems Research Lab, General Motors R&D, Oakland, USA
e-mail: hanho.yun@gm.com

RCCI	Reactivity-Controlled Compression Ignition
GDCI	Gasoline Direct-Injection Compression Ignition
HCCI	Homogeneous Charge Compression Ignition
LTC	Low Temperature Combustion
FTP	Federal Test Procedure
SULEV	Super Ultra Low Emissions Vehicle
NVO	Negative Valve Overlap
PVO	Positive Valve Overlap
RI	Ringing Index (Intensity)
CA10	Crank Angle for 10% mass fraction burned
CA50	Crank Angle for 50% mass fraction burned
CA90	Crank Angle for 90% mass fraction burned
A/F ratio	Air/Fuel ratio.
COV	Coefficient of Variation
IMEP	Indicated Mean Effective Pressure
NMEP	Net Mean Effective Pressure
BMEP	Brake Mean Effective Pressure
TDC	Top Dead Center
aTDC, bTDC	After/before Top Dead Center
BSFC	Brake Specific Fuel Consumption
HRR	Heat Release Rate
SPK	Spark
IGN	Ignition
PASS	Passive SCR System
TWC	Three-Way Catalyst
SCR	Selective Catalytic Reduction
WRAF	Wide Range Air Fuel
TP	Tail-Pipe

1 Introduction

Gasoline SI (Spark-ignition) engines have been the primary power source for US passenger vehicles due to their numerous benefits such as low emissions, relatively simple control, and low cost. However, with evolving CO_2 emissions legislation in the world, there is a growing need to extract more fuel efficiency from the gasoline SI engine. Charge boosting and engine downsizing can be used to reduce parasitic losses and is an industry-wide step in the direction of increased efficiency, but doesn't address heat losses and/or work extraction losses. Gasoline low temperature combustion (lean and exhaust gas recirculation dilution, i.e., EGR), on the other hand, directly targets these losses. Gasoline low temperature combustion provides a significant opportunity for decreasing fuel consumption, with estimated system and emission control costs less than half of those associated with modern diesel engines.

High efficiency is achieved through an increase in the compression ratio as well as an increase in the specific-heat ratio of the lean mixture resulting in improved work extraction and a reduction in engine heat losses [1, 2]. As such, the successful integration of downsizing and low temperature combustion allows for the simultaneous reduction of all major efficiency losses.

Most gasoline low temperature combustion technologies rely on auto-ignition of the mixture in the cylinder [3–6]. Auto-ignition characteristics are strongly dependent on the reactivity of the mixture and prior approaches in this space such as Reactivity-Controlled Compression Ignition (RCCI) and Gasoline Direct-Injection Compression Ignition (GDCI), are briefly reviewed here. RCCI combustion allows the reactivity of the combustion mixture to be tailored to the speed-load condition through fuel delivery and in-cylinder mixing of fuels of differing reactivity. By premixing the low-reactivity (and higher volatility) fuel to a lean overall mixture, auto-ignition is avoided prior to the diesel fuel injection (high-reactivity). The early diesel injection then provides a timed and distributed ignition source [3, 4]. The GDCI concept has a high compression ratio with a 200-bar peak pressure capability and two boost systems, leading to high parasitic losses and implementation costs. Furthermore, since GDCI relies on auto-ignition for all operating conditions, combustion noise (ringing) issues at high load and combustion robustness at light load conditions are challenging [5, 6]. In addition, since both RCCI and GDCI rely on lean operation for all operating conditions the specific output of these engines is relatively low, leading to large engine displacement requirements for equivalent vehicle level performance.

To address these vehicle level implementation issues, the engine concept proposed here, differs from these existing low temperature combustion engines by operating with higher specific output, lower parasitic losses, lower engine/vehicle mass, and reduced cost through synergistic integration. The proposed engine concept relies on engine downsizing to reduce parasitic losses and lean, low temperature combustion over the majority of the driving cycle to minimize heat losses and maximize work extraction. Since the proposed engine concept relies on simple, fixed geometry turbocharger technology as opposed to complex multi-stage and/or high-pressure boost devices there will be significant implementation cost and complexity reductions along with lower peak pressure requirement resulting in reduced mass and frictional losses.

It is generally accepted that the main obstacles to commercial implementation of gasoline low temperature combustion (LTC) are limited operating ranges as shown in Fig. 1 and difficulty in combustion timing control. The operating range is restricted by misfiring at low load and limited by a high rate of pressure rise at high load. Over the past 15 years, GM has developed and demonstrated the potential to solve these obstacles with model-based engine controls and novel mixture control strategies. This capability was demonstrated publicly in 2008/09 and has since been augmented by the intelligent integration of advanced fuel injection strategies [7–9]. These systems allow robust low temperature combustion and precise combustion phasing control across a wide range of operating conditions while relying on a single readily available fuel and simple low-cost solenoid injectors. By applying gasoline

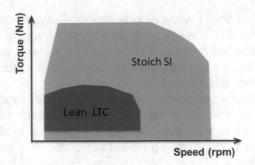

Fig. 1 Representation of the limited operating range of robust LTC operation in a conventional 4-stroke engine

low temperature combustion over typical driving cycle conditions and traditional stoichiometric spark-ignition combustion over the high speed and high load operating conditions, reduced peak pressures and low friction losses can be maintained across the entire operating range. The seamless transition between low temperature combustion and homogeneous spark-ignition combustion is achieved through valve event matching and precise model-based control as demonstrated publicly with the GM HCCI (homogeneous charge compression ignition) demonstration vehicles [10].

The combined effects of such advanced combustion methods with various levels of electrification/hybridization have also been investigated by various researchers. Lawler et al. [11] have explored the synergies between powertrain electrification and HCCI. They found that HCCI combustion offered a 17% improvement in fuel economy over a conventional engine, when coupled with mild hybridization in the form of a 5 kW integrated starter/motor, this benefit jumped to 35%. Burke [12] has explored the benefits of a series hybrid arrangement relative to a parallel hybrid arrangement. His simulation results for hybrid/electric vehicles using ultracapacitors to load level the engine, show a large potential improvement in fuel economy, by allowing engine down-sizing and reduced specific power requirements. Shahed [13] presents a study that shows that engine downsizing, assisted turbocharging and light hybridization with limited launch assist can possibly provide the same benefits (fuel consumption reduction and launch torque restoration) as full hybrid system without the added weight and cost.

The primary contribution of the present research is the development and demonstration of a downsized boosted, lean, low temperature gasoline combustion engine system, showing a significant fuel economy improvement relative to a contemporary naturally aspirated stoichiometric combustion engine, while meeting SULEV30 emission requirements using marketplace gasolines. This engine may be used in a conventional powertrain, as well as in electrified drivelines with various levels of hybridization.

2 Experimental Setup

The enabling technologies selected for evaluation in this research were lean low temperature combustion, engine downsizing and boosting to minimize friction, a low-cost lean after-treatment system, and physics-based control. These technologies were integrated into a multi-cylinder engine with four valves per cylinder and centrally located direct fuel injectors The compression ratio was 12.0:1. The configuration of piston, spark plug position and injector position were designed to enable stratified charge combustion. The specifications of the engine are given in Table 1. Full engine assembly installed in the dynamometer is shown in Fig. 2.

Table 1 Comparison of engine specification between baseline and proposed engine

	Baseline engine	Proposed engine
Displacement volume	2.2L	2.2L as a surrogate of 1.4L
Compression ratio	12:1	12:1
Bore/Stroke	86 mm/94.6 mm	86 mm/94.6 mm
Intake cam	9.3 mm*280°CA	9.3 mm*280°CA; 6.0 mm*160°CA (2-step)
Exhaust cam	9.3 mm*280°CA	9.3 mm*280°CA; 5.0 mm*150°CA (2-step)
Ignition system	Conventional spark plug system	Conventional spark plug system
# of Injector holes	8 holes	10 holes
Spray cone angle	60°	2 holes  60 8 holes  90
Injection pressure	12 MPa	12 MPa

Fig. 2 Full engine assembly installed in dynamometer

For negative valve overlap (NVO) operation, the peak valve lift of intake and exhaust valves was 5.0 mm and 6.0 mm, respectively, and the opening duration of intake and exhaust valves was 150 and 160 crank angle degrees (°CA), respectively. The low lift valve profile was designed considering the compatibility with a two-step mechanism. For positive valve overlap (PVO) operation, the same valve profiles as conventional SI operation were used, which are 9.3 mm of peak valve lift and 280°CA of valve duration for both intake and exhaust valves. The injectors used in these experiments were a 60° 8-hole solenoid actuated injector for the baseline and a 60/90 combination, 10-hole injector for proposed engine's low-temperature operation. The fuel injection pressure was maintained at 12 MPa throughout the experiment. The fuel used was a fully blended gasoline with octane number (RON(90.9) + MON(83.2))/2 of 87. The engine coolant and engine oil were set to 95 °C. The cylinder pressure was measured with a Kistler 6125A pressure transducer. Exhaust emissions were sampled from the exhaust system and analyzed by a Horiba Mexa 7000 exhaust gas analyzer. The air fuel ratio in the exhaust was measured with a wide-range air–fuel sensor.

The heat release calculation was performed according to the first law of thermodynamics. The heat transfer was accounted for whereas blow-by is assumed to be negligible and not taken into account for the heat release calculation. The heat transfer calculation used the heat transfer coefficient correlation proposed by Woschni [14] with a slight modification of the velocity term. The residual gas fraction was estimated using the correlation published by Yun and Mirsky [15]. To evaluate combustion noise, the ringing index concept is used in this study [16]. Ringing index is defined as

$$RI = 2.88 \times 10^{-8} \times \frac{(MPRR \times RPM)^2}{PP}$$

where $MPRR$ is the maximum pressure rise rate (kPa/deg), RPM is the engine speed, and PP is the peak pressure (kPa).

3 Results and Discussion

3.1 Homogeneous Stoichiometric Spark-Ignition Combustion (Baseline)

The baseline homogeneous stoichiometric calibrations to cover the FTP cycles were developed by optimizing cam timing and combustion phasing (CA50, crank angle where 50% mass fraction is burned) to build a baseline. The optimal cam timing and combustion phasing were selected, considering efficiency and combustion stability. These baseline data serve as the reference data for comparisons with the lean low temperature combustion (LTC) data from our proposed novel concept.

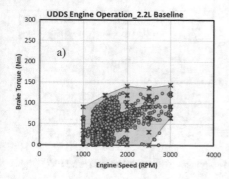
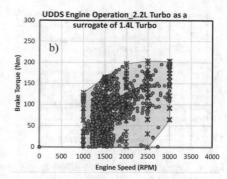

Fig. 3 Comparison of calibration points to cover FTP cycle for the baseline (left) and LTC (right) engines

Since stoichiometric SI operation would be used to run portions of the FTP cycle for our proposed LTC engine too, due to the downsized displacement effect, additional steady-state calibrations were obtained from the LTC engine by optimizing cam timing, CA50 and EGR rates. Figure 3 shows the comparison of calibrations obtained from the baseline engine and LTC engine to cover FTP cycle.

3.2 Low Temperature Combustion (LTC) Strategy Development

To extend the LTC operating range, two different valving strategies were employed. One is negative valve overlap, which is used at low load operation. The other is positive valve overlap which is used at medium to high load operation.

3.2.1 Low Temperature Combustion Characteristics During NVO and PVO

In NVO operation, exhaust valve closing (EVC) timing determines the internal residual (hot residual gas) mass fraction. The burned gases trapped inside the cylinder are recompressed and allowed to expand before the intake valve opens. The decreasing residual gas fraction means that more air is inducted in the cylinder, which changes the mixture air/fuel (A/F) ratio. The basic principle of LTC in this operating condition is that the air and fuel are premixed prior to ignition and then ignited by the compression from the piston motion. The ignition is provided in multiple points and therefore the charge gives a parallel energy release. This results in a uniform and simultaneous auto-ignition throughout the charge without flame propagation, which is relatively fast and more akin to an idealized combustion event. Therefore, more efficient engine operation with lower emissions can be realized. Combustion timing

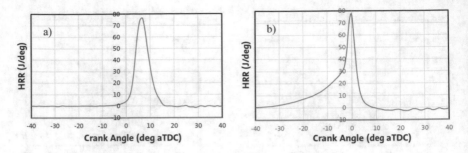

Fig. 4 Typical heat release rate of low temperature combustion during NVO (left) and PVO (right) operation

control and combustion noise have been known as main obstacles for commercialization of this concept. A representative plot of heat release rate of low temperature combustion during NVO operation is shown in Fig. 4a.

On the other hand, in PVO operation, the amount of PVO determines the trapped residual mass. Therefore, as engine load varies (and injected fuel mass changes), the amount of PVO or EGR rate should be adjusted to change the mixture A/F ratio. A plot of typical heat release rate of low temperature combustion during PVO operation is presented in Fig. 4b. Combustion starts with flame-burn and is followed by auto-ignition later in the cycle. The trade-off between NOx emissions and combustion stability is the main challenge to be resolved. When combustion executes by flame-propagation, it becomes stable, but NOx emissions are increased. When combustion is dominated by auto-ignition, NOx emissions are decreased, but combustion becomes unstable due to a lack of internal hot residuals.

Seamless mode switching between NVO and PVO plays an important role in improving combustion robustness. The LTC engine is equipped with two-step valve lifters which can instantaneously switch from low-lift valve profiles (for NVO mode) to high-lift valve profiles (for PVO and SI modes) and vice versa. When intake and exhaust valve timings with the low-lift valve profiles reach minimum NVO, the desired mode changes from NVO to PVO. The cam lobes of the low-lift and high-lift profiles are designed such that the geometrical cylinder volumes are closely matched before and after mode change from NVO to PVO and vice versa, so that cylinder charge has minimal disruptions during mode transitions. Once valve profile switching is successfully completed, intake and exhaust valve timings are continuously controlled to meet the desired geometrical cylinder volume with necessary restrictions.

3.2.2 Ignition Timing Control Methodology

One of the known barriers to LTC is the difficulty in ignition timing control. Figure 5 explains why the optimal combustion phasing is important for gasoline low temperature combustion. The most advanced CA50 (crank angle for 50% mass fraction

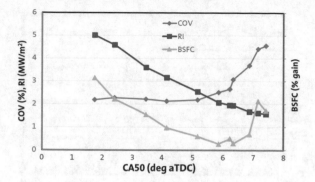

Fig. 5 Effect of combustion phasing on COV, ringing and efficiency

burned) is limited by the combustion noise (ringing intensity is larger than 5 MW/m²), and the most retarded CA50 is limited by combustion stability (COV of IMEP is larger than 3%). The optimum CA50 should be chosen within this range, considering the best efficiency at each operating condition.

For successful combustion phasing control, two important aspects of the control parameters need to be considered. One is real time response to the combustion and the other is independence of other combustion parameters. Intake air temperature, valve timing and EGR are not appropriate parameters to use as control knobs for these reasons. The methodology of combustion phasing control using injection timing and/or spark timing was developed and demonstrated as part of the present effort. Figure 6 shows the effect of the injection timing on the combustion phasing at the 2000 rpm, 2 bar BMEP condition. As the injection timing is advanced, the combustion phasing is advanced and the total burn duration (defined as CA10 to CA90) is shortened. The corresponding variation of heat release rate is shown in Fig. 7. Injection timing was chosen as a control parameter to change combustion phasing at this operating condition.

As engine load increases, injection timing is no longer a strong parameter to influence combustion phasing. Instead, spark timing was used to control the combustion phasing combined with a multiple injection strategy at this operating condition.

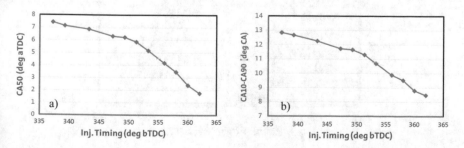

Fig. 6 Effect of injection timing on the combustion phasing (left) and the total burn duration (right)

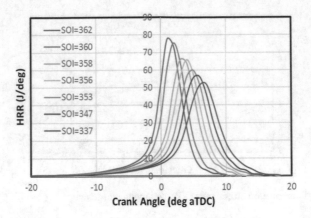

Fig. 7 Comparison of heat release rate as a function of injection timing

Figure 8 shows the effect of spark timing on the combustion phasing and combustion noise at 2000 rpm, 4 bar BMEP. The corresponding heat release rate for each spark

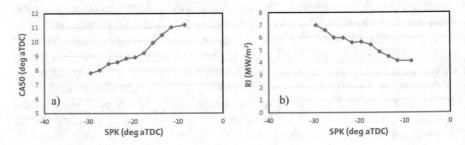

Fig. 8 Effect of spark timing on the combustion phasing (left) and combustion noise (right)

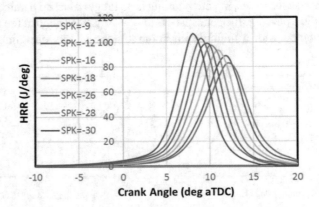

Fig. 9 Comparison of heat release rate as a function of spark timing

timing is presented in Fig. 9. As shown in Fig. 8b, combustion noise can be reduced by retarding combustion phasing, by suitably adjusting the spark timing.

3.2.3 Lean Limit Extension During NVO Operation (Effect of Ignition on LTC)

In LTC engine, the ignition system is needed for high load SI operation for a high specific output. However, the combustion characteristic of LTC during NVO operation is auto-ignition. Then it is fair to ask what the role of the ignition system is in LTC during NVO operation? Figure 10 shows a comparison of lean limit extension with ignition and without ignition at the 2000 rpm, 2 bar BMEP condition, with combustion phasing fixed at 5° aTDC. More stable lean operation is achieved with the ignition system than without it. Figure 11 presents a comparison of cyclic variability of IMEP, obtained from the conditions indicated by the dotted circle in

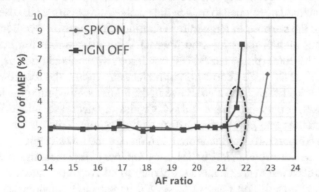

Fig. 10 Comparison of AF ratio sweep at 2000 rpm, 2 bar BMEP condition (CA50 = 5° aTDC)

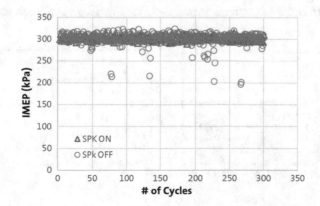

Fig. 11 Comparison of cyclic variability of IMEP at 2000 rpm, 2 bar BMEP condition (AF ratio = 21.5)

Fig. 10. As shown in the plot, spark ignition provides stability by limiting problematic cycles. It was found that the ignition system does not drive combustion, but it provides stability in LTC mode.

3.2.4   Combustion Noise Reduction During NVO Operation

In this section, the strategy developed to reduce the combustion noise at high load LTC operation is presented. The effect of EGR on combustion noise at the 2000 rpm, 3.3 bar BMEP, CA50 7° aTDC condition, is shown in Fig. 12a. Lower ringing was obtained with higher EGR rate, due to the decrease of combustion temperature (chemical effect of EGR). For a clear explanation, the corresponding heat release rates are compared and presented in Fig. 12b. Since EGR suppresses the auto-ignitability of the mixture, combustion begins with flame burn and volumetric auto-ignition takes place later in the cycle, which leads to lengthening of the total burn duration. This plays a key role in decreasing combustion noise at high load NVO operation.

Another method developed to reduce the combustion noise was the use of multiple injections for the purpose of temperature stratification. Figure 13 presents the effect of the amount of fuel mass injected near TDC on the combustion noise (ringing intensity) at the 2000 rpm, 4 bar BMEP condition, with a fixed combustion phasing (CA50 at 9.5° aTDC). The corresponding NOx emissions, the total burn duration, and combustion efficiency are also shown in the figure. As the fuel mass injected near TDC increases, combustion noise and NOx emissions decrease, while the total burn duration increases and combustion efficiency decreases. This clearly explains the effect of the temperature stratification. Even though the lowest ringing intensity and NOx emissions were obtained with the largest amount of fuel mass injected near TDC, a significant decrease in combustion efficiency should be anticipated. Therefore, the optimal fuel mass injected near TDC should be selected considering combustion noise, NOx emissions and combustion efficiency at each operating condition.

This finding was also applied to the reduction of combustion noise during transient operation. When the LTC engine is in a rapid load transient condition starting from

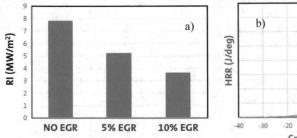 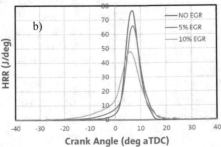

Fig. 12 Effect of EGR on combustion noise (left) at 2000 rpm, 3.3 bar BMEP condition (NVO operation, AF = 17:1) and corresponding heat release rate (right)

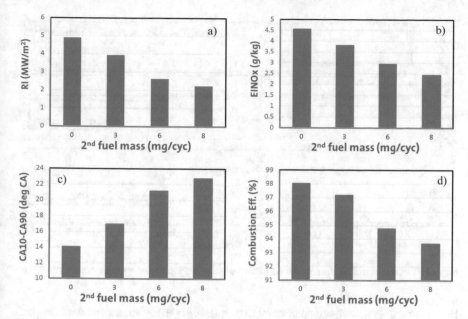

Fig. 13 Effect of the fuel mass injected near TDC on combustion noise, NOx, total burn duration and combustion efficiency at 2000 rpm, 4 bar BMEP condition (CA50 = 9.5° aTDC)

a light load, a burst of combustion noise can occur as the temperature of the cylinder charge may be too hot during mode transition, due to EGR transport delay. When the cylinder charge is too hot, combustion tends to be auto-ignited with rapid pressure rise, causing combustion noise and ringing. To reduce the combustion noise during mode transitions, an extra amount of fuel is temporarily injected at around spark timing, immediately after mode-switching from NVO to PVO, until the exhaust gas reaches the intake manifold through the EGR pipe. The total amount of fuel injected in the cylinder is still maintained at the desired value to meet the torque demand. The extra fuel helps the combustion to slow down by fortifying spark-assisted auto-ignition, but this is at the cost of NOx emissions, as the cylinder charge is more stratified. Thus, the amount of extra fuel must be optimized to decrease emissions while maintaining combustion stability during mode transitions.

Figure 14 shows experimental data collected from the LTC engine at 2000 rpm, when a step change of 4 bars of NMEP (load demand) was requested. The figure shows that the burst combustion noise is successfully suppressed during the mode transition with the variable injection strategy in the presence of EGR transport delay.

3.2.5 Stability Improvement During PVO Operation

To improve the trade-off between combustion stability and NOx emissions in low temperature combustion with PVO operation, a double injection strategy was applied.

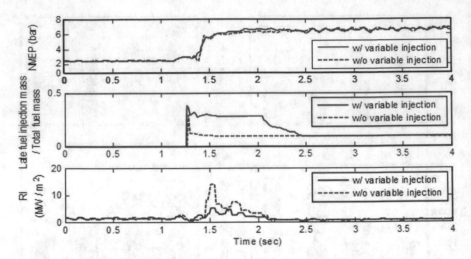

Fig. 14 Variable injection strategy for combustion noise reduction during mode transition

Figure 15 shows a comparison of combustion stability using single and double injection, at a 2000 rpm, 4.5 bar BMEP condition, during PVO operation. The combustion phasing (CA50) was maintained at 4.5° aTDC. A significant improvement in combustion stability was demonstrated without increasing NOx emissions. It was found that robust flame kernel development by a small amount of fuel near the spark plug is critical to ensuring auto-ignition, later in the cycle. This can be clearly seen in Fig. 16, which presents the corresponding heat release rate. For single injection, there is a large cyclic variability of heat release generated by flame burn. Weak flame burn leads to weak auto-ignition later in the cycle, which contributes to instability of the combustion. Providing repeatable and consistent flame burn at each cycle is a key enabler for enhancing the robustness of low temperature combustion during PVO operation.

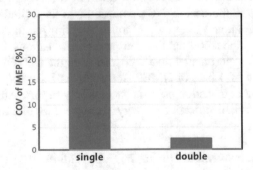

Fig. 15 Effect of the fuel mass injected near TDC on combustion stability during PVO operation at 2000 rpm, 4.5 bar BMEP condition (CA50 = 4.5° aTDC)

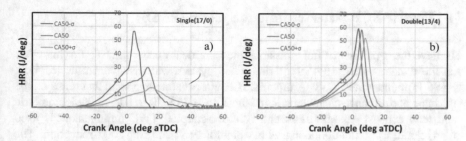

Fig. 16 Comparison of HRR between single injection (left) and double injection (right) during PVO operation at 2000 rpm, 4.5 bar BMEP condition

3.2.6   LTC Operating Range Extension

One of the known drawbacks of LTC is a limited operating range. Figure 17 explains why high load LTC operation is limited. At a given load, the operable CA50 range is determined by the combustion noise and stability. The most advanced CA50 is limited by the combustion noise and the most retarded CA50 is limited by combustion stability. As the engine load increases, CA50 should be retarded to reduce combustion noise. As the engine load increases further, we reach a singular limiting point, determined by the stable combustion and low combustion noise criteria. This is effectively, the high load limit point of LTC operation. However, by applying EGR and the multiple injection strategy, this operable range of CA50 can be enlarged, thereby extending the operating range of LTC. In addition, the range of LTC may be further extended by employing the PVO valving strategy, thus eliminating the combustion noise problem.

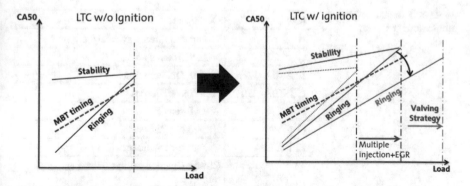

Fig. 17 Conceptual diagram of the reason for the limited operating range of LTC and methodology to show how to extend the operating range of LTC

4   Passive SCR Aftertreatment System (PASS)

To have the capability of full emissions measurement during the FTP driving cycle on the dynamometer, a PASS aftertreatment system was installed in the test cell. PASS is an integrated TWC-Ammonia SCR aftertreatment system that relies on the intrinsic performance characteristics of the TWC and SCR to address stoichiometric and lean exhaust gas aftertreatment. A close-coupled TWC arrangement is used for HC and CO oxidation along with stoichiometric and rich NOx reduction. The engine needs to operate rich periodically to generate NH_3 on the TWC and store it on the SCR. During lean operating conditions, the stored NH_3 on the SCR may be used for NOx conversion. Figure 18 shows the schematic diagram of the PASS aftertreatment system installed in the test cell. Table 2 summarizes the details of the PASS aftertreatment system.

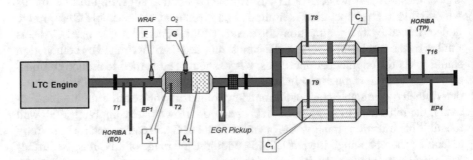

Fig. 18 Schematic diagram of the PASS aftertreatment system

Table 2 Summary of overall aftertreatment system

Item	Description
A1 & A2	Production three-way catalyst
C1 & C2	NOx SCR catalyst
HORIBA (EO)	Engine-Out emissions
HORIBA (TP)	Tail-Pipe emissions
F	WRAF sensor
G	Heated oxygen sensor
EP1 & EP4	Exhaust gas static pressure
T1	Closed-Coupled catalyst inlet gas temperature
T2	TWC catalyst brick temperature—1" from front face
T8 & T9	SCR catalyst brick temperature—1" from front face
T16	Exhaust system outlet gas temperature

5 Federal Test Procedure (FTP) Drive Cycle Test Results

One may recall from a prior section, that a hypothetical 1.4L turbo engine is the natural "equivalent" of the baseline 2.2L naturally-aspirated engine, for comparison purposes. However, our dynamometer test property is a 2.2L turbo LTC engine. Hence the desired torque profile for the FTP test is scaled up by 2.2/1.4. Once the driving test is completed, the experimental data collected from the LTC engine, such as fuel consumption, emissions, torque measurement, etc., are scaled down by the ratio of 2.2/1.4, for comparison with the baseline 2.2L naturally-aspirated SI engine.

The dynamometer test was repeated three times for each engine, and experimental data collected from the engine were averaged for consistency. The experimental results of the FTP driving cycle test of the LTC engine (scaled as described above) are plotted in Fig. 19. The orange line indicates NVO LTC operation, the blue line presents PVO LTC operation, and the black line shows stoichiometric SI operation. It was confirmed that 86% of the FTP driving cycle was operated at lean operation. This contributes to the fuel economy gain.

To generate NH_3 on the TWC for NOx reduction, the LTC engine is periodically operated with enriched fuel and air mixture, when the engine is in the lean PVO combustion mode. The level of enrichment is a function of engine load. The duration and frequency of enrichment are set to optimal values, determined by a trade-off between emissions and fuel consumption over the FTP cycle. The whole process taken to reduce tail-pipe emissions is presented in Fig. 20. The red-dotted rectangular box indicates the SULEV30 emissions target. Initially, without the implementation of any enrichment strategy, unacceptably high levels of NOx emissions were measured. By applying a more sophisticated enrichment strategy, NOx, HC and CO emissions were significantly reduced. Finally, the project team met the SULEV30 emissions target while still achieving fuel economy gains, as shown in Fig. 21. These measurements are from the hot (engine warmed up) FTP cycle operation, A detailed comparison of LTC emissions data versus the stoichiometric baseline emissions is shown in Table 3. A 20.5% fuel economy gain was obtained for the hot FTP75 driving cycle test over the baseline, while meeting SULVE30 emissions regulation (15.37 mg of NOx+HC; 0.16 g/mile of CO).

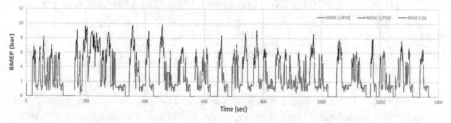

Fig. 19 FTP driving cycle test results of 1.4L LTC engine

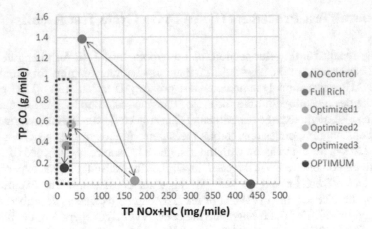

Fig. 20 History of optimization process for tail-pipe emissions

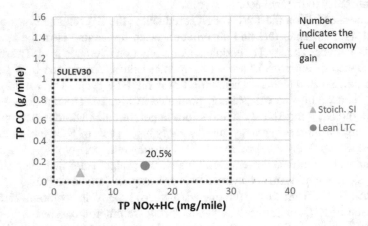

Fig. 21 Comparison of tail-pipe emissions (baseline vs LTC)

Table 3 Comparison of emissions results obtained from stoichiometric Hot FTP75

	NOx (mg/mile)	HC (mg/mile)	CO (g/mile)	NOx+HC (mg/mile)	Fuel economy (mpg)	% Gain
Baseline	1.81	2.70	0.09	4.51	30.53	
LTC	5.74	9.63	0.16	15.37	36.80	20.5%

The objective of this project was to develop and demonstrate the application of key enabling technologies such as downsizing, boosting, low temperature combustion, ignition, and advanced control. These enabling technologies have been shown to enable a lean low temperature combustion down-sized boosted spark ignition gasoline engine operating with EGR dilution to achieve a significant fuel economy benefit relative to conventional naturally aspirated gasoline engines. By combining this technology with hybridization, additional synergistic benefits may be expected in the mid-term light-duty market.

6 Conclusions

This research describes an LTC combustion engine, integrating a low temperature combustion system with enabling technologies, including downsizing, boosting, aftertreatment, and physics-based control. Driving cycle fuel economy improvement (hot FTP cycle) of 20.5% relative to a naturally aspirated 2.2L baseline engine, was obtained while meeting current and anticipated future emissions standards (SULEV30). Key accomplishments in the project may be summarized as follows.

Homogeneous stoichiometric SI combustion development

- The baseline homogeneous stoichiometric calibrations to cover the FTP driving cycle have been developed by optimizing cam timing and combustion phasing (CA50). These data served as the reference baseline for the subsequent lean low temperature combustion system development.

Low Temperature Combustion (LTC) Strategy Development

- Ignition Timing Control Methodology—The methodology of combustion phasing control was successfully developed and demonstrated using injection timing and spark timing.
- Lean limit extension during NVO operation—Improved stable lean operation was achieved via the ignition system, because spark ignition provides stability by limiting problematic cycles. It was found that the ignition system does not drive combustion, but it provides stability in LTC mode.
- Combustion noise reduction during NVO operation—The use of EGR and temperature stratification helped to reduce combustion noise during LTC operation.
- Stability improvement during PVO operation—A double injection strategy was applied to improve the combustion stability during PVO operation. It was found that robust flame kernel development by a small amount of fuel near the spark plug, is critical to ensuring auto-ignition later in the cycle.
- LTC operating range extension—One of the known drawbacks of LTC is a limited operating range. By applying EGR and a multiple injection strategy, the range of CA50 (limited by stability and ringing) can be enlarged. This leads to the extension of the operating ranges of LTC. In addition, the range of LTC can be further extended by employing a new valving strategy during PVO operation.

FTP Test Results

- To allow full emissions measurement on the dynamometer, the project team installed a PASS aftertreatment system. PASS consists of a production TWC and an underfloor NOx SCR catalyst.

- By applying a sophisticated enrichment strategy, sub-SULEV30 emissions (15.37 mg/mile of NOx + HC; 0.16 g/mile of CO) were demonstrated, while still achieving a 20.5% fuel economy gain on the hot FTP cycle.

References

1. Yun, H., Wermuth, N., Najt, P.: Extending the high load operating limit of a naturally-aspirated gasoline HCCI combustion engine. SAE Int. J. Engines. **3**(1), 681–699 (2010). https://doi.org/10.4271/2010-01-0847
2. Yun, H., Idicheria, C., Najt, P.: The effect of advanced ignition system on gasoline low temperature combustion. Int. J. Engines. Res. (2019).https://doi.org/10.1177/1468087419867543
3. Kokjohn, S., Hanson, R., Splitter, D., Kaddatz, J., et al.: Fuel reactivity controlled compression ignition (RCCI) combustion in light- and heavy-duty engines. SAE Int. J. Engines. **4**(1), 360–374 (2011). https://doi.org/10.4271/2011-01-0357
4. Kokjohn, S.L., Hanson, R.M., Splitter, D.A., Reitz, R.D.: Fuel reactivity controlled compression ignition (RCCI): a pathway to controlled high-efficiency clean combustion. Int. J. Engine Res. **12**, 209 (2011). https://doi.org/10.1177/1468087411401548
5. Sellnau, M., Foster, M., Hoyer, K., Moore, W., Sinnamon, J., Husted, H.: Development of a gasoline direct injection compression ignition (GDCI) engine. SAE Int. J. Engines. **7**(2), 2014. https://doi.org/10.4271/2014-01-1300
6. Sellnau, M., Moore, W., Sinnamon, J., Hoyer, K., Foster, M., Husted, H., GDCI multi-cylinder engine for high fuel efficiency and low emissions. SAE Int. J. Engines. **8**(2), 2015. https://doi.org/10.4271/2015-01-0834
7. Yun, H., Wermuth, N., Najt, P.: Development of robust gasoline HCCI idle operation using multiple injection and multiple ignition strategy. SAE 2009-01-0499 (2009). https://doi.org/10.4271/2009-01-0499
8. Wermuth, N., Yun, H., Najt, P.: enhancing light load HCCI combustion in a direct injection gasoline engine by fuel reforming during recompression. SAE Int. J. Engines **2**(1), 823–836 (2009). https://doi.org/10.4271/2009-01-0923
9. Yun, H., Wermuth, N., Najt, P.: High load HCCI operation using different valving strategies in a naturally-aspirated gasoline HCCI engine. SAE Int. J. Engines **4**(1), 1190–1201 (2011). https://doi.org/10.4271/2011-01-0899
10. Alt, M., Grebe, U.D., Dulzo, J.R., Ramappan, V.A., Kafarnik, P., Najt, P.M.: HCCI—from lab to the road. In: International Vienna Motor Symposium (2008)
11. Lawler, B., Ortiz-Soto, E., Gupta, R., Peng, H., Filipi, Z.: Hybrid electric vehicle powertrain and control strategy optimization to maximize the synergy with a gasoline HCCI engine. SAE 2011-01-0888
12. Burke, A.: Hybrid/electric vehicle design options and evaluations. SAE 920447
13. Shahed, S.M.: An analysis of assisted turbocharging with light hybrid powertrain. SAE 2006-01-0019
14. Woschni, G.: A universally applicable equation for the instantaneous heat transfer coefficient in the internal combustion engine. SAE 670931 (1967). https://doi.org/10.4271/670931
15. Yun, H.J., Mirsky, W.: Schlieren-streak measurements of instantaneous exhaust gas velocities from a spark ignition engine. SAE 741015 (1974). https://doi.org/10.4271/741015
16. Eng, J.A.: Characterization of pressure waves in HCCI combustion. SAE 2002-01-2859 (2002). https://doi.org/10.4271/2002-01-2859

Simulation of Multistage Autoignition in Diesel Engine Based on the Detailed Reaction Mechanism of Fuel Oxidation

S. M. Frolov[ID], S. S. Sergeev, V. Ya. Basevich, F. S. Frolov, B. Basara, and P. Priesching

Abstract Three-dimensional numerical simulations of mixture formation, autoignition, and combustion processes in a cylinder of a sample Diesel engine using the detailed reaction mechanism of fuel oxidation are performed. Particular attention is paid to the autoignition process. The three-stage nature of fuel autoignition in a Diesel engine characterized by the successive appearance of cool, blue, and hot flame exothermic centers has been observed computationally for the first time. The specific features of each of the three stages of autoignition and their interaction with each other are revealed. The location of the first centers of autoignition is identified. The influence of the parameters of numerical procedure on the calculated characteristics of multistage autoignition is investigated. The influence of the fuel Cetane number on the engine operation process is discussed.

Keywords Diesel engine · Computational fluid dynamics · Detailed reaction mechanism · Multistage autoignition · Exothermic centers

1 Introduction

As stated in [1], the theory of two-stage (multistage) autoignition as applied to piston engines was first formulated by Sokolik, who worked in the famous research group of Semenov at the Institute of Chemical Physics of the USSR Academy of Sciences. The two-stage nature of autoignition of injected fuel under Diesel engine conditions with the necessary preliminary formation of cool flame was proved in 1954 by means

S. M. Frolov (✉) · S. S. Sergeev · V. Ya. Basevich · F. S. Frolov
Semenov Federal Research Center for Chemical Physics of the Russian Academy of Sciences (FRC), 4 Kosygin Str., Moscow 119991, Russian Federation
e-mail: smfrol@chph.ras.ru

S. M. Frolov · F. S. Frolov
Federal State Institution "Scientific Research Institute for System Analysis of the Russian Academy of Sciences", 36-1 Nakhimovskii Prosp., Moscow 117218, Russian Federation

B. Basara · P. Priesching
AVL LIST GmbH, 1 Hans-List-Platz, Graz, 8020, Austria

© The Author(s), under exclusive license to Springer Nature Switzerland AG 2022
T. Parikyan (ed.), *Advances in Engine and Powertrain Research and Technology*,
Mechanisms and Machine Science 114,
https://doi.org/10.1007/978-3-030-91869-9_6

of direct experiments in an adiabatic compression machine designed by Voinov for reproducing a single operation cycle of Diesel engine. The time history of pressure as well as the glow in the cylinder with a transparent cover were registered in [2]. During fuel injection, the ranges of temperature and pressure were 575–790 K and 10–25 bar, respectively. Later, in [3], based on the analysis of experiments [2] the intermediate stage of blue flame was assumed to be present. It is worth noting that the concept of blue flame is identical with the concept of "intermediate temperature heat release" (ITHR) [4], because the ITHR spectrum is predominantly blue-light emission with a peak at 450 nm. In review [5], Diesel ignition is treated as an applied example of intermediate-temperature ignition, whereas in [6], the blue flame exothermic centers are detected experimentally during fuel spray autoignition. Multistage autoignition is a series of successive flames associated with certain stages of conversion of the initial hydrocarbon. In the cool, blue, and hot flames, this is mainly a conversion to formaldehyde (H_2CO), CO, and final products, respectively [3, 7]. Formaldehyde formation during autoignition of fuel sprays (i.e., the cool flame stage) was recently investigated in [8], but the blue flame stage was not discussed therein.

The objective of this work is to find out whether all the stages of multistage autoignition are present in a Diesel engine operation process using the three-dimensional (3D) numerical simulation with the detailed reaction mechanism (DRM) of fuel oxidation. Moreover, it is worth to evaluate the contribution of each of the stages to the autoignition process and how these stages interact with each other.

2   Short Description of DRM Under Consideration

Over the past 30 years, the Federal Research Center for Chemical Physics has been working on the construction of DRMs of oxidation and combustion of hydrocarbons [9–16]. The algorithm of DRM construction is based on two assumptions: (1) Low-temperature branching is described by a group of reactions with a single addition of oxygen:

$$R + O_2 = RO_2,$$
$$RO_2 + RH = RO_2H + R,$$
$$RO_2H = RO + OH,$$

(here, RH stands for the initial hydrocarbon and R stands for the hydrocarbon radical), which are followed by other reactions of decomposition and oxidation of radicals and molecules. It does not include reactions of so-called double oxygen addition (first to the peroxide radical, and then to its isomeric form), i.e., the first addition turns out to be sufficient; and (2) The channels of oxidation via isomeric forms are excluded because they are slower than the oxidation through nonisomerized species.

In 2013, using this algorithm, a DRM for the oxidation of n-hexadecane (n-$C_{16}H_{34}$) [16] was developed and validated against available experimental data. In particular, the characteristics of the autoignition and combustion of homogeneous mixtures of air with gaseous and liquid (droplets) hydrocarbons from n-undecane to n-hexadecane over a wide range of initial conditions were calculated and the calculation results were compared with the experimental data, which turned out to be in satisfactory agreement. In 2014–2016, the mechanism of [16] has been extended by adding the DRMs of toluene [17] and iso-octane [18] oxidation and combustion. The resultant DRM can be applied to modeling both low-temperature oxidation with cool and blue flames and high-temperature hot explosion and combustion of the realistic Diesel oil containing normal alkanes, iso-alkanes, and aromatic hydrocarbons [19]. The DRM is relatively short and contains only several thousand reactions between several hundred species. More precisely, the DRM of n-tetradecane oxidation includes 1944 reactions between 161 species. For comparison, the DRMs of n-heptane oxidation proposed in [20, 20] include 2300 reactions between 650 species and 1567 reactions between 100 species, respectively, and the DRM of n-hexadecane oxidation proposed in [22] includes 8130 reactions between 2116 species. None of the mechanisms in [20–22] considers the blue flame appearance. Despite its compactness, the mechanism of [16–18] correctly describes the phenomena accompanying low-temperature oxidation of heavy hydrocarbons, for example, the region of the Negative Temperature Coefficient (NTC) of the reaction rate, where the ignition delay increases with temperature [1–3]. In the DRM under consideration, acceleration of the reaction in a cool flame is mainly a consequence of branching during the decomposition of alkyl hydroperoxide ($C_{2n}H_{2n-1}O_2H$) to oxyradical ($C_{2n}H_{2n-1}O$) and hydroxyl (OH). The blue flame appears due to the branching caused by the decomposition of hydrogen peroxide H_2O_2 with the formation of OH ($H_2O_2 + M = 2OH + M$, where M denotes some third species [7, 23]. Finally, the rapid acceleration of the reaction in a hot explosion occurs due to a well-known branching reaction between atomic hydrogen and molecular oxygen ($H + O_2 = 2OH + O$). The DRM was implemented in AVL FIRE, a CFD code for simulating the operation process in internal combustion engines [24].

3 Object of Investigations and Mathematical Model

The object of numerical investigation is the Diesel engine of a passenger car [25] with a common-rail injection system. The operation mode is described by the following parameters and conditions: crankshaft speed $n = 4000$ rpm (cycle time is 30 ms, duration of crankshaft rotation by a crank angle (CA) of $1°$ is ~ 0.042 ms); the number of nozzle holes in the injector is 7; the maximum droplet velocity in the fuel spray is 490 m/s; the initial temperature of droplets in the spray is $T_{d0} = 373$ K; the overall air-to-fuel equivalence ratio $\alpha = 1.28$; fuel injection starts at $701.8°$ CA; and the mean in-cylinder temperature at the start of injection is 1100 K.

Assuming that all cylinders of the Diesel engine operate identically, we consider only a single cylinder in the calculations. Furthermore, assuming that the combustion process in the cylinder is axisymmetric, we consider only a segment of the cylinder, which is the 1/7th part of the combustion chamber in accordance with the number of nozzle holes in the injector. The compensation volume accounting for the various technological recesses in the piston and in the cylinder head is placed above the piston so that the portion of the air charge, which is concentrated in the recesses, is guaranteed to participate in the combustion process. With this arrangement of the compensation volume, the calculated operation mode of the engine is reproduced more correctly than when the compensation volume is placed along the cylinder wall [26].

In the calculations, the moving computational mesh is used (Fig. 1). The bottom wall (piston firing surface) with a profiled cavity moves back and forth from the Bottom Dead Center (BDC) to the Top Dead Center (TDC). To check the influence of mesh resolution, a mesh with a double reduction in the average cell size in the spray block is also constructed, which leads to an increase in the total number of cells in the TDC from 274,000 to 975,000. The spray block is a nonmoving trapezoidal cell block in the vertical section of the combustion chamber, in which the cells are oriented along the motion of the spray to reduce the effect of numerical diffusion.

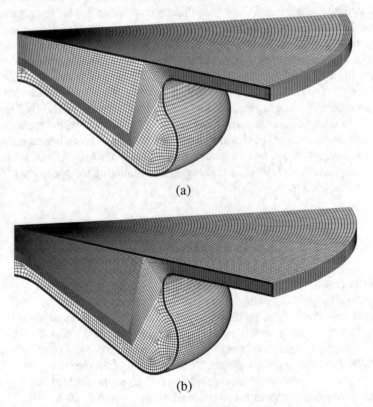

(a)

(b)

Fig. 1 Geometry of the computational domain (TDC): **a** base mesh and **b** mesh refined along the spray axis

The intake and exhaust strokes are not modeled. Calculations begin when the intake valve closes (35° CA after the BDC) and ends when the exhaust valve is opened (40° CA before the BDC). The initial rotation of the intake air is given by the law of rotation of the solid body, and its value is obtained from the experiment by the static blowing of the cylinder head with the maximum lift of the intake valves. An experimental injection rate is used, considering the time delay between the electrical signal to the nozzle control valve and the nozzle needle stroke.

The calculations of the 3D reacting flow are based on the Reynolds-averaged Navier–Stokes equations using the k-ξ-f-model of turbulence [27]. The choice of the turbulence model is based on the results of the preliminary aerodynamic calculations (without combustion), where the k-ξ-f-model showed the best agreement with the results of LES simulation and experiment (in terms of spray penetration).

For combustion simulation, a quasi-laminar model is used. For the sake of simplicity, the effect of turbulent pulsations on the mean rates of chemical reactions is not considered. In the combustion model, each computational cell is treated as a perfectly stirred reactor. The equations of chemical kinetics are solved by the CVODE-solver incorporated in the CFD code [28]. For decreasing the computational costs, the multizone approach [29] available in the code is used. In this approach the cells which have similar thermodynamic conditions (temperature, air-to-fuel equivalence ratio, etc.) are identified and grouped to clusters. The equations of chemical kinetics are solved only for the mean of each cluster. After the results from the solver are acquired, the species vector and the energy source are mapped back to the cells contained in the cluster.

To calculate the motion, evaporation, and break-up of liquid fuel droplets, the standard Lagrangian-droplet method is used. The initial diameter of the droplets is assumed to be equal to the diameter of nozzle holes (120 μm).

The integration time step Δ varies depending on the current crankshaft angle value: at the interval 575–700° CA, $\Delta = 0.5°$ CA (0.021 ms), at the interval 700–780° CA, $\Delta = 0.1°$ CA (0.0042 ms), and at the interval 780–860° CA, $\Delta = 0.5°$ CA (0.021 ms).

The equations for the mean flow variables are solved in Cartesian coordinates using collocated variable arrangements. For the solution of a linear system of equations, a conjugate gradient (CG) and an algebraic multigrid (AMG) types of solvers are used (AMG is used by default for the pressure and CG for all other equations). The overall solution procedure is iterative and is based on the SIMPLE algorithm. A more detailed description of the numerical integration algorithm is given in [30].

4   Results of Calculations

The model fuel for research is n-tetradecane (n-$C_{14}H_{30}$), which is the closest to the realistic Diesel oil in terms of the molecular mass. At the first stage of this study, the calculation model was verified by comparing the calculated and experimental time histories of pressure (Fig. 2). The agreement between the calculation and experiment

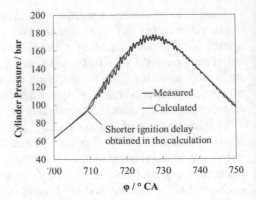

Fig. 2 Comparison between the calculated and measured time histories of the in-cylinder pressure; fuel is n-tetradecane

is generally reasonable, but an earlier autoignition is obtained in the calculation. The latter is explained by the fact that the Cetane number (CN) of n-tetradecane (CN = 96) is higher than the CN of realistic Diesel oil (CN ≈ 53). This drawback can be corrected by replacing n-tetradecane by a proper surrogate fuel (Fig. 3, see Sect. 5).

At the second stage of the study, the specific features of the autoignition process of a fuel spray were investigated. Figure 4 compares the mean (T) and the maximum (T_{max}) in-cylinder temperatures as functions of the crank angle. Firstly, these temperatures are seen to differ from each other. Secondly, the curve $T_{max}(\varphi)$ indicates that autoignition occurs in two stages. Thirdly, the autoignition event is not visible on the $T(\varphi)$ curve at all. The possible reason for the latter is that the mass of evaporated fuel during the induction period of autoignition $(\Delta \varphi = 701.8\text{–}708.8° \text{ CA})$ was only 6% of the entire fuel mass injected.

This uncertainty can be resolved if one averages the flow parameters of interest over the variable volume of fuel spray (V_{spray}) rather than over the entire cylinder volume. To separate the cells occupied by the fuel spray from all others cells the following criterion is proposed: the cells are identified as located inside the fuel spray if their local value of the air-to-fuel equivalence ratio $\alpha < 1.0$ (the local air-to-fuel equivalence ratio is calculated based solely on the concentration of fuel vapor, i.e.,

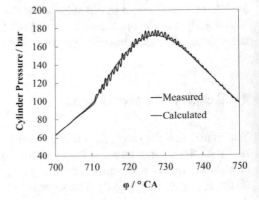

Fig. 3 Calculated and measured time histories of the in-cylinder pressure; fuel is the surrogate containing n-hexadecane (56%wt.) and toluene (44%wt.)

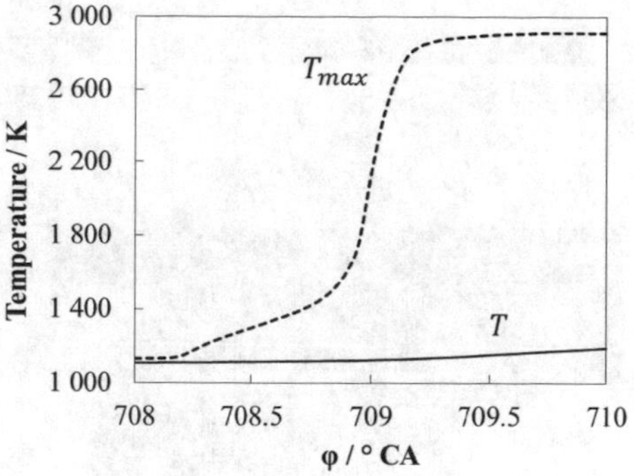

Fig. 4 Calculated
dependencies of the mean, T,
and maximum, T_{max},
temperature in the engine
cylinder on the crank angle

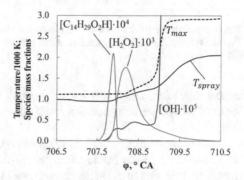

Fig. 5 Calculated curves for
temperature and mass
fractions of hydroxyl,
hydrogen peroxide and alkyl
hydroperoxide obtained by
averaging over the variable
volume of the fuel spray
where $\alpha \leq 1.0$

fuel in the liquid phase is not considered). This means that the fuel spray occupies
the volume covered by the isosurface of the stoichiometric mixture composition.
With the use of such an approach the mean gas temperature in the spray volume,
T_{spray}, becomes closer to T_{max} (Fig. 5). Moreover, the curve $T_{max}(\varphi)$ still indicates
that autoignition occurs in two stages, however the curve $T_{spray}(\varphi)$ now exhibits three
stages of the temperature rise. The first rise of the mean temperature in the spray
volume (from 950 to 1030 K) is nothing else as the cool flame, because it corresponds
to the decomposition of alkyl hydroperoxide ($C_{14}H_{29}O_2H$). The heat released in the
cool flame promotes the second rise in the mean temperature that is the thermal
decomposition of hydrogen peroxide (blue flame) accumulated in the spray volume
during the induction period. The heat released in the blue flame further increases the
gas temperature in the spray volume up to 1270 K and gives rise to the hot flame.
Thus, despite the cool flame stage is not visible on the $T_{max}(\varphi)$ curve, the autoignition
of fuel is a three-stage process with the successive cool, blue, and hot flames. The
three-stage nature of autoignition is clearly reflected by the kinetic curve for OH.

In the calculation, the cool flame appears when the mean gas temperature in
the spray volume attains a value of $T_{spray} = 950$ K. According to [18], at such a
temperature there is no accumulation of alkyl hydroperoxide in homogeneous fuel–
air mixtures. However, when looking at the calculated flow fields (Fig. 6), one can
see that the local temperature at the tip of the fuel spray is lower than the mean

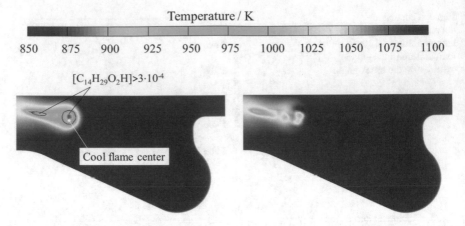

Fig. 6 Appearance of cool flame: calculated fields of the mass fraction of alkyl hydroperoxide $C_{14}H_{29}O_2H$ (isolines) and temperature at two instants of time: **a** $\varphi = 707.9°$ CA and **b** $\varphi = 708.1°$ CA

temperature and is about 800–850 K (Fig. 6a). These latter values are well appropriate for the $C_{14}H_{29}O_2H$ accumulation and its decomposition with cool flame appearance (Fig. 6b).

Figure 7 shows the calculated distributions of the local air-to-fuel equivalence ratio α and the local Sauter Mean Diameter (SMD) of fuel droplets for two instants of time corresponding to the appearance of cool flame ($\varphi = 707.9°$ CA) and subsequent appearance of blue flame ($\varphi = 708.4°$ CA). The exothermic center of cool flame with the maximum concentration of $C_{14}H_{29}O_2H$ is formed in front of the fuel spray in the region with $\alpha \sim 0.3$ (shown by the filled circle in Fig. 7a). In this center, there are fuel droplets with a diameter of 10–25 μm by $\varphi = 707.9°$ CA. The exothermic

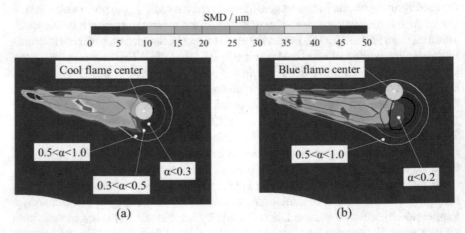

Fig. 7 Calculated fields of the local air-to-fuel equivalence ratio and SMD of fuel droplets at two instants of time: **a** $\varphi = 707.9°$ CA; **b** $\varphi = 708.4°$ CA

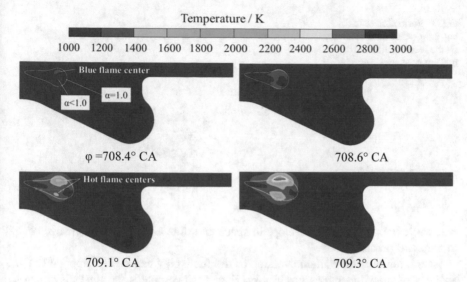

Fig. 8 Time evolution of the exothermic centers of blue flame and hot explosion

center of blue flame originates in the region shown by the circle in Fig. 7b. In this region $\alpha \sim 0.2$–1.0. By $\varphi = 708.4°$ CA, there are no fuel droplets in this region.

Figure 8 shows the calculated temperature fields at four instants of time after the appearance of blue flame. The temperature fields contain isolines of the local air-to-fuel equivalence ratio $\alpha = 1.0$, which can be approximately interpreted as the boundaries of the fuel spray. Inside the isolines the mixture in the gas phase is fuel rich: $\alpha < 1$.

The exothermic centers of blue flame appear at $\varphi \sim 708.1°$ CA and rapidly develop at the periphery of the front part of the fuel spray due to the localized decomposition of accumulated hydrogen peroxide. At $\varphi > 709.0°$ CA, the exothermic centers of hot explosion occur and rapidly develop inside the regions filled with the products of the blue flame. The exothermic center of hot explosion occurs near the top wall of the combustion chamber (the fire surface of the cylinder head) on the leeward side of the fuel spray relative to the rotational motion of the flow around the cylinder axis.

Figure 9 shows the formation of a hot-flame exothermic center in the form of a volume with local temperatures exceeding 2500 K inside. The mean value of the local air-to-fuel equivalence ratio in the hot flame center is ~ 0.45–0.60, i.e., hot explosion occurs in the regions enriched with fuel vapor. This result agrees well with the findings reported in [31] where the autoignition of a monodisperse cloud of n-decane droplets in heated air was studied computationally and a hot explosion was found to occur in regions enriched with fuel vapor. After the occurrence, the region with the maximum rate of heat release in hot explosion shifts to the stoichiometric isoline (see Fig. 8). The common property of the exothermic centers of cool and blue flames, and hot explosion is that they appear in the region of the fuel spray where α

Fig. 9 Formation of the
hot-flame exothermic center
on the leeward side of the
fuel spray at $\varphi = 709°$ CA

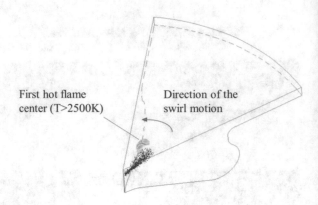

First hot flame
center (T>2500K)

Direction of the
swirl motion

< 1, while the region itself develops in space and time and its volume changes due
to transport processes.

Averaging over the variable volume of the fuel spray can be also applied to the
analysis of in-cylinder pressure history. Figure 10 compares the time histories of
in-cylinder pressure (solid curve) and the mean pressure in the spray volume (dashed
curve). The first visible deviation of curves at 707.8° CA corresponds to the onset of
cool flame. Heat release in the cool flame results in the rise of overpressure in the spray
volume by about 1 bar. After decomposition of accumulated alkyl hydroperoxides
(mainly $C_{14}H_{29}O_2H$), the rate of energy release decreases and the overpressure in
the spray volume starts to decrease due to the dominating role of thermal expansion
in the cool-flame exothermic centers.

The second increase of the overpressure in the spray volume at 708.1° CA by
about 1 bar corresponds to the onset of blue flame. After decomposition of accumulated hydrogen peroxide (H_2O_2), the rate of energy release decreases causing the

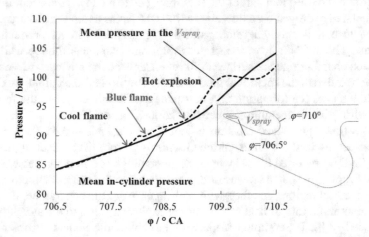

Fig. 10 Time histories of in-cylinder pressure (solid curve) and the mean pressure in the variable
volume of the fuel spray (dashed curve)

Fig. 11 Time histories of various temperatures during ignition

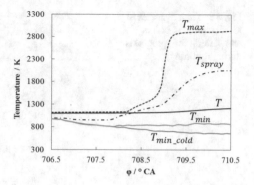

decrease in the overpressure due to the dominating role of thermal expansion in the blue-flame exothermic centers. The subsequent noticeable localized pressure rise starting from 708.9° CA corresponds to the onset of hot explosion. The hot explosion increases the overpressure in the selected volume by 3–4 bar. Thereafter the hot-explosion exothermic center undergoes thermal expansion leading to the inertial decrease in the local pressure below the instantaneous mean in-cylinder pressure at 709.9° CA. Further evolution of the pressure in the variable spray volume exhibits oscillations around the mean in-cylinder pressure indicating the existence of reverberating pressure waves in the cylinder.

Another possibility of demonstrating the multistage nature of autoignition in Diesel engine is to plot time histories of various gas temperatures, like it is done in Fig. 11. This figure compares the mean in-cylinder temperature, T; the maximum in-cylinder temperature, T_{max}; the gas temperature obtained by averaging over the variable volume of the fuel spray, T_{spray}; the minimum in-cylinder temperature obtained by averaging over the entire engine cylinder except for near-wall regions, T_{min}; as well as the minimum in-cylinder temperature obtained like T_{min} but with deactivated chemical energy release, T_{min_cold}. The onset of cool flame can be readily attributed to $\varphi = 707.9°$ CA, when the curve $T_{min}(\varphi)$ deviates from the curve $T_{min\_cold}(\varphi)$. This happens when the temperature in the spray core drops to ~ 800 K. The onset of the blue flame can be attributed to $\varphi = 708.2$–$708.3°$ CA, when the curve $T_{max}(\varphi)$ sharply deviates from the curve $T(\varphi)$. The time of this deviation matches with the time when the curve $T_{spray}(\varphi)$ exhibits the beginning of the second rise. The onset of hot explosion should be attributed to $\varphi = 709°$ CA, when the curve $T_{spray}(\varphi)$ exhibits the beginning of the third rise and the curve $T_{max}(\varphi)$ exhibits the beginning of the second rise.

When analyzing the local flow temperatures on the spray axis (points 1, 2 and 3 in Fig. 12) one observes remarkably interesting local phenomena. In point 1, only cool and blue flames are seen to arise whereas hot explosion is missing here during the selected time interval. In point 2, there is only blue flame, while hot explosion is also missing. The absence of hot explosion in these points is caused by continuing evaporation of spray droplets, which absorbs heat and makes the local mixture composition extremely fuel rich. As for point 3, there is a prominent temperature

Fig. 12 Time histories of temperatures in points 1, 2, and 3 along the spray axis during ignition

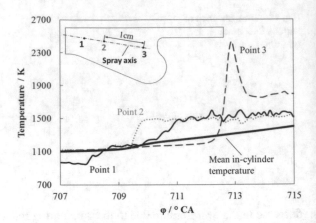

peak at about 713° CA up to 2400 K followed by a temperature drop down to ~ 1700 K. This peak is caused by the high-temperature flame arising on the spray tip and traversing through this point. The subsequent temperature drop in this point is caused by arrival of relatively cool fuel-rich mixture in the spray core.

The findings described above imply that calculations of autoignition in Diesel engine must be overly sensitive to the spatial and temporal resolution of zones and processes occurring in the exothermic centers of cool and blue flames and hot explosion. To verify this implication, we have made additional calculations with different values of the integration time step. It turned out (Fig. 13a) that with an increase in the time step, the maximum value of $C_{14}H_{29}O_2H$ mass fraction decreases and at a time step $\Delta = 0.5°$ CA this species virtually does not form at all. This means that at $\Delta = 0.5°$ CA autoignition occurs without the stage of cool flame inherent in the conditions under consideration. Regarding the blue flame stage, because of its longer duration in comparison with the cool flame stage, the influence of numerical time

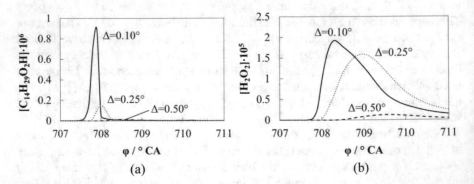

Fig. 13 Influence of the integration time step on the calculated mass fractions of alkyl hydroperoxide $C_{14}H_{29}O_2H$ (**a**) and hydrogen peroxide H_2O_2 (**b**) in the autoignition induction period; Averaging is made over the entire volume of the combustion chamber

Fig. 14 Influence of the integration time step on the calculated maximum gas temperature during the autoignition induction period; Averaging over the entire volume of the combustion chamber

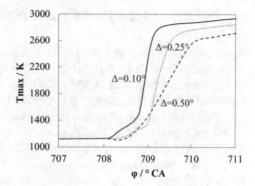

step is not that critical, nevertheless with the time step $\Delta = 0.5°$ a blue flame stage becomes very weak (Fig. 13b).

It can be expected that the "loss" of cool flame in the calculation with the increased value of the integration time step will influence the autoignition induction period. As a matter of fact, Fig. 14 shows the dependences of the maximum temperature $T_{max}(\varphi)$ obtained with different integration time steps while averaging over the entire volume of the combustion chamber.

The use of relatively small integration time steps ensuring the temporal resolution of the processes of accumulation and decomposition of alkyl hydroperoxide provides a lower value of the ignition delay period, in contrast to the use of relatively large time steps. At the time step $\Delta = 0.5°$ CA, there is no cool flame, and the blue-flame stage is very weak, therefore autoignition appears as a single-stage process with an increased induction period.

Calculations on a finer mesh showed that with a commensurate decrease in the time step (Fig. 15), the calculated kinetic curves virtually repeat the curves obtained on the base mesh with the initial value of the time step, except for alkyl hydroperoxide, which, however, does not affect the induction period. Based on this, it can be concluded that the use of a basic computational mesh is quite acceptable for modeling the formation of centers of cool and blue flame.

Fig. 15 Calculated time histories of temperature and mass fractions of hydroxyl, hydrogen peroxide and alkyl hydroperoxide on the mesh with the refined fuel spray block and the reduced time step ($\Delta = 0.05°$)

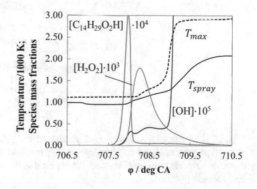

5   Influence of the Fuel Cetane Number

As mentioned above, the CN of n-tetradecane (CN = 96) is higher than the CN of Diesel oil (CN ≈ 53). In the calculations this leads to earlier autoignition. According to in-house investigations at AVL LIST GmbH, the most suitable surrogate of the realistic Diesel oil, considering all its physical and chemical properties, is a mixture of two individual liquid hydrocarbons: n-hexadecane (56%wt.) and toluene (44%wt.).

Figure 3 referred earlier in this paper compares the calculated and measured time histories of in-cylinder pressure when the DRM of n-tetradecane oxidation is replaced by the DRM of oxidation of the surrogate fuel containing 56%wt. n-hexadecane and 44%wt. toluene. The calculation with the surrogate fuel shows better agreement with the experiment, both in terms of the ignition induction period (the ignition delay increases) and the subsequent pressure rise.

Figure 16 shows the calculated time histories of the mass fractions of alkyl hydroperoxide $C_{16}H_{33}O_2H$, hydrogen peroxide, and hydroxyl averaged over the variable volume of fuel spray. This figure should be compared with Fig. 4 plotted for n-tetradecane. As seen, with the surrogate fuel the accumulation of $C_{16}H_{33}O_2H$ is started somewhat later and the cool flame stage is more extended. These factors shift the timing of the onset of the other two stages of autoignition.

When n-tetradecane is replaced by the surrogate fuel, the maximum concentration of hydroxyl (OH) in the cool and blue flames decreases approximately by a factor of 2 indicating the decrease in the overall reaction rate in the engine cylinder. Due to increase in the autoignition induction period with the use of the surrogate fuel, the minimum temperature in the fuel spray core is getting lower due to longer evaporation of droplets in the preignition period. As a result, the maximum mass fraction of alkyl hydroperoxide $C_{16}H_{33}O_2H$ prior to the onset of cool flame with the use of the surrogate fuel is higher than that of alkyl hydroperoxide $C_{14}H_{29}O_2H$ with the use of n-tetradecane, while the maximum mass fractions of hydrogen peroxide in both

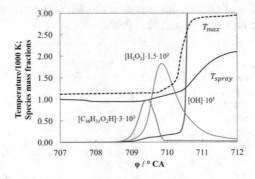

Fig. 16 Calculated time histories of the maximum gas temperature and mean gas temperature, and mass fractions of hydroxyl, hydrogen peroxide, and alkyl hydroperoxide obtained by averaging over the variable volume of the fuel spray; Fuel is the surrogate containing n-hexadecane (56%wt.) and toluene (44%wt.)

cases show an opposite trend. The effect of addition of the aromatic component on chemical reactions during the induction period deserves a separate study.

6 Conclusions

Three-dimensional numerical simulations of mixture formation, autoignition and combustion processes in a cylinder of Diesel engine using the detailed reaction mechanism of fuel oxidation are performed. Special attention is paid to the autoignition process. For the first time, the three-stage nature of fuel autoignition characterized by the successive appearance of foci of cool, blue, and hot flames, has been computationally observed. Relatively low temperatures in the core of the fuel spray create preconditions for the formation and accumulation of alkyl hydroperoxides $(C_{2n}H_{2n-1}O_2H)$. Further thermal decomposition of the accumulated alkyl hydroperoxides leads to the appearance of cool flame. In turn, the appearance of the cool flame initiates the decomposition of hydrogen peroxide accumulated in the fuel spray and the appearance of blue flame. Further heating of the mixture due to the blue flame leads to the occurrence of hot explosion. The first center of hot explosion arises near the upper wall of the combustion chamber (firing surface of the cylinder head) on the leeward side of the fuel spray relative to the rotational motion of the flow around the cylinder axis. The common property of the foci of cool and blue flames and hot explosions is found: they form in the region of the fuel spray, where $\alpha < 1.0$.

A new method for the analysis of physicochemical processes during the induction period by averaging flow properties over the cells located inside the fuel spray volume is proposed. In view of the fact that the induction period in high-speed engines can be very short (especially at full load conditions), for the correct modeling of the autoignition delay, it is necessary to choose the time integration step and the spatial resolution of the corresponding flow regions in such a way that chemical reactions in the centers of spontaneous ignition pass through all the characteristic stages. It has been established that the addition of the aromatic component to obtain the realistic value of the Cetane Number of the fuel affects mainly the cool flame stage.

Acknowledgements The work was supported by the subsidies given to the Semenov Federal Research Center for Chemical Physics of the Russian Academy of Sciences to implement the state assignment on the topic No. 0082-2019-0006 and to the Federal State Institution "Scientific Research Institute for System Analysis of the Russian Academy of Sciences" to implement the state assignment on the topic No. 0580-2021-0005.

References

1. Kondratiev, V.N., Nikitin, E.E.: Kinetics and Mechanism of Gas-Phase Reactions. Nauka Publ., Moscow (1974)

2. Sokolik, A.S., Basevich, V.Ya.: About kinetic nature of autoignition in diesel engine conditions. Zh. Fiz. Khim. 28(11) (1954)
3. Sokolik, A.S.: Autoignition, Combustion, Flame and Detonation in Gases. USSR Acad. Sci. Publ., Moscow (1960)
4. Hwang, W., Dec, J., Sjöberg, M.: Spectroscopic and chemical-kinetic analysis of the phases of HCCI autoignition and combustion for single- and two-stage ignition fuels. Combust. Flame 154, 387–409 (2008)
5. Westbrook, C.K.: Chemical kinetics of hydrocarbon ignition in practical combustion systems. Proc. Combust. Inst. 28, 1563–1577 (2000)
6. Lee, C., et al.: The experimental investigation on the impact of toluene addition on low-temperature ignition characteristics of diesel spray. Fuel 254, 1–12 (2019)
7. Grajetzki, P., et al.: A novel reactivity index for SI engine fuels by separated weak flames in a micro flow reactor with a controlled temperature profile. Fuel 245, 429–437 (2019)
8. Kundu, P., et al.: Importance of turbulence-chemistry interactions at low temperature engine conditions. Combust. Flame 183, 283–298 (2017)
9. Basevich, V.Y., Vedeneev, V.I., Frolov, S.M., Romanovich, L.B.: Nonextensive principle for construction of oxidation and combustion mechanisms for normal alkane hydrocarbons: transition from C1–C2 to C_3H_8. Russ. J. Phys. Chem. 25(11), 87–96 (2006)
10. Basevich, V.Y., Belyaev, A.A., Frolov, S.M.: The mechanisms of oxidation and combustion of normal alkane hydrocarbons: the transition from C1–C3 to C_4H_{10}. Russ. J. Phys. Chem. B 2(5), 477–484 (2007)
11. Basevich, V.Y., Belyaev, A.A., Frolov, S.M.: Mechanisms of the oxidation and combustion of normal akanes: passage from C1–C4 to C_5H_{12}. Russ. J. Phys. Chem. B 3(4), 629–635 (2009)
12. Basevich, V.Y., Belyaev, A.A., Frolov, S.M.: Mechanisms of the oxidation and combustion of normal alkanes: transition from C1–C5 to C_6H_{14}. Russ. J. Phys. Chem. B 4(4), 634–640 (2010)
13. Basevich, V.Y., Belyaev, A.A., Posvyanskii, V.S., Frolov, S.M.: Mechanism of the oxidation and combustion of normal paraffin hydrocarbons: transition from C1–C6 to C_7H_{16}. Russ. J. Phys. Chem. B 4(6), 985–994 (2010)
14. Basevich, V.Y., Belyaev, A.A., Medvedev, S.N., Posvyanskii, V.S., Frolov, S.M.: Oxidation and combustion mechanisms of paraffin hydrocarbons: transfer from C1–C7 to C_8H_{18}, C_9H_{20}, and $C_{10}H_{22}$. Russ. J. Phys. Chem. B 5(6), 974–990 (2011)
15. Frolov, S.M., Medvedev, S.N., Basevich, V.Y., Frolov, F.S.: Self-ignition of hydrocarbon–hydrogen–air mixtures. Int. J. Hydrogen Energy 38, 4177–4184 (2013)
16. Basevich, V.Y., Belyaev, A.A., Posvyanskii, V.S., Frolov, S.M.: Mechanisms of the oxidation and combustion of normal paraffin hydrocarbons: transition from C1–C10 to C11–C16. Russ. J. Phys. Chem. B 7(2), 161–169 (2013)
17. Basevich, V.Ya., Belyaev, A.A., Frolov, F.S., Frolov, S.M., Medvedev, S.N.: Detailed chemistry of heavy alkane hydrocarbon fuel oxidation: application to combustion and detonation of gaseous and liquid fuels. Transient Combustion and Detonation Phenomena: Fundamentals and Applications, pp. 14–25. TORUS PRESS, Moscow (2014)
18. Basevich, V.Y., Belyaev, A.A., Medvedev, S.N., Posvyanskii, V.S., Frolov, F.S., Frolov, S.M.: A detailed kinetic mechanism of multistage oxidation and combustion of isooctane. Russ. J. Phys. Chem. B 10(5), 801–809 (2016)
19. Merker, G., Schwarz, C.: Grundlagen Verbrennungsmotoren. Simulation der Gemischbildung, Verbrennung, Schadstoffbildung und Aufladung. Praxis. 6 Auflage. Vieweg+Teubner Verlag, Wiesbaden (2012)
20. Chevalier, C., Warnatz, J., Melenk, H.: Automatic generation of reaction mechanisms for description of oxidation of higher hydrocarbons. Ber. Bunsenges. Phys. Chem. 94 (1990)
21. Stagni, A., Cuoci, A., Frassoldati, A., Faravelli, T., Ranzi, E.: Lumping and reduction of detailed kinetic schemes: an effective coupling. Ind. Eng. Chem. Res. 53, 9004–9016 (2014)
22. Westbrook, C.K., Pitz, W.J., Herbinet, O., et al.: A comprehensive detailed chemical kinetic reaction mechanism for combustion of n-alkane hydrocarbons from n-octane to n-hexadecane. Combust. Flame 156, 181–199 (2009)

23. Basevich, V.Y., Frolov, S.M.: Kinetics of "blue" flames in the gas-phase oxidation and combustion of hydrocarbons and their derivatives. Russ. Chem. Rev. **76**(9), 867–884 (2007)
24. AVL FIRE®—Computational Fluid Dynamics for Conventional and Alternative Powertrain Development. https://www.avl.com/fire. Accessed 21 Feb 2021
25. Blanchard, E., Visconti, J., Coblence, P., et al.: Der neue dCi 130 1,6l Dieselmotor von RENAULT, pp. 247–273. 19 Aachener Kolloquium Fahrzeug- und Motorentechnik (2010)
26. Sergeev, S.S., Frolov, S.M., Basara, B.: Numerical modeling of combustion and pollutants formation in cylinder of diesel using a detailed kinetic mechanism of n-heptane oxidation. Goren. Vzryv (Mosk.)—Combust. Explos. **10**(2), 26–34 (2017)
27. Hanjalic, K., Popovac, M., Hadziabdic, M.: A robust near wall elliptic relaxation eddy-viscosity turbulence model for CFD. Int. J. Heat Fluid Flow **25**, 897–901 (2004)
28. Cohen, S.D., Hindmarsch, A.C.: CVODE User Guide (1994)
29. Liang, L., Stevens, J.G., Farell, J.T.: A dynamic multi-zone partitioning scheme for solving detailed chemical kinetics in reactive flow computations. Combust. Sci. Technol. **181**, 1345–1371 (2009)
30. Przulj, V., Basara, B.: Bounded convection schemes for unstructured grids. AIAA paper, 2001-2593 (2001)
31. Frolov, S.M., Basevich, V.Y., Frolov, F.S., Borisov, A.A., Smetanyuk, V.A., Avdeev, K.A., Gots, A.N.: Correlation between drop vaporization and self-ignition. Russ. J. Phys. Chem. B **3**(3), 333–347 (2009)

Towards Lower Engine-Out Emissions with RCCI Combustion

Urban Žvar Baškovič⬤, Tine Seljak⬤, and Tomaž Katrašnik⬤

Abstract Future emission standards and renewable energy targets present a significant challenge for internal combustion engine manufacturers, requiring continuous improvements in efficiency and use of newly developed, preferably carbon neutral, fuels. As an answer to the future challenges, Reactivity Controlled Compression Ignition (RCCI) concept was developed with the aim to attain low emissions of NO_x and particulate mass (PM) simultaneously. In the recent years, the RCCI concept is becoming even more attractive, since recently up-scaled renewable fuels exhibit a wide interval of different reactivities, making them ideal for obtaining the desired combustion parameters in RCCI combustion concept. In the present chapter, an application of direct injection of renewable Hydrotreated Vegetable Oil (HVO) and port injection of Methane rich natural gas in RCCI concept is discussed. The application features low local combustion temperatures preventing local formation of high NO_x concentrations, with a high degree of charge homogenization, which is crucial for achieving low emissions of particulate matter. The presented results and discussion focus on the influence that different engine control strategies have on the main combustion process and emission indicators. Relations between different engine control strategies, including variation of energy shares of utilized fuels, direct injection timing of the direct injection and gas path control, engine thermodynamic parameters and engine-out emissions were deeply investigated and benchmarked with conventional diesel combustion. The results show that with an innovative combination of alternative fuels, significant simultaneous reduction of NO_x and PM engine-out emissions can be achieved.

Keywords RCCI · Renewable fuels · Dual-fuel combustion · Light-duty engine · Emissions

U. Žvar Baškovič · T. Seljak · T. Katrašnik (✉)
Faculty of Mechanical Engineering, University of Ljubljana, Ljubljana, Slovenia
e-mail: tomaz.katrasnik@fs.uni-lj.si

© The Author(s), under exclusive license to Springer Nature Switzerland AG 2022
T. Parikyan (ed.), *Advances in Engine and Powertrain Research and Technology*,
Mechanisms and Machine Science 114,
https://doi.org/10.1007/978-3-030-91869-9_7

Abbreviations

CO	Carbon monoxide
HC	Hydrocarbons
HVO	Hydrotreated vegetable oil
LHV	Lower heating value
LTC	Low temperature combustion
NG	Natural gas
NO_x	Nitrogen oxides
PM	Particulate mass
RCCI	Reactivity controlled compression ignition
ROHR	Rate of heat release
ROPR	Rate of pressure rise
SOC	Start of combustion

1 Introduction

Internal combustion engines have been in the recent decades subjected to various emission standards with an objective to minimize their negative impact on the environment and to increase a quality of living, especially in densely populated areas. Their public acceptance is, in the electrification era, tightly related to their overall environmental influence thus forcing producers to provide advanced solutions leading to mitigated engine-out emissions. Among the regulated emissions, particulate mass (PM) and nitrogen oxides (NO_x) present a major concern due to their adverse effects on the environment and health. To mitigate emissions of pollutants, either advanced combustion concepts or improved exhaust after-treatment systems have to be employed. As the latter can lead to production costs increase of the powertrain, substantial effort has been directed into development of advanced combustion concepts, which can achieve comparable engine efficiencies to the conventional combustion concepts and significantly lower engine-out emissions. In contrast to conventional compression ignition combustion concept, which features fuel rich and high-temperature lean in-cylinder regions leading to high soot and high NO_x formation rates respectively, advanced combustion concepts feature low combustion temperatures and a high degree of charge homogenization. Low in-cylinder temperatures prevent local formation of high NO_x concentrations, whereas a high degree of charge homogenization is crucial for achieving low emissions of particulate matter. In contrast to conventional compression ignition engines, low-temperature combustion concepts (LTC) can thus simultaneously reduce NO_x and particulate emissions.

1.1 RCCI and Low-Temperature Combustion Concepts

Several approaches towards low temperature combustion with high degree of charge homogenization are available and in general they can be divided into:

- Homogeneous charge compression ignition (HCCI) concept, e.g. [1];
- Partially premixed combustion (PPC) concept, e.g. [2];
- Reactivity-controlled compression ignition (RCCI) combustion concept, e.g. [3].

Among these, RCCI features the highest fuel-flexibility as well as ability to operate across the entire engine operating space [4] and can be therefore considered as a suitable concept for utilization in real engine applications where a high specific power and, consequently, economic viability are important aspects.

The RCCI concept employs two fuels with significantly different reactivities to create local reactivity pockets and combustion mixture reactivity gradient governing combustion characteristics of the mixture. In a study, which introduced RCCI concept [3], it was concluded that global fuel reactivity enhances control of combustion phasing and fuel reactivity stratification allows to control rate of the heat released during engine load increase. In RCCI concept, low reactivity fuel with high volatility is usually injected in the intake manifold, whereas high-reactivity fuel is injected directly into the combustion chamber during the compression stroke leading to high degree of charge homogenization. Due to reactivity and equivalence ratio stratification in combustion chamber, sequential combustion process from high reactivity areas towards low reactivity areas within flammability limits is established. Combining stratification of fuel reactivity and retention of control over the combustion event, which can be achieved with suitable injection and gas path control strategies, RCCI concept makes it possible to achieve extended combustion duration leading to low local combustion temperatures and low rate of pressure rise (ROPR) and rate of heat release (ROHR) values, which work in favour of low wear of engine components. Due to a combination of low air–fuel equivalence ratio as a result of high degree of charge homogenization and low combustion temperatures that remain below the NO_x formation threshold, RCCI possesses potential to significantly decrease NO_x and PM emissions compared to conventional compression ignition engines. An RCCI concept thus presents a viable approach towards simplified PM and NO_x exhaust after treatment systems or towards even lower tailpipe emissions.

Due to specific characteristics of RCCI combustion concept, its drawback can be seen in higher CO and HC emissions, which can be substantially higher compared to the conventional diesel combustion process, especially at low engine loads [5]. Whereas HC emissions can be ascribed to a high volume of very lean regions outside flammability limits containing low reactivity fuel, injected early in the cycle, CO formation rate is a function of the HC availability and the local charge temperature, which in combination control the rate of HC decomposition and oxidation. Due to low local temperatures, RCCI concept is prone to combustion reaction quenching leading to intermediate species in exhaust gasses, such as CO. Increased CO and HC emissions can be mitigated in an oxidation catalyst, which is already an essential part of modern compression ignition engines and is cheaper to produce compared to exhaust aftertreatment systems for NO_x emissions.

In the past, various studies, e.g. [4, 6, 7], were carried out to establish cause and effect relationships between engine control parameters, thermodynamic parameters describing combustion characteristics and engine-out emissions, whereas most of them were performed in heavy-duty engines. Due to pronounced focusing on hybrid powertrain and downsized engines, this chapter analyses potentials of introducing RCCI concept with alternative fuels to small swept volume light-duty engines thus allowing to reduce PM and NO_x emissions while potentially targeting mass market.

1.2   Fuels in RCCI Combustion Concept

To maximize potential of RCCI operation and achieve best output, it is in general considered that fuels with a wide range of reactivities should be used in RCCI concept [8]. RCCI concept can make use of a combination of various gaseous and liquid fossil fuels or biofuels, such as diesel, gasoline, natural gas (NG), ethanol, methanol and biodiesel [9]. A typical combination of fossil fuels featuring a wide range of reactivity is represented by gasoline and diesel [6, 10–12], where gasoline represents a low-reactivity fuel with high volatility and diesel fuel represents a high reactivity fuel. At higher engine loads it is recommended to achieve lower reactivity of the combustion mixture [13], whereas at lower engine loads low reactivity of the combustion mixture can lead to significant increase of HC and CO emissions [14]. Therefore, to enable engine operation across the entire engine operating map, it is desirable to utilize fuels covering high range of reactivities.

With a goal to extend engine operating map, fuels with very high and very low reactivities should be used in RCCI combustion concept. A promising representative of low-reactivity fuels is natural gas, which features higher octane number than gasoline thus leading to larger reactivity gradient in combustion chamber and therefore to a better control over combustion propagation and ROPR values [15]. In the past, several RCCI studies have been performed, in which NG was used in combination with conventional diesel fuel—e.g. [16–19]. Although control parameters vary significantly between performed studies, reduction in NO_x and PM emissions compared to conventional compression ignition engine operation was observed on an account of increased HC and CO emissions. As determined in [19], an increased energy share supplied by NG significantly reduces soot emissions due to the absence of C–C bond in the fuel's chemical structure thus proving that NG possesses high potential for utilization in RCCI concept.

To meet future environmental standards defining required shares of biofuels in stationary and transport sectors, which would lead to low carbon footprint and sustainable environment, RCCI concept should be used in combination with suitable biofuels. Among various gaseous and liquid low-reactivity fuels, biomethane possesses high potential due to its high reactivity, which is even higher than the one of conventional natural gas. In terms of bio-derived high-reactivity fuels, various fuels can be used in RCCI concept. One of them is waste based fish oil, which was used in RCCI study in combination with natural gas [20] and resulted in more stable cycle-to-cycle operation and lower HC emissions compared to NG/diesel RCCI operation due to higher percentage of in-fuel bound oxygen and higher reactivity, compared to diesel fuel. Another biofuel featuring potential for utilization in RCCI concept is hydrotreated vegetable oil (HVO), which is fully paraffinic renewable fuel with higher cetane number than conventional diesel fuel.

1.3 Scope of the Study

Within the framework of the future emission standards and renewable fuel regulations, this study aims to experimentally explore influence of start of direct injector energizing (SOE), energy share of low-reactivity fuel and exhaust gas recirculation (EGR) rate on the engine operation in low-temperature RCCI concept while utilizing NG as a low-reactivity fuel and HVO as a high-reactivity fuel in a small swept volume light duty engine. The results obtained, consisting of analyses of in-cylinder thermodynamic parameters and engine-out emission response lead to demonstration of suitable settings of control parameters for engine operation in RCCI mode while using innovative gaseous and liquid fuels.

2 Methodology

Research methodology was developed in a way to establish the interrelation between engine control parameters, combustion characteristics and engine emission response during engine operation in the RCCI mode. It comprises variation of three crucial control parameters, which significantly influence combustion characteristics and emission formation rates. Varied control parameters, presented in Fig. 1, cover injection characteristics as well as composition characteristics of intake gasses and can be divided into SOE timing of HVO, Energy share supplied by NG, and EGR rate.

SOE timing of the directly injected HVO and energy share, supplied by NG, enable control over charge reactivity stratification thus governing start of combustion and the rate of combustion reactions. EGR rate increase on the other hand allows to

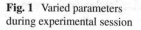

Fig. 1 Varied parameters during experimental session

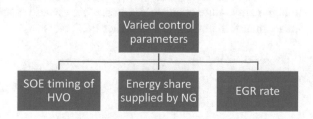

Table 1 Values of varied control parameters

Operating points	SOE timing of the HVO °CA	Energy share of the NG %	EGR rate %
01	−30	~92	0
02	−40	~92	0
03	−45	~92	0
04	−50	~92	0
05	−55	~92	0
06	−50	93	0
07	−60	90	0
08	−120	86	0
09; 10; 11	−50	~94	0
10	−50	~94	23
11	−35	~94	43

alter combustion process by supressing rate of combustion reactions due to reduced oxygen share and increased share of inert species with high specific heats.

Values of the varied control parameters were determined on a basis of an initial screening process with a goal to present an influence of the varied parameters to combustion thermodynamic parameters comprising in-cylinder pressure, ROPR, ROHR and mean in-cylinder temperature as well as gaseous and particulate engine-out emissions in a wide range of parameters' levels at a constant speed and torque values. Overall, eleven operating points with characteristics, presented in Table 1, were defined, measured and analysed. Other operating parameters, which are not listed in Table 1 were kept constant at the following approximate levels: Engine speed was 1500 rpm, IMEP was 9 bar, Intake manifold pressure was 2.2 bar, NG SOE timing was −320 °CA and HVO injection pressure was 900 bar.

Values, presented in bold text in the table represent the main variation for each set of operating points, however, at operating points 06 to 08, SOE timing of the HVO was also adjusted with an aim of maintaining a constant SOC timing thus allowing to see relations between energy share of the NG, SOE timing of the HVO and combustion event in terms of injection event timing limits and combustion event coupling. At operating point 11, SOE timing of the HVO was also adjusted with the aim to enable stable combustion. Intake manifold temperature was kept constant at 35 °C for all operating points except 09 to 11, where it was kept constant at 50 °C due to high temperature of exhaust gasses, which would otherwise lead to an increase in intake manifold temperature at higher EGR rates.

3 Utilized Fuels

In the presented study, liquid and gaseous fuels with different characteristics were utilized to establish favourable conditions for RCCI combustion concept. Low-reactivity fuel is represented with the natural gas (NG) featuring a very high octane number due to its composition as it mostly consists of methane, which achieves research octane number above 127 according to [21]. Utilized methane-rich natural gas could be mixed or substituted with a biomethane [22], which represents one of the promising renewable fuels. Main properties of the utilized natural gas are presented in Table 2.

High-reactivity fuel in this study is represented by a Hydrotreated vegetable oil (HVO), which can be classified as a waste-derived biofuel and can be produced from various feedstocks including vegetable oils, waste cooking oils and animal fats [24]. In terms of a chemical composition and physical characteristics it is very similar to conventional diesel fuel. Additionally, it is free of aromatics [25], which leads to its very high reactivity being beneficial to establish fuel reactivity stratification in the combustion chamber during operation in RCCI mode. To establish local zones with different reactivities within the reactivity gradient in the combustion chamber, HVO was mixed in ratio 75 vol% to 25 vol% with gasoline. Characteristics of the mixture are presented in Table 3.

In the following sections, high-reactivity fuel mixture will be denoted as a HVO due to its significantly higher share compared to gasoline.

Table 2 Main properties of natural gas

Property/fuel	NG
Lower heating value on mass basis [MJ/kg]	48.557
Density [kg/m^3] [23]	0.718
CH_4 [mol%]	95.216
C_2H_6 [mol%]	2.698
H:C ratio [% m/m]	0.33

Table 3 Main properties of HVO/gasoline mixture

Property/fuel	HVO/gasoline 75 vol%:25 vol%
Density [kg/L] [26]	0.77
LHV on mass basis [MJ/kg] [27]	43.65
Stoichiometric ratio	14.7
H:C ratio [% m/% m]	0.165
95% (V/V) recovered at [°C] [28]	298.2
Calculated cetane index [29]	90.3

4   Experimental Setup

Experimental setup consists of a light-duty 4-stroke 4-cylinder turbocharged compression ignition engine with characteristics, presented in Table 4, advanced control system and exhaust emission measurement system.

It can be seen from the experimental setup scheme (Fig. 2) that an original compression ignition engine was adapted to enable full control of gas path and injection parameters in the first cylinder where RCCI combustion was established by separating the entire gas path and fuel supply system from the other three cylinders.

To enable full control over thermodynamic parameters and injection parameters, separate control system was developed based on the embedded real-time controller.

Table 4 Characteristics of experimental engine

Engine	PSA DV6ATED4
Cylinders	4, inline
Engine type	4-stroke
Displacement	1560 cm^3
Compression ratio	18:1
Fuel injection system	Common rail, up to 1600 bar
Maximum power	66 kW @ 4000 rpm
Maximum torque	215 Nm @ 1750 rpm
Bore × stroke	75 mm × 88.3 mm

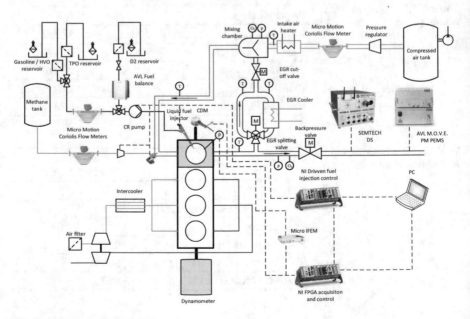

Fig. 2   Experimental setup [30]

While engine speed was kept constant by conventional operation of the three cylinders, RCCI operation was established in the separated cylinder by injecting NG into the intake port and HVO directly into the cylinder.

Engine was coupled with an Eddy-current dynamometer, which was controlled by a real-time control system. In-cylinder pressure was measured with a piezo-electric pressure transducer every 0.1 °CA. Crank angle was measured with an optical crankshaft encoder and the reference to the top dead centre (TDC) position of the piston was determined with a capacitive TDC sensor during motored operation of the separated cylinder. Fuel and air mass flows were measured with a gravimetric balance measurement method and Coriolis mass flow meters. During engine operation, both gaseous and particulate engine-out emissions were measured with flame ionization detector (total hydrocarbons—THC), non-dispersive infrared analyser (CO and CO_2), non-dispersive ultraviolet analyser (NO_x) and combination of micro soot sensor/gravimetric filter module (particulate mass—PM). 0D thermodynamic analysis was performed with averaged and/or filtered measured parameters in a state-of-the-art commercial software.

5 Results

Experimental results comprising in-cylinder thermodynamic parameters and engine-out emissions are presented in the following section for all variations of control parameters.

5.1 Variation of HVO SOE Timing

To evaluate influence of HVO SOE timing on in-cylinder thermodynamic parameters and engine-out emissions, which is crucial to set the basis for RCCI concept control strategy, HVO SOE timing was varied between −30 °CA and −55 °CA. In-cylinder thermodynamic parameters are presented in Fig. 3.

From the presented results, it is evident that advanced SOE timings lead to longer ignition delay, which can be attributed to an increased homogenization rate of high-reactivity HVO and, therefore, to a lower reactivity of the combustion mixture. Ignition delay was calculated as a delay between SOE timings and 5% of burned fuel mass fraction and is presented in Table 5.

As a result of an increased ignition delay and consequently retarded SOC timing, leading to higher share of heat being released after TDC, lower in-cylinder pressures and mean in-cylinder temperatures are achieved when advancing SOE timing. Additionally, from the ROHR results, it can be concluded that an increased ignition delay leads not only to an increase of in-cylinder pressure and mean in-cylinder temperatures but also to a change in combustion process characteristics. It namely influences

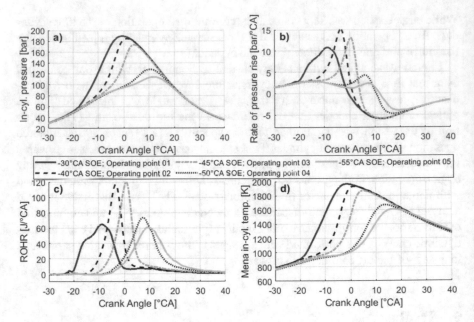

Fig. 3 In-cylinder pressure (**a**), ROPR (**b**), ROHR (**c**) and mean in-cylinder T (**d**) at operating points 01 to 05

Table 5 Ignition delay depending on SOE timing

Operating point	01	02	03	04	05
SOE timing [°CA]	−30	−40	−45	−50	−55
Ignition delay [°CA]	12.4	30	39.1	49.2	55.1

the rate of homogenization of the high-reactivity fuel and, therefore, reactivity stratification of the combustion mixture. By advancing SOE timings, kinetically-controlled phenomena start prevailing over the mixing-controlled phenomena. In addition, by retarding SOE timing, a distinct ROHR peak characteristic for premixed combustion can be observed at operating point 01, which resembles characteristics of the conventional diesel engine operation, whereas other operating points move towards low-temperature kinetically-controlled combustion.

Engine-out emissions for analysed operating points are presented in Fig. 4.

It is evident that NO_x emissions increase with an increase of in-cylinder temperatures when SOE timing is retarded and they are subjected to an even more rapid increase at operating point 01, at which premixed combustion can be observed. PM emissions tend to resemble NO_x emission trend due to an increasing degree of charge homogenization when advancing SOE timing, which leads to reduced rich in-cylinder regions, characterised by high PM production rates. CO and HC emissions are on the other hand subjected to an inverse trend when compared to NO_x and PM emissions, which is characteristic for low-temperature combustion concepts.

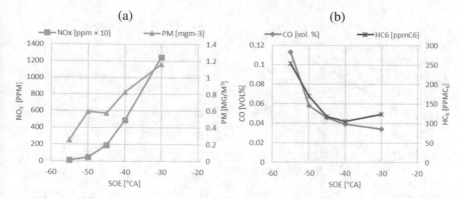

Fig. 4 NO_x and PM engine-out emissions (**a**) and CO and HC_6 engine-out emissions at operating points 01 to 05 (**b**)

CO emissions constantly increase during SOE advancing due to CO to CO_2 conversion reaction quenching as a result of mixture dilution into cylinder areas outside of flammability limits. HC emissions increase when increasing ignition delay mainly due to a higher dilution rate of hydrocarbons into areas outside flammability limits. Additionally, at operating point 01, HC emissions increase can be attributed to a reduction of areas within flammability limits as combustion occurs in the vicinity of injection sprays due to a mixing-controlled nature of the combustion process thus avoiding distant regions with high NG compositions.

5.2 Variation of Energy Share Supplied by the NG

NG energy share was varied between approximately 93% (Operating points 06) and 86% (Operating point 08) by adapting injection durations of high reactivity HVO and low reactivity NG. Simultaneously, SOE timing of the HVO was advanced while decreasing NG energy share with a goal to determine dependency of ignition and combustion events. As presented in Fig. 5, in-cylinder pressure, ROPR, ROHR and mean in-cylinder temperatures are similar for operating points 06 and 07, whereas values of the analysed thermodynamic parameters significantly differ for operating point 08.

To retain similar trends of thermodynamic parameters for operating points 06 and 07, HVO SOE timing was advanced from −50 °CA to −60 °CA while decreasing NG energy share from approximately 93–90%. Advancing of the SOE timing of the HVO and an increase in the energy share of the NG lead to opposite effects in terms of charge reactivity, which influences SOC timing. While advancing SOE of the HVO leads to a reduced reactivity of the combustion mixture due to a higher degree of charge homogenization, decreasing energy share of the NG leads to an increase in charge reactivity due to a higher energy share of highly reactive HVO in

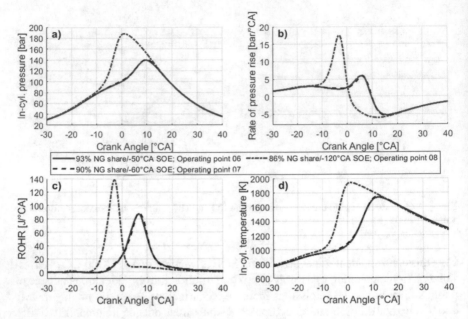

Fig. 5 In-cylinder pressure (**a**), ROPR (**b**), ROHR (**c**) and mean in-cylinder T (**d**) at operating points 06 to 08

the combustion mixture. By simultaneously altering combustion timing and share of injected fuels it is possible to actively control combustion process and consequently influence engine emission response.

Further decrease of energy share of the NG from approximately 90–86% leads to decoupling of the direct injection timing of HVO and SOC timing. Decoupling event was observed during initial engine operation screening process when energy share of the NG was approximately 86% and SOC timing was constant at approximately −13 °CA independently of HVO SOE timing. Based on the SOC and SOE timing trends, presented at operating points 01 to 05, advanced SOE timing of the HVO should lead to retarded SOC timing. However, when moving from operating point 07 to operating point 08, SOC timing was advanced from −8 to −13 °CA, although SOE timing was advanced from −60 to −120 °CA. Observed decoupling event can be attributed to an increase in local reactivity of combustion mixture as a result of high energy share of highly-reactive HVO leading to a spontaneous ignition at lower in-cylinder temperatures. Earlier ignition event in combination with high reaction rates contribute to the majority of the heat released before TDC and therefore, compared to operating points 06 and 07, peak values of all analysed thermodynamic parameters are significantly higher at operating point 08. Increased in-cylinder pressure and ROPR values can have a significant influence on the wear of engine components, and the rate of heat release peak before TDC leads to negative torque, which reduces engine efficiency. Due to the listed aspects, it is advisable to retain dependency

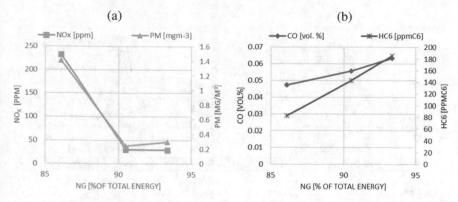

Fig. 6 NO_x and PM engine-out emissions (**a**) and CO and HC_6 engine-out emissions at operating points 06 to 08 (**b**)

between ignition and combustion events in RCCI combustion concept by maintaining sufficiently high energy share of low reactivity fuel, in this study presented with NG.

Engine-out emissions for operating points 06 to 08 are presented in Fig. 6.

When comparing engine-out emission at operating points 06 (93 energy % of NG) and 07 (90 energy % of NG), it can be observed that NO_x emissions resemble similar values due to similar in-cylinder temperatures, whereas at operating point 08 NO_x emissions achieve significantly higher values due to high combustion reaction rates and the most heat released before TDC, which increase local in-cylinder temperatures and consequently increase production of thermal NO_x emissions. CO and HC emissions, on the other hand, decrease with a decrease in NG energy share independently of the ROHR characteristics as presented in Fig. 6b, which can be ascribed to lower share of the NG in volumes outside flammability limits or volumes in which combustion reactions take place but the conversion to CO_2 is not complete resulting in high CO emissions.

5.3 Variation of the EGR Rate

At the operating points 09 to 11, EGR rate is being gradually increased from 0 to 23% and to 43%. As one can see in Fig. 7c, increased EGR rate results in a decreased rate of combustion reactions. In combination with a relatively high specific heat of recirculated exhaust gases, lower local combustion temperatures and mean in-cylinder temperatures are achieved as presented in Fig. 7d.

An increase of EGR rates lead to a decrease of thermal NO_x formation rates and therefore to a reduction of NO_x emissions, as presented in Fig. 8. PM, HC and CO emissions, on the other hand, increase with increasing the EGR rate due to a decrease of local air–fuel equivalence ratio, which increases production of PM and due to an increase of regions outside the flammability limits as a result of lower

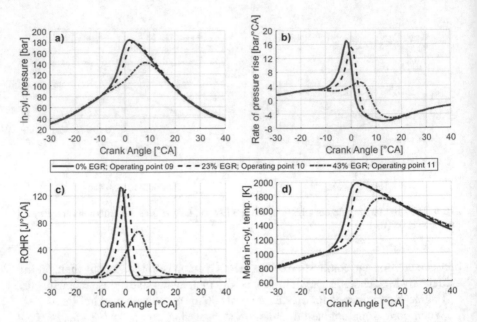

Fig. 7 In-cylinder pressure (**a**), ROPR (**b**), ROHR (**c**) and mean in-cylinder T (**d**) at operating points 09 to 11

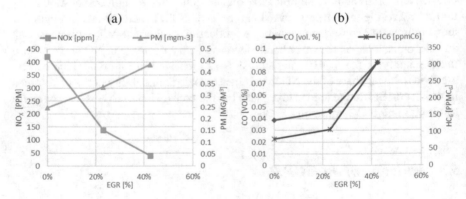

Fig. 8 NO$_x$ and PM engine-out emissions (**a**) and CO and HC$_6$ engine-out emissions at operating points 09 to 11 (**b**)

local temperatures and lower oxygen concentrations, which influences CO and HC emissions.

When evaluating engine-out emissions, it should be, however, considered that an introduction of EGR reduces exhaust mass flow into the environment. Therefore, to determine impact on the environment, total engine-out emissions in unit "g/kWh" were calculated and compared with volumetric emissions, presented by "ppm", "vol%" or "mg/m^3" in Table 6. Calculated increase in % suggests that partial

Table 6 Absolute and relative engine-out emissions for various EGR rates

Operating point	EGR [%]	HC_6 [ppm]	HC_6 [g/kWh]	CO [vol%]	CO [g/kWh]	PM [mg/m^3]	PM [g/kWh]	NO_x [ppm]	NO_x [g/kWh]
09	0	79.00	2.07	0.04	3.12	2.57	0.028	420.00	3.90
11	43	308.00	4.50	0.09	3.99	4.47	0.027	40.00	0.22
Increase [%]	–	290	117	132	28	74	−3	−90	−94

recirculation of exhaust gases results in significantly lower increase of CO and HC_6 emissions and significantly higher decrease of PM and NO_x values when evaluating total emissions compared to volumetric emissions.

Although introduction of EGR leads to 117% increase in HC_6 emissions and 28% increase in CO emissions while keeping PM emission almost constant, significant reduction of NO_x emissions in the range of −94% makes introduction of EGR one of the crucial means to establish a low PM and NO_x engine operation.

5.4 Summary

To establish RCCI combustion in compression ignition engine, SOE timing of directly injected high-reactivity fuel, energy shares of utilized fuels, and EGR rate have to be adjusted at the same time. General influences of analysed control parameters on combustion parameters and engine emission response are summarized in Table 7. The symbols in the table stand for parameter increasing (+), decreasing (−), retarding (>) or advancing (<). It can be summarized that in terms of engine-out total emissions, advanced SOE timing of the HVO, increased energy share of the NG and increased EGR rate have similar effect as they increase CO and HC emissions and decrease NO_x and PM emissions although altering combustion event in different ways. The same variations also lead to lower ROPR values and retarded or unchanged SOC timing.

Comparison between NO_x and PM emissions of analysed operating points and operating points, measured during engine operation in original compression ignition

Table 7 Interrelation between control parameters, thermodynamic parameters and engine-out emissions

Control parameter	Variation	SOC	ROPR max	CO	HC	NO_x	PM
SOE timing of the HVO	<	>	−	+	+	−	−
Energy share of the NG	+	>	−	+	+	−	−
EGR rate	+	Negligible eff.	−	+	+	−	−

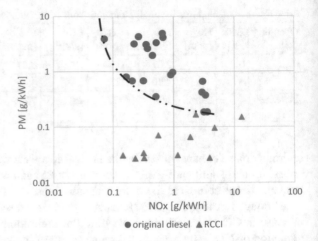

Fig. 9 OEM and measured operating points' NO_x and PM emissions

configuration, is presented in Fig. 9. The results are presented for the analysed small swept volume light duty compression ignition engine. It should be emphasized that with further optimization of control parameters and/or use of heavy-duty engines instead of light-duty engine, further decrease of RCCI NO_x and PM emissions could be achieved.

It can be seen that original diesel operating points achieve up to 100 times higher PM emissions and up to 10 times higher NO_x emissions than RCCI operating points. Furthermore, the original diesel PM and NO_x emissions are bounded by a Pareto front, which is shown with a black dashed line. The Pareto front is characteristic for a conventional compression ignition combustion process due to a stratified combustion mixture leading to inability to simultaneously reduce PM and NO_x emissions. The RCCI combustion concept, however, allows to achieve a simultaneous reduction of NO_x and PM emissions due to low local in-cylinder temperatures, which is crucial to reduce NO_x formation rates and high degree of charge homogenization leading to low PM emissions.

6   Conclusions

RCCI combustion possesses a potential to simultaneously achieve significantly lower engine-out emissions of NO_x and PM due to low local combustion temperatures and high degree of charge homogenization compared to conventional compression ignition engine operation. To evaluate interrelation between engine control parameters, in-cylinder thermodynamic parameters and engine-out emissions, variations of crucial control parameters were performed. The main findings are summarized as follows:

- Advance of the **SOE timing** of high-reactivity HVO leads to a higher degree of charge homogenization and therefore at early injection timings, combustion is reactivity controlled. At late injection timings, combustion characteristics tend to be also mixing-controlled and additionally resemble conventional diesel engine operation with a distinct premixed combustion. By advancing SOE timings of the HVO, NO_x and PM emissions decrease, whereas CO and HC emissions increase due to a combination of underlying parameters, presented in results section.
- An increase in the **energy share of the NG** leads to a decrease in charge reactivity due to a lower share of low-reactivity fuel, which is reflected in lower peak values for ROHR and ROPR values as well as mean in-cylinder temperature. Low energy shares of the NG can result in decoupling the direct injection and combustion events and consequently in advanced SOC timing. Increasing the energy share of the NG, in general, leads to lower NO_x and PM emissions, while CO and HC emissions increase.
- Increasing the **EGR rate** leads to a prolonged combustion duration due to a decreased rate of combustion reactions caused mainly by a reduction of oxygen levels in combustion zones and addition of species with relatively high specific heats. Prolonged combustion duration results in significantly reduced ROHR and ROPR peak values, whereas the SOC is only slightly retarded, which is expected as combustion rate is reduced. Introduction of EGR leads to a reduction in NO_x emissions and an increase of CO and hydrocarbon emissions, while PM emissions stay almost constant when observing total engine-out emission.

Based on the findings, a combination of presented measures, summarized in following suggestions should be taken into account when establishing RCCI concept in internal combustion engines:

- SOE timing of high-reactivity fuel should be sufficiently advanced to achieve high degree of charge homogenization and suitable combustion mixture reactivity gradient leading to kinetically-controlled combustion;
- Sufficient energy share of low-reactivity fuel should be introduced to retain dependency between ignition and combustion events;
- Sufficient EGR rate should be utilized to supress rate of combustion reactions and decrease local in-cylinder temperature.

The presented results indicate that RCCI combustion can lead to up to 100-fold reduction in PM emissions and up to tenfold reduction in NO_x emissions compared to conventional compression ignition engine operation, whereas it should be taken into consideration that a small swept volume light-duty compression ignition engine was used and the presented control parameters were not rigorously optimized. To achieve optimized engine operating results, design space should be further explored.

References

1. Najt, P.M., Foster, D.E.: Compression-ignited homogeneous charge combustion. SAE Technical Papers 830264 (1983). http://doi.org/10.4271/830264
2. Noehre, C., Andersson, M., Johansson, B., Hultqvist, A.: Characterization of partially premixed combustion. SAE Technical Papers 2006-01-3412 (2006). http://doi.org/10.4271/2006-01-3412
3. Kokjohn, S.L., Hanson, R.M., Splitter, D.A., Reitz, R.D.: Experiments and modeling of dual-fuel HCCI and PCCI combustion using in-cylinder fuel blending. SAE Int. J. Engines **2**, 24–39 (2009). https://doi.org/10.4271/2009-01-2647
4. Benajes, J., Pastor, J.V., García, A., Monsalve-Serrano, J.: The potential of RCCI concept to meet EURO VI NOx limitation and ultra-low soot emissions in a heavy-duty engine over the whole engine map. Fuel **159**, 952–961 (2015). https://doi.org/10.1016/j.fuel.2015.07.064
5. Abdelaal, M.M., Hegab, A.H.: Combustion and emission characteristics of a natural gas-fueled diesel engine with EGR. Energy Convers. Manag. **64**, 301–312 (2012). https://doi.org/10.1016/j.enconman.2012.05.021
6. Benajes, J., Molina, S., García, A., Belarte, E., Vanvolsem, M.: An investigation on RCCI combustion in a heavy duty diesel engine using in-cylinder blending of diesel and gasoline fuels. Appl. Therm. Eng. **63**, 66–76 (2014). https://doi.org/10.1016/j.applthermaleng.2013.10.052
7. Nieman, D.E., Dempsey, A.B., Reitz, R.D.: Heavy-duty RCCI operation using natural gas and diesel. SAE Int. J. Engines **5**, 270–285 (2012). https://doi.org/10.4271/2012-01-0379
8. Imtenan, S., Varman, M., Masjuki, H.H., Kalam, M.A., Sajjad, H., Arbab, M.I., et al.: Impact of low temperature combustion attaining strategies on diesel engine emissions for diesel and biodiesels: a review. Energy Convers. Manag. **80**, 329–356 (2014). http://doi.org/10.1016/j.enconman.2014.01.020
9. Reitz, R.D.: Directions in internal combustion engine research. Combust. Flame **160**, 1–8 (2013). https://doi.org/10.1016/j.combustflame.2012.11.002
10. Gross, C.W., Reitz, R.: Investigation of steady-state RCCI operation in a light-duty multi-cylinder engine using "dieseline." SAE Technical Paper 2017-01-0761 (2017). http://doi.org/10.4271/2017-01-0761
11. Benajes, J., García, A., Monsalve-Serrano, J., Boronat, V.: An investigation on the particulate number and size distributions over the whole engine map from an optimized combustion strategy combining RCCI and dual-fuel diesel-gasoline. Energy Convers. Manag. **140**, 98–108 (2017). https://doi.org/10.1016/j.enconman.2017.02.073
12. Heuser, B., Ahling, S., Kremer, F., Pischinger, S., Rohs, H., Holderbaum, B., et al.: Experimental investigation of a RCCI combustion concept with in-cylinder blending of gasoline and diesel in a light duty engine (2015). https://doi.org/10.4271/2015-24-2452
13. Kokjohn, S.L., Hanson, R.M., Splitter, D.A., Reitz, R.D.: Fuel reactivity controlled compression ignition (RCCI): a pathway to controlled high-efficiency clean combustion. Int. J. Engine Res. **12**, 209–226 (2011). https://doi.org/10.1177/1468087411401548
14. Liu, H., Yao, M., Zhang, B., Zheng, Z.: Effects of Inlet pressure and octane numbers on combustion and emissions of a homogeneous charge compression ignition (HCCI) engine. Energy Fuels **22**, 2207–2215 (2008). https://doi.org/10.1021/ef800197b
15. Kakaee, A.-H., Rahnama, P., Paykani, A.: Influence of fuel composition on combustion and emissions characteristics of natural gas/diesel RCCI engine. J. Nat. Gas Sci. Eng. **25**, 58–65 (2015). https://doi.org/10.1016/j.jngse.2015.04.020
16. Zoldak, P., Sobiesiak, A., Bergin, M., Wickman, D.D.: Computational study of reactivity controlled compression ignition (RCCI) combustion in a heavy-duty diesel engine using natural gas. SAE Technical Paper 2014-01-1321 (2014). http://doi.org/10.4271/2014-01-1321
17. Poorghasemi, K., Saray, R.K., Ansari, E., Irdmousa, B.K., Shahbakhti, M., Naber, J.D.: Effect of diesel injection strategies on natural gas/diesel RCCI combustion characteristics in a light duty diesel engine. Appl. Energy **199**, 430–446 (2017). https://doi.org/10.1016/j.apenergy.2017.05.011

18. Jia, Z., Denbratt, I.: Experimental investigation of natural gas-diesel dual-fuel RCCI in a heavy-duty engine. SAE Int. J. Engines **8**, 797–807 (2015). https://doi.org/10.4271/2015-01-0838
19. Li, J., Yang, W., An, H., Zhou, D.: Soot and NO emissions control in a natural gas/diesel fuelled RCCI engine by φ-T map analysis. Combust. Theor. Model. **21**, 309–328 (2017). https://doi.org/10.1080/13647830.2016.1231936
20. Gharehghani, A., Hosseini, R., Mirsalim, M., Jazayeri, S.A.: An experimental study on reactivity controlled compression ignition engine fueled with biodiesel/natural gas. Energy **89**, 1–10 (2015). https://doi.org/10.1016/j.energy.2015.06.014
21. Kubesh, J., King, S.R., Liss, W.E.: Effect of gas composition on octane number of natural gas fuels. SAE Technical Paper Series 1 (1992). http://doi.org/10.4271/922359
22. Tuner, M.: Review and benchmarking of alternative fuels in conventional and advanced engine concepts with emphasis on efficiency, CO_2, and regulated emissions. SAE Technical Paper (2016). https://doi.org/10.4271/2016-01-0882
23. ISO 6976:2016: Natural gas—calculation of calorific values, density, relative density and Wobbe indices from composition (2016)
24. Rantanen, L., Linnaila, R., Aakko, P., Harju, T.: NExBTL—biodiesel fuel of the second generation. SAE Technical Paper Series 2005-01-3771 (2005). http://doi.org/10.4271/2005-01-3771
25. Bezergianni, S., Dimitriadis, A.: Comparison between different types of renewable diesel. Renew. Sustain. Energy Rev. **21**, 110–116 (2013). https://doi.org/10.1016/j.rser.2012.12.042
26. ISO 12185:1996: Crude petroleum and petroleum products—determination of density—oscillating U-tube method, p. 9 (1996)
27. ASTM D4868-00(2010): Standard test method for estimation of net and gross heat of combustion of burner and diesel fuels (2010)
28. ISO 3405:2011: Petroleum products—determination of distillation characteristics at atmospheric pressure, p. 36 (2011)
29. ISO 4264:2007: Petroleum products—calculation of cetane index of middle-distillate fuels by the four-variable equation (2012)
30. Žvar Baškovič, U.: Advanced combustion concepts with innovative waste derived fuels. Ph.D. thesis (2019)

Investigation of Charge Mixing and Stratified Fuel Distribution in a DISI Engine Using Rayleigh Scattering and Numerical Simulations

Stina Hemdal and Andrei N. Lipatnikov 📖

Abstract The stratified fuel distribution and early flame development in a firing spray-guided direct-injection spark-ignition (DISI) engine are characterized applying optical diagnostics. The goal is to compare effects of single and double injections on the stratified air–fuel mixing and early flame development. Vaporized in-cylinder fuel distributions resulting from both single and double injections before, during and after ignition are selectively visualized applying Rayleigh scattering. Reynolds-averaged Navier–Stokes (RANS) simulations are performed to facilitate interpretation of the obtained experimental data. Two hypotheses are tested. First, injecting the fuel as a closely coupled double injections can improve mixing. Second, the better mixing putatively associated with double injections is mainly due to either a longer mixing time or higher mixing rate (driven by turbulence generated by the injections). The optical investigation of the in-cylinder fuel distributions and early flame propagation corroborated the better mixing, showing that double injections are associated with more evenly distributed fuel, fewer local areas with high fuel concentrations, faster initial flame spread and more even flame propagation (more circular flame spreading). The results from both the experiments and the simulations support the hypothesis that delivering fuel in closely coupled double injections results in better mixing than corresponding single injections. According to the simulations, the improved mixing stems from the longer time available for mixing of the air and fuel in double injection events, which has stronger effects than the higher computed peak bulk mixing rate for single injections.

Keywords Gasoline direct injection · Multiple injection · Stratified charge · Optical diagnostics · Rayleigh scattering · RANS simulations

S. Hemdal
Volvo Group Trucks Technology, 41715 Gothenburg, Sweden

A. N. Lipatnikov (✉)
Department of Mechanics and Maritime Sciences, Chalmers University of Technology, 41296 Gothenburg, Sweden
e-mail: andrei.lipatnikov@chalmers.se

1   Introduction

Burning of globally (i.e., volume-averaged) lean stratified-charge in a Spray-Guided (SG) Direct-Injection (DI) Spark Ignition (SI) engine is a promising technological solution capable for substantially reducing fuel consumption [1–3] when compared to combustion of homogeneous stoichiometric mixture in a Port-Fueled (PF) SI engine. In the stratified operation mode of a SG DI SI engine, the spark is triggered shortly after late injection of a fuel into the engine cylinder during the compression stroke. Accordingly, due to a short time allowed for turbulent mixing, the highly inhomogeneous cloud of the fuel and air is confined to the center of the cylinder volume at the ignition time, with the local mixture composition being ignitable around the spark electrodes even if the global mixture composition is well beyond the lean flammability limit after the end of the fuel injection. Such a technology allows the engine to operate with highly diluted fuel–air charge, thus, resulting in an increase in the specific heat ratio, reduction of pumping and heat losses, and, finally, gains in the efficiency and fuel consumption [1–3].

However, side by side with these benefits, there are also issues that should be resolved in order to fully utilize this potential of the SG stratified-charge combustion for substantial reduction of fuel consumption. For instance, locally rich zones of a stratified charge can be a source of particles [4]. Moreover, local mixture inhomogeneities around the spark make growth of the early flame kernel less regular, thus, contributing to cycle-to-cycle variations. Furthermore, if the time between the end of injection and the start of ignition is too long, the local mixture composition can become too lean even in the vicinity of the spark (the so-called "overmixing"), thus, reducing combustion stability and causing misfires [5].

The above negative phenomena are mainly controlled by the local fuel distribution, which in its turn is controlled by the fuel injection and turbulent mixing. Moreover, interaction between the fuel distribution and turbulence just after spark ignition is also of great importance for stability of a SG DI SI engine [6]. For instance, too intense turbulence can impede igniting the mixture and growing the flame kernel [7]. Moreover, the fluctuating velocity field can move the early flame kernel outside the boundaries of a flammable mixture, thus, further increasing combustion instability [8, 9].

Thus, characteristics of the local fuel–air mixture around the spark are of vital importance for stable and efficient operation of a SG DI SI engine. Accordingly, fuel injection and mixing were in the focus of a number of experimental and numerical investigations. However, there were two important gaps in such studies. First, the majority of them dealt with multi-hole injectors, whereas piezo-actuated outward-opening injectors, which are installed in most SG DI SI engines produced by the car industry, have yet been investigated substantially less. The latter injectors are also of great interest, because they enable precise deliveries of small amounts of fuel, thus, offering a technological solution such as multiple (split) injections separated by short delay times. Split-injection strategies were already shown to yield a higher Indicated Mean Efficient Pressure (IMEP) at the same mass of injected fuel, better combustion stability, a larger ignition window, and lower engine-out emissions [5, 10–12]. Moreover, split injections are known to reduce spray penetration, thus, impeding impinging the fuel on the piston crown [10, 12]. Accordingly, the probability of piston pool fires is decreased and, consequently, less amount of soot is formed in the cylinder [13]. Furthermore, split injection strategy was noted [14] to improve combustion stability by reducing cycle-to-cycle variability of the gas flow around the spark when compared to a single injection of the same fuel mass. Split injections were also hypothesized [5, 10, 12, 15] to enhance convection of fuel back towards the injector position by upward flow associated with vortexes documented [15–18] in the central regions of hollow cone sprays.

Second, the majority of advanced experimental investigations of fuel injection followed by turbulent mixing, performed by applying laser diagnostic techniques, have yet examined these phenomena in constant-volume spray chambers under a constant pressure and a constant temperature. However, applications of advanced laser diagnostic techniques to exploring fuel injection and turbulent mixing processes in revolving and firing engines have yet been limited, especially in SG DI SI engines with outward-opening injectors.

The present study aims at filling these gaps by characterizing both liquid and vaporized fuel distributions using Laser-Induced Fluorescence (LIF) of diacetyl [19] added to the fuel and Rayleigh scattering, respectively, in a firing SG DI SI engine. Moreover, single and double injection strategies are compared by analyzing the visualized fuel distributions and the early flame kernel growth. The goal of this comparison is to explore whether or not the double injections yield better fuel distribution at the ignition instant. Furthermore, to facilitate interpretation of the obtained images and to elucidate the mixing process, results of unsteady 3D Reynolds-Averaged Navier–Stokes (RANS) simulations are reported and discussed. The simulations aim, in particular, at identifying the major cause (different mixing times or different mixing rates) of differences between fuel distributions imaged in the cases of single and double injections.

In the next section, the experimental setup and conditions are reported, and the diagnostic techniques are discussed. Since the Rayleigh-scattering-based visualization technique adopted in the present study was in the focus of another recent paper by the first author [20], we will restrict ourselves to a brief discussion of this method of imaging fuel distribution and the experimental setup. The reader interested in more details is referred to Ref. [20]. In the third section, numerical simulations are briefly summarized. Experimental and numerical results are discussed in the fourth section, followed by conclusions.

2 Experimental Setup

2.1 Engine and Conditions

Experiments were performed using a single-cylinder engine AVL 5411.018 equipped with an elongated piston. The piston crown contains a flat quartz plate with diameter of 63 mm, thus providing optical access from below via a mirror inclined under the crown (Bowditch design), see Fig. 1. Moreover, optical access is available through the upper part of the cylinder liner and the pent roof.

The engine characteristics are reported in Table 1, where abbreviations TDC, aTDC, CA, IVO, IVC, EVO, and EVC mean: top dead center, after TDC, crank angle, intake valve opens, intake valve closes, exhaust valve opens, and exhaust valve closes, respectively. A fuel (*iso*-octane with addition of 10 volume percent of diacetyl adopted as a tracer in LIF measurements) is delivered (at a pressure of 200 bar) and injected using a Haskel pump and a Bosch outward-opening piezo-actuated injector, respectively. Combustion is initiated adopting a BMW triple electrode spark plug. Both the injector and the spark plug are centrally mounted on the combustion chamber roof on a line between the exhaust and intake valves. The spark plug and injector are inclined by 10^0 and 20^0, respectively, with respect to the vertical axis.

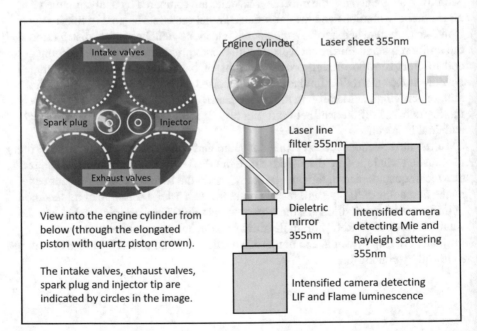

Fig. 1 The view of the engine cylinder through the optical piston and a sketch of the optical set-up with laser sheet forming lenses, dielectric mirror as beam splitter, laser line filter and two intensified cameras

Table 1 Engine characteristics and operating parameters

Characteristic	Value	Parameter	Value
Bore	83 mm	Single injection duration	600 μs
Stroke	90 mm	Single injection SOI	− 24° CA aTDC
TDC volume	53 cm^3	Single injection IGN	− 16° CA aTDC
Displacement volume	488 cm^3	Double injection duration 1	300 μs
Compression ratio	10.1	Dwell	400 μs
IVO/IVC	340/− 120° CA aTDC	Double injection duration 2	300 μs
EVO/EVC	105/− 355° CA aTDC	Double injection SOI	− 29° CA aTDC
Engine speed	1500 rpm	Double injection IGN	− 16° CA aTDC
IMEP	3.6 bar	1° CA	111 μs

In the present experiments, the engine speed was equal to 1500 rpm. Accordingly, 1° CA corresponds approximately to 111 μs. The engine was operated in skip-fire mode, i.e. 100 fired cycles were followed by about 300 cycles without a fire. To set appropriate Start of Ignition (SOI) and Ignition timing (IGN) for the major measurements, these two parameters were preliminarily varied in order to find the maximum engine load for a fixed amount of injected fuel (either gasoline or *iso*-octane, with major results being similar for both fuels). The load was calculated by processing in-cylinder pressure traces measured using a fast piezo pressure transducer.

Such a pre-study was performed in the cases of (i) a single injection during 600 μs and (ii) double injections. In the latter case, the fuel was injected during 300 μs, followed by a delay of 400 μs and the second injection of the fuel during 300 μs. Obtained dependencies of the Coefficient of Variation (COV) of IMEP on the time between SOI and IGN instants show that the use of the double injections reduces COV and increases the injection timing and ignition window [20]. Based on results of this pre-study, subsequent optical measurements were performed by setting (i) the injection timing equal to − 24 and − 29° CA for the single and double injections, respectively, and (ii) the ignition timing equal to − 16° CA in both cases. Henceforth, the values of CA degrees are reported after TDC, i.e. a negative CA degree corresponds to an instant before the TDC and 0° CA corresponds to the TDC. As a result, the end of injection (EOI) and SOI timings were similar for both injection strategies (more specifically, the EOI timing was equal to − 19 and − 20° CA for the single and double injections, respectively). While IMEP was maximal for ignition at the EOI or slightly before it, in the present experiments, ignition was started after the EOI, in particular, to perform Rayleigh scattering measurements, which would be disturbed by liquid droplets otherwise.

2.2 Optical Diagnostics, Image Acquisition and Post-Processing

LIF and scattering (Mie or Rayleigh) measurements were performed using the third harmonic (355 nm) of a Nd:YAG laser (Spectra Physics Quanta Ray lab 120) with mean intensity of 150 mJ. To transform the laser beam into a laser sheet, a set of cylindrical lenses (f − 100, f + 200, f − 200, f + 500) was adopted. The sheet width and thickness were equal to 45 mm and 0.6 mm, respectively. The laser sheet passed into the engine cylinder through the pent roof windows. The sheet was located below the spark plug electrodes, with the vertical distance between the sheet and the top of the cylinder pent being about 12 mm. Figure 1 displays a sketch of the optical set-up and the obtained view into the cylinder.

Scattered light, LIF signal, or flame luminosity passed through the quartz piston crown, passed through or was redirected by a dielectric mirror (355 ± 10 nm), passed through a laser line filter (355 ± 2 nm), and was finally detected with one of two Intensified Charge Coupled Device (ICCD) cameras (DynaMight and LaVision) with a resolution of 1024×1024 pixels. In these measurements, a dynamic range of 16-bits was used. Since acquisition of images was limited by the pulse repetition rate of the laser (10 Hz), the images were acquired as single-shot images with 10 repetitions at each step from SOI until $- 10°$ CA.

One of the ICCD cameras detected the Mie or Rayleigh scattering light. Simultaneously, the other ICCD camera captured either the LIF signal from diacetyl in the fuel or flame luminosity. More specifically, the former camera captured the Mie scattering from the liquid fuel during injection. When the Mie scattering light disappeared due to evaporation of the injected fuel, the gain of the intensifier on this camera was increased from 35 to 65 to capture the Rayleigh scattering from the vaporized fuel. The latter camera detected LIF signal since injection until ignition. Subsequently, the LIF signal was drowned by flame luminosity, and the camera captured the flame images. Thus, the two cameras either simultaneously captured Mie scattering and LIF signal during an early stage of the studied process or simultaneously captured Rayleigh scattering and flame luminosity during a late stage.

Due to the high level of background noise, results obtained using Rayleigh scattering were only analyzed from the qualitative perspective. Flame kernel growth was explored by finding probabilities of flame presence in the image pixels. For this purpose, each image was binarized after subtraction of background noise. Subsequently, images obtained at the same CA degree were averaged pixel by pixel. Finally, the number of pixels that contained flame was counted to evaluate the aforementioned probability.

The reader interested in a more detail discussion of the optical diagnostics, image acquisition and post processing tools used in the present study is referred to Ref. [20].

3 Numerical Simulations

Unsteady multidimensional RANS simulations of the experiments discussed above were performed using *Converge* Computational Fluid Dynamics (CFD) software [21]. The simulations were started at $-440°$ CA and modelled air intake and turbulence generation, the *iso*-octane hollow-cone spray discharged by the piezo-actuated outward-opening injector, and evaporation of the fuel, followed by turbulent mixing in the cylinder of the optical GDI SI engine used in the experiments. The simulations were performed invoking the Renormalization Group (RNG) k-ε turbulence model [22], Kelvin–Helmholtz Rayleigh–Taylor (KHRT) spray model [23–29], No Time Counter (NTC) droplet collision model [30], and Frossling correlation model of fuel evaporation [31]. All these models were used as implemented in *Converge*, but default values of two constants of the KHRT model were changed. First, the value of the fraction of the parent parcel mass that contributed to the child parcel mass ("*shed_factor*") was set equal to 0.1 (the default value is 0.25). Second, the constant C_1 that affected the maximum time of the growth of RT waves on a droplet surface was increased from 1 to 3. The two changes slightly improved agreement between measured and computed images, but effects of such variations in the two constants were quite moderate. It is also worth noting that the KHRT spray model was earlier validated [32] by simulating experiments with gasoline and ethanol hollow-cone sprays discharged at the same injection pressure by a similar piezo-actuated outward-opening injector into Chalmers' spray rig [33], but those simulations were performed using another CFD software (*OpenFOAM*) and slightly different implementation of the model. Spark ignition and combustion were beyond the scope of the numerical work presented here.

4 Results and Discussion

Figures 2 and 3 show images captured simultaneously by the two cameras at subsequent time steps of a single injection event. Figure 2 reports images acquired by capturing Mie scattering from the liquid fuel (the first and third row) and LIF from the fuel (the second and fourth row) during the early stage (until EOI). Figure 3 shows images acquired by capturing Rayleigh scattering from the vaporized fuel (first and third row) and flame luminosity (second and fourth row) after the EOI. After the EOI ($-18°$ CA), the Mie scattering light vanished, thus, indicating that all injected fuel had been vaporized. Accordingly, a significantly less intense Rayleigh scattering light was detectable starting from $-18°$ CA. As soon as the spark became visible, the LIF signal was obscured by the flame luminosity. Since this happened shortly after $-18°$ CA, the most left image in the second row in Fig. 3 still shows the captured LIF signal. Since all injected fuel had been vaporized before $-18°$ CA, as noted earlier, comparison of brightness of the discussed (left bottom cell) and the previous ($-19°$ CA) LIF images indicates that the LIF signal from the vaporized diacetyl

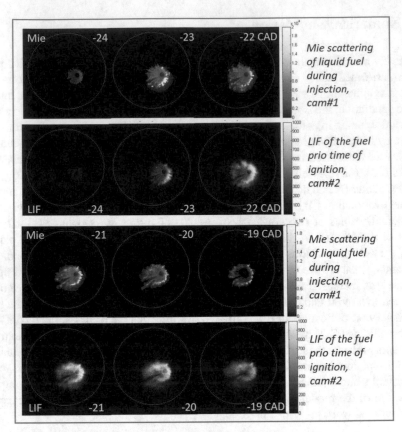

Fig. 2 Single-shot images acquired in the case of a single injection started and ended at − 24° CA and − 19° CA, respectively. Mie scattering light from the liquid fuel, LIF signal the liquid and vaporized fuel

(− 18° CA) was weak when compared to the LIF signal from the liquid diacetyl (− 19° CA) under elevated pressures and temperatures in the engine combustion chamber. A detailed discussion of these images can be found in Ref. [20]. Here, we restrict ourselves to noting a few points.

First, while the Mie scattering and LIF images correspond closely to one another, the latter images show a wider fuel cloud when compared to the former images. The point is that the Mie scattering occurs only on the fuel droplets, whereas both the liquid and vaporized fuel contributes to the LIF signal. Accordingly, the difference between the two kinds of images increases with time as more fuel evaporates.

Second, the Rayleigh scattering images show that the fuel vapor concentrated initially (after the EOI) in a circular arc of the hollow cone spray. Subsequently, the fuel spread mostly inwards, i.e. to the center of the circular arc., thus, making the fuel cloud compact and close to the center of the cylinder. Within the cloud, the fuel was unevenly distributed with a high degree of stratification (brighter local areas

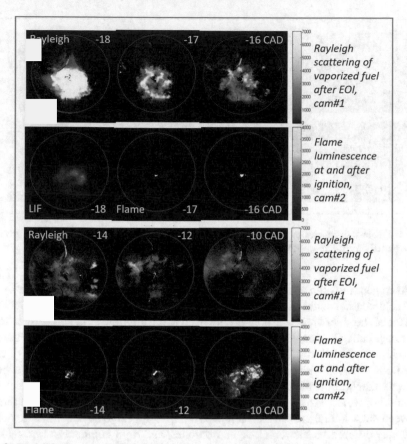

Fig. 3 Rayleigh scattering light from the vaporized fuel and flame luminescence are shown in the first and third and second and fourth rows, respectively (with the exception of the most left image in the bottom row, which still shows LIF signal from the fuel)

in the images indicate zones with a higher fuel concentrations). With time, the fuel distribution became more even, but the fuel cloud was still close to the center of the cylinder.

Third, the shape of the flame kernel was uneven, thus, indicating differences in the flame propagation speeds in different directions. At − 10° CA, the flame shape was still far from circular and was highly wrinkled at the edges. The uneven flame expansion is associated with significant spatial variations in the local air–fuel ratio in the highly stratified fuel cloud.

Single-shot images of Rayleigh scattering light from the vaporized fuel and flame luminosity, captured after the end of the double injections, are reported in Fig. 4. When compared to the single-shot images obtained in the case of a single injection (Fig. 3), Fig. 4 indicates a less uneven fuel distribution and a faster initial growth of the flame kernel.

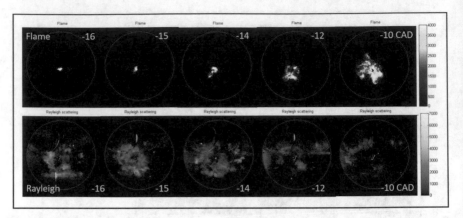

Fig. 4 Single-shot images of the vaporized fuel (Rayleigh scattering, bottom row) and growing flame kernel (flame luminosity, top row) after the end of double injections

Averaged Rayleigh scattering images of fuel distributions are compared for single (left column) and double (right column) injection events in Fig. 5. In the center of the fuel cloud, brighter areas are more pronounced in the left images (a single injection). In the right images (double injections), the fuel cloud occupies a larger area. Figure 6 also shows a faster growth of the area occupied by the flame in the case of double injections (red line with circles). These differences imply that the double injections yield weaker stratification of the fuel distribution, i.e. better mixing of the fuel and air in other words. At the same time, it is worth noting that significant differences between flame kernel growth rates reported in Fig. 6 appear to be in contrast to rather moderate differences between images in the left and right columns in Fig. 5. This apparent inconsistency will be explained later using results of numerical simulations.

Before using numerical results for interpretation and explanation of the experimental findings, it is necessary to assess the RANS simulations qualitatively. For this purpose, the calculated mass of the evaporated fuel is compared with the experimentally obtained images of the fuel distributions in selected injection events in Fig. 7. The computed evaporated fuel mass is shown as a function of CA degrees for a single (black solid line) or a double (red dashed line) injection event, together with experimentally obtained images of the injection event at selected time steps, indicated with capital letters (A-E) near the curves.

In the simulated single injection event, see black solid line, the vaporized fuel mass starts to increase shortly after SOI at $-24°$ CA and then increases throughout the rest of the event. Almost all the fuel is vaporized shortly after the EOI at $-20°$ CA. The experimentally obtained images of the liquid and vaporized fuel show the same patterns. The liquid fuel is visualized by Mie scattering from the SOI (image E in Fig. 7) until the EOI. Shortly after the EOI ($-19°$ CA) the signal from the liquid disappears and the vaporized fuel is visualized using Rayleigh scattering (image F).

In the simulated double injection event, see red dashed line, the evaporated fuel mass starts to increase shortly after the SOI ($-29°$ CA) and increases until the end of

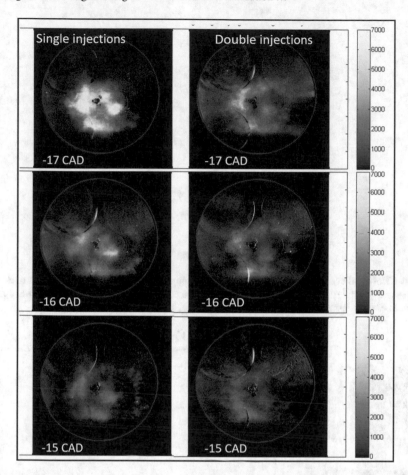

Fig. 5 Average images of the Rayleigh scattering signal at time of ignition for single injections (left column) and double injections (right column)

the first injection at $-24°$ CA. Experimentally the liquid fuel is detectable from just after the SOI (image A) and throughout the first injection pulse. Shortly after the end of the first injection ($-26°$ CA) the liquid signal vanishes, and the vaporized fuel is detected (image B). During the delay between the first and second injection pulses, the vaporized fuel mass is almost constant since no more fuel is injected, but the previously injected fuel fully evaporates. Accordingly, the experimentally obtained images show only vaporized fuel until the second injection starts. The vaporized fuel mass increases from shortly after start of the second injection until the EOI. Image C shows the liquid fuel at the beginning of the second injection. Shortly after the EOI, the vaporized fuel mass becomes constant and, in the experiments, the vaporized fuel is detected (image D). Thus, the calculated vaporized fuel mass shows qualitative agreement with the experimentally obtained images.

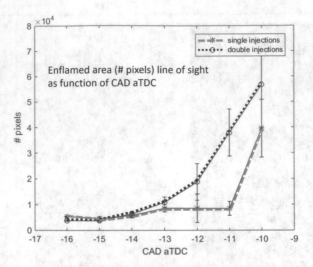

Fig. 6 Growth of average flame area measured in the number of pixels in images. Results obtained in the cases of single and double injections are shown in blue dashed line with stars and in red dotted line with circles, respectively

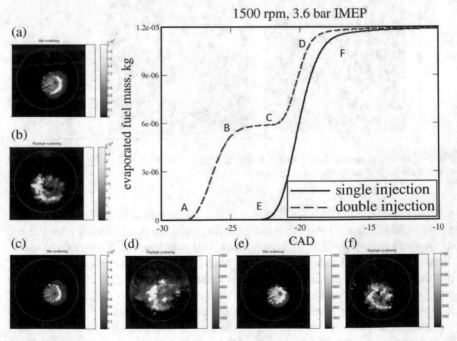

Fig. 7 Calculated evaporated fuel mass as a function of CA degrees for single injection (black solid line) and double injection (red dashed line) events and experimentally obtained images of the liquid or vaporized fuel distribution at selected time steps indicated with capital letters (A–E) near curves

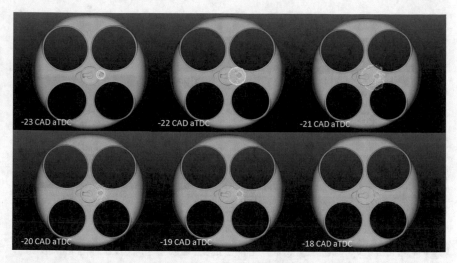

Fig. 8 Calculated distributions of droplets, shown as white particles, in the cylinder volume during a single injection of 12 mg of *iso*-octane with SOI at − 24° CA. The image matrix shows droplet distributions at consecutive time steps from − 23 (upper left image) to − 18° CA (lower right image)

Images of the simulated droplet distribution inside the cylinder volume during the injection event for a single injection of 12 mg of *iso*-octane at − 24° CA are presented in Fig. 8. The droplet distribution is displayed as white particles in the images. Droplets appear shortly after the SOI and, at − 23° CA, a small hollow cone can be seen close to the injector position. During the subsequent time steps, as more fuel is injected, the hollow cone grows. The injection ends at around − 19° CA and the quantity of droplets then falls due to vaporization of the fuel. The calculated droplet distribution shows the same pattern as the experimentally obtained images of the Mie scattering (Fig. 3). The numerical images show droplets from the whole cylinder volume whereas the experimental images are cut throughs obtained using a laser sheet, and thus mainly show droplets in the intersection of the laser sheet. However, in the experiments, the multiple scattering by the droplets was so strong (relative to signals in their surroundings), that the whole hollow cone was visible, although less intensely than the arc-shaped intersection with the laser sheet. The calculated droplet distribution agrees qualitatively with the experimentally obtained images. When comparing the experimental and CFD results, it is also worth remembering that the numerical images presented in Figs. 8, 9, 10, and 11 show the entire cross-section of the engine cylinder, whereas the experimental images reported in Figs. 2, 3, 4, and 5 show a central area of the cylinder, with a diameter just 70% of the cylinder diameter.

Computed spatial distributions of the equivalence ratio in a horizontal plane (with thickness of 1 mm) located 2 mm above the TDC position of the piston crown and corresponding to the position of the laser sheet during the experiments are shown in Fig. 9 at consecutive time steps from − 21° to − 16° CA. In the first image (− 21° CA), a small arc is visible to the right. This agrees well with the experimental

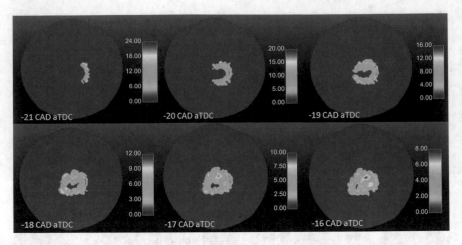

Fig. 9 Calculated equivalence ratios for a single injection event in a 2D plane 2 mm above the TDC position of the piston at consecutive time steps from − 23° to − 16° CA

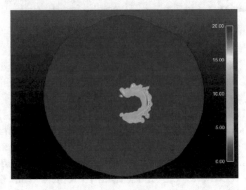

Fig. 10 Calculated overall equivalence ratios for the single injection event in a 2D plane 2 mm above the TDC position of the piston at − 20° CA

observations and is due to the inclination of the injector. The leading edge of the spray at the right reaches the position of the cut-through (in the simulations) or the laser sheet (in the experiments) first, due to the shorter travel distance from the injector orifice. In the subsequent time steps, the rest of the circular arc, formed by the hollow cone spray, becomes visible.

The experimentally obtained fuel distribution provides detail, but only qualitative, information on the local fuel distribution, while the simulations of the same system provide quantitative data. They indicate that the local equivalence ratios are highest in the circular arc during the injection event, as also observed in the experiments. After the EOI (− 19° CA), the fuel cloud still forms a circular arc, but it spreads out somewhat and the local equivalence ratios decrease as the fuel mixes with the air. From about − 18° CA, the fuel is drawn into the center of the arc, forming a

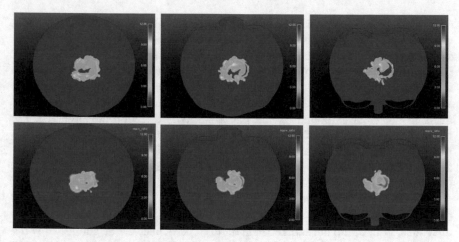

Fig. 11 2D fields of the equivalence ratio, computed in the cases of single (top row) and double (bottom row) injections at − 18° CA. Cuts made 1, 3, and 5 mm above the TDC position of the piston are reported in the left, middle, and right columns, respectively

circular cloud in the center of the cylinder. The size of this cloud agrees well with the size of the fuel cloud observed in the LIF image obtained at − 18° CA (Fig. 3). Local areas or spots of high equivalence ratios are found in the arc and the center of the fuel cloud. The resulting fuel cloud at the time of ignition (− 16° CA) is well confined to the center of the cylinder. It has the same spread and basic features, such as inhomogeneity with local areas with high equivalence ratios and areas with much lower ratios between them, like the experimentally observed fuel distribution. The highest equivalence ratios (around 8.0) are found in spots on the inside of the circular arc close to the center of the circular area. This is in good agreement with the experimentally observed fuel distribution (Figs. 2 and 3). The values of the estimated equivalence ratios are high. For the early time steps during the injection event, this is easily explained by the presence of liquid fuel. However, both experiments and simulations show that, at the time of ignition, all the fuel was vaporized. In the experiments, this was confirmed by disappearance of the Mie scattering signal and, in the calculations, the liquid fraction approached zero before the time of ignition.

Results of the present simulations facilitate interpretation of the experimental images. For instance, first, when comparing Mie scattering and LIF images (Fig. 3), a larger size of the fuel cloud in the latter images was already noted and attributed to partial evaporation of the fuel. The simulations further support such an explanation. Figure 10 shows an image of the overall equivalence ratio computed at − 20° CA, whereas the corresponding image of the equivalence ratio in the gas phase is reported in Fig. 9 (middle column, top row). At first glance, the two images look similar, but a red dotted arc (droplets) is solely observed in Fig. 10. Thus, comparison of the two images indicates that the cloud of the vaporized fuel is much larger (at that instant) than the arc of liquid droplets, in line with earlier discussion of the experimental images.

Second, since LIF and Rayleigh-scattering techniques show fuel distribution in a single plane, representativeness of such images is worth investigating. Figure 11 shows that 2D fields of the equivalence ratio, simulated in various horizontal planes parallel to the piston, are similar, especially in the case of double injections (bottom row). Accordingly, experimental images obtained from a single plane appear to be sufficiently representative. Nevertheless, some non-uniformities of the computed fields in the vertical (normal to the piston surface) direction are observed. For instance, in the single-injection case, the central spot of a very lean mixture rotates with the vertical coordinate (top row), whereas a similar spot is pronounced less and only far (right column) from the piston surface in the case of double injections (bottom row).

Comparison of images reported in the two rows in Fig. 11 also indicates that the simulated cloud of the vaporized fuel is more uniform in the case of double injections, in line with the experimental data. In particular, the magnitude of variations in the equivalence ratio is higher in the case of a single injection, whereas the aforementioned lean spot is weakly pronounced in the case of double injections.

In addition to local effects, the simulations offer an opportunity to estimate the bulk influence of injection strategy on fuel–air mixing in the chamber of the studied engine. For instance, the left plot in Fig. 12 shows that, at ignition time ($- 16°$ CA), the computed volume-averaged root mean square (rms) of spatial variations of the equivalence ratio (Φ) is lower in the case of double injections (red dashed line), thus, indicating better mixing, in line with the experiments. It is worth remembering that numerical results obtained at $- 15°$ CA or later do not straightforwardly relate to the experimental data, because spark ignition and combustion were not addressed in the simulations, i.e. spark discharge was not activated.

Shown in the right plot in Fig. 12 is the time-derivative $d\langle\Phi'\rangle/dt$. Positive values of $d\langle\Phi'\rangle/dt$ indicate time intervals during that the bulk effect of fuel injection overwhelms the bulk effect of evaporation and mixing. Accordingly, in the case of double (single) injections, there are two (one, respectively) peaks of $d\langle\Phi'\rangle/dt$. Negative

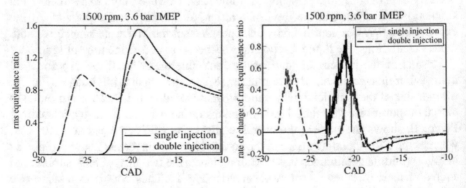

Fig. 12 Volume-averaged rms $\langle\Phi'\rangle$ of spatial variations of the equivalence ratio (left) and its rate of change, $d\langle\Phi'\rangle/dt$ (right)

values of this derivative indicate time intervals during that the bulk effect of evaporation and mixing overwhelms the bulk effect of fuel injection. In particular, the negative peaks of $d\langle\Phi'\rangle/dt$ are observed either after the single injection (solid black line, $-18°$ CA) or between two injections ($-24°$ CA) and after ($-19°$ CA) the second injection (dashed red line). It is of interest to note that the minimal values of $d\langle\Phi'\rangle/dt$ (i.e. negative values with the largest magnitude) are comparable in the two cases, but the minimal value of $d\langle\Phi'\rangle/dt$ is characterized by a larger magnitude $|d\langle\Phi'\rangle/dt|$ in the case of a single injection. Accordingly, in the case of double injections, the highest bulk rate of mixing is lower than the counterpart rate simulated in the case of a single injection. These numerical results imply that the better mixing associated with the double injections is mainly due to a longer mixing time, rather than enhancement of the bulk mixing rate due to eventual generation of turbulence during the first injection.

Indeed, in the case of double injections, fuel supplied during the first pulse can mix with the air over a significantly longer time interval when compared to the case of a single injection. Even if the second injection could accelerate the in-cylinder motion, such an effect appears to be stronger in the case of a single injection. For instance, turbulent diffusivity simulated during a time interval from $-23°$ till $-19°$ CA (the second injection pulse in the case of double injections) was lower (higher) in the case of double (single, respectively) injections.

Finally, while both experimental and numerical data indicate better mixing in the case of double injections, the magnitude of the documented effect appears to be sufficiently low (Fig. 5) when compared to the strong difference in the rates of growth of the average flame area, measured during an early stage (from -13 till $-11°$ CA) of flame kernel growth (Fig. 6). The simulations could give a clue to explaining significantly different magnitudes of effects shown in Figs. 5 and 6. Indeed, Fig. 13

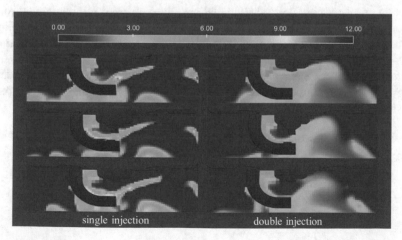

Fig. 13 Two-dimensional fields of the equivalence ratio computed in the cases of single (left column) and double (right column) injections at $-18°$ CA. In each column, three vertical cuts separated by a distance of 1 mm are shown

indicates that, in the vicinity of the spark plug, the fields of Φ, simulated in the case of double injections (right column), are much more uniform than the fields of Φ, simulated in the case of a single injection (left column). Therefore, the strong difference between two measured dependencies plotted in Fig. 6 does not seem to be surprising. It is associated with significant differences between local fields of the equivalence ratio in the vicinity of the spark plug. Unfortunately, this very important zone is not accessible in optical measurements.

5 Conclusions

The aim of this study was to compare in-cylinder fuel distributions and mixing processes in single and double injection events using advanced optical diagnostics (LIF, Mie and Rayleigh scattering, as well as direct flame luminosity) and unsteady three-dimensional CFD simulations.

The Rayleigh images indicate that double injections improve mixing, i.e. generate fewer local areas with high intensity, when compared to a single injection. Images of liquid fuel obtained using Mie scattering support this finding by showing fewer local areas with high fuel concentrations and more evenly distributed fuel in the case of double injections. Optical investigation of early flame propagation supports this finding also by showing faster initial flame spread and more even flame propagation in the case of double injections.

Results of unsteady RANS simulations of fuel injection, evaporation, and mixing in the engine combustion chamber corroborate the above findings also by showing that the computed volume-averaged rms of spatial variations of the equivalence ratio Φ is smaller in the case of double injections when compared to the single injection event. According to the simulations, the improved mixing stems from a longer time available for mixing of the air and fuel in the case of double injections. On the contrary, the computed peak bulk mixing rate is higher in the case of a single injection.

Thus, both experimental and numerical results presented here indicate that delivering fuel in closely coupled double injections results in better mixing when compared to a single injection of the same amount of the fuel.

Acknowledgements This work was supported by Combustion Engine Research Center (CERC). The second author (AL) is grateful to *Converge CFD* for granting a license for using their software.

References

1. van Basshuysen, R., Spicher, U.: Gasoline Engine with Direct Injection: Processes, Systems, Development, Potential. Vieweg+Teubner, Leipzig (2009)
2. Fansler, T.D., Reuss, D.L., Sick, V., Dahms, R.N.: Combustion instability in spray-guided stratified-charge engines: A review. Int. J. Engine Res. **16**(3), 260–305 (2015)
3. Alkidas, A.C.: Combustion advancements in gasoline engines. Energy Conserv. Manage. **48**, 2751–2761 (2007)
4. Hemdal, S., Andersson, M., Dahlander, P., Ocheterena, R., Denbratt, I.: In-cylinder soot imaging and emissions of stratified combustion in a spark-ignited spray-guided direct-injection gasoline engine. Int. J. Engine Res. **12**(6), 549–563 (2011)
5. Oh, H., Bae, C.: Effects of the injection timing on spray and combustion characteristics in a spray-guided DISI engine under lean stratified conditions. Fuel **107**, 225–235 (2013)
6. Dahms, R.N., Drake, M.C., Fansler, T.D., Kuo, T.-W., Peters, N.: Understanding ignition processes in spray-guided gasoline engines using high-speed imaging and the extended spark ignition model Spark CIMM Part B: Importance of molecular fuel properties in early flame front propagation. Combust. Flame **158**, 2245–2260 (2011)
7. Dahms, R.N., Drake, M.C., Fansler, T.D., Kuo, T.-W., Peters, N.: Understanding ignition processes in spray-guided gasoline engines using high-speed imaging and the extended spark ignition model Spark CIMM Part A: Spark channel processes and the turbulent flame front propagation. Combust. Flame **158**, 2229–2244 (2011)
8. Petersen, B., Sick, V.: Simultaneous flow field and fuel concentration imaging at 4.8 kHz in an operating engine. Appl. Phys. B **97**, 887–895 (2009)
9. Petersen, B., Reuss, D.L., Sick, V.: On the ignition and flame development in a spray-guided direct-injection spark-ignited engine. Combust. Flame **161**, 240–255 (2014)
10. Johansen, L.C.R., Hemdal, S.: In cylinder visualization of stratified combustion of E85 and main sources of soot formation. Fuel **159**, 392–411 (2015)
11. de Francqueville, L., Effects of ethanol addition in RON 95 gasoline on GDI stratified combustion. SAE technical Paper 2011-24-0055 (2011)
12. Oh, H., Bae, C., Park, J., Jeon, J.: Effect of multiple injection on stratified combustion characteristics in a spray-guided DISI engine under lean stratified operation. SAE Technical Paper 2011-24-0059 (2011)
13. Johansen, L.C.R., Hemdal, S., Denbratt, I.: Comparison of E10 and E85 spark ignited stratified combustion and soot formation. Fuel **205**, 11–23 (2017)
14. Zeng, W., Sjöberg, M., Reuss, D.: Using PIV measurements to determine the role of the in-cylinder flow field for stratified DISI engine combustion. SAE Int. J. Engines **7**(2), 615–632 (2014)
15. Stiehl, R., Schorr, J., Kruger, C., Dreizler, A., Böhm, B.: In-cylinder flow and fuel spray interactions in a stratified spray-guided gasoline engine investigated by high-speed laser imaging techniques. Flow Turbul. Combust. **91**, 431–450 (2013)
16. Aleferis, P.G., Serras-Pereira, J., van Romunde, Z.R., Caine, J., Wirth, M.: Mechanisms of spray formation and combustion from a multi-hole injector with e85 and gasoline. Combust. Flame **157**, 735–756 (2010)
17. Aleferis, P.G., van Romunde, Z.R.: An analysis of spray development with *iso*-octane, *n*-pentane, gasoline, ethanol and *n*-butanol from a multi-hole injector under hot fuel conditions. Fuel **105**, 143–168 (2013)
18. Stiehl, R., Bode, J., Schorr, J., Kruger, C., Dreizler, A., Böhm, B.: Influence of intake geometry variations on in-cylinder flow and flow-spray interactions in a stratified direct-injection spark-ignition engine captured by time-resolved particle image velocimetry. Int. J. Engine Res. **17**(9), 983–997 (2016)
19. Smith, J.D., Sick, V.: Crank-angle resolved imaging of biacetyl laser-induced fluorescence in an optical internal combustion engine. Appl. Phys. B **81**, 579–584 (2005)
20. Hemdal, S.: Characterization of stratified fuel distribution and charge mixing in a DISI engine using Rayleigh scattering. Combust. Flame **193**, 218–228 (2018)

21. Converge CFD Homepage. https://convergecfd.com/. Last accessed 21 Jan 2021
22. Yakhot, V., Orszag, S.A., Thangam, S., Gatski, T.B., Speziale, C.G.: Development of turbulence models for shear flows by a double expansion technique. Phys. Fluids A **4**, 1510–1520 (1992)
23. Reitz, R.D.: Modeling atomization processes in high-pressure vaporizing sprays. Atomization Spray Technol. **3**, 309–337 (1987)
24. Kong, S.-C., Han, Z., Reitz, R.D.: The development and application of a diesel ignition and combustion model for multidimensional engine simulation. SAE Paper No. 950278 (1995)
25. Xin, J., Ricart, L., Reitz, R.D.: Computer modeling of Diesel spray atomization and combustion. Combust. Sci. Technol. **137**, 171–194 (1998)
26. Patterson, M.A., Reitz, D.R.: Modeling the effects of fuel spray characteristics on diesel engine combustion and emission. SAE Technical Paper 980131 (1998)
27. Beale, J.C., Reitz, R.D.: Modeling spray atomization with the Kelvin-Helmholtz/Rayleigh-Taylor model. Atom. Sprays **9**, 623–650 (1999)
28. Senecal, R.K., Schmidt, D.P., Nouar, I., Rutland, C.J., Reitz, R.D., Corradini, M.L.: Modeling high-speed viscous liquid sheet atomization. Int. J. Multiph. Flow **25**, 1073–1097 (1999)
29. Schmidt, D.P., Nouar, I., Senecal, R.K., Rutland, C.J., Reitz, R.D., Hoffman, J.A.: Pressure-swirl atomization in the near field. SAE Technical Paper 1999-01-0496 (1999)
30. Schmidt, D.P., Rutland, C.J.: A new droplet collision algorithm. J. Comput. Phys. **164**, 62–80 (2000)
31. Amsden, A.A., O'Rourke, P.J., Butler, T.D.: KIVA-II: A computer program for chemically reactive flows with sprays. Los Alamos National Laboratory Report No. LA-11560-MS (1989)
32. Huang C., Lipatnikov, A.: Modelling of gasoline and ethanol hollow-cone sprays using OpenFOAM. SAE Technical Paper 2011-01-1896 (2011)
33. Hemdal, S., Denbratt, I., Dahlander, P., Warnberg, J.: Stratified cold start sprays of gasoline-ethanol blends. SAE Int. J. Fuels Lubr. **2**(1), 683–696 (2009)

Evolution of Fuels with the Advancement of Powertrains

Stamatis Kalligeros

Abstract The primary source of energy in all engines and powertrains is the "Fuel". This term includes a range of different types from gas to distillate or synthetics and to heavy liquid fuels. The primary function of a fuel is to offer the requested amount of energy to the powertrain in all conditions and in all domains (land, sea, and air). Additionally, fuel is an important source of emissions that strongly impact air quality but also a pollutant for land and marine environments. The continuous effort to impose stricter limits led to the alteration of the fuel characteristics. Simultaneously, compatible advance alternative fuels (ex. biofuels such as biomethane, biodiesel, bioethanol, etc.) and renewable and synthetic fuels entering the market. This chapter presented an overview of Fuel characteristic associated with the development of different standards (both civil and military) for different kind of fuels which are associated with different propulsion systems. The continuous development of the fuel standards for compressed ignition engines in association with the biofuel standards will be analyzed. For spark ignition engines, unleaded petrol in association with ethanol fuel as blending components for petrol will be examined. The development of marine fuels including the biofuels and the new synthetic fuels from synthesized hydrocarbons will be presented. The Liquefied Petroleum Gas (LPG) will be examined. An important analysis will be presented for the aviation turbine fuels through the development of synthesized hydrocarbons.

Keywords Diesel · Biodiesel · Petrol · Ethanol · Marine fuel · Aviation fuel · Synthetic hydrocarbons

Abbreviations

ASTM	American Society for Testing and Materials
ATJ-SPK	Alcohol-To-Jet synthetic paraffinic kerosene

S. Kalligeros (✉)
Fuels & Lubricants Laboratory, Hellenic Naval Academy, 18539 Piraeus, Athens, Greece
e-mail: sskalligeros@hna.gr

© The Author(s), under exclusive license to Springer Nature Switzerland AG 2022
T. Parikyan (ed.), *Advances in Engine and Powertrain Research and Technology*,
Mechanisms and Machine Science 114,
https://doi.org/10.1007/978-3-030-91869-9_9

BTL	Biomass-to-Liquid
CCAI	Calculated Carbon Aromaticity Index
CFPP	Cold Filter Plugging Point
CHJ	Catalytic Hydrothermolysis Jet
COS	Carbonyl Sulfide
cSt	Centistokes (1cSt $= 1$ mm^2/s)
CTL	Coal-to-Liquid
DPFs	Diesel Particulate Filters
DNV	Det Norske Veritas
EGD	European Green Deal
FAME	Fatty Acid Methyl Esters
FT	Fischer–Tropsch
CEN	European Committee for Standardization
GHG	Green House Gas
GND	Green New Deal
GTL	Gas-to-Liquid,
HC	Hydrocarbons
HEFA	Hydroprocessed Esters and Fatty Acids
HRD	Hydroprocessed Renewable Diesel
HVO	Hydrotreatment Vegetable Oils
IMO	International Maritime Organization
ISO	International Organization for Standardization
JP	Jet Propulsion
l	Liter
LNG	Liquefied Natural Gas
LPG	Liquefied Petroleum Gas
Max	Maximum
Min	Minimum
mg	Milligram
MON	Motor octane number
NCV	Net Calorific Value
NOx	Nitrogen Oxides
OECD	Organization for Economic Cooperation and Development
PAH	Polycyclic aromatic hydrocarbons
PGND	Pact for a Green New Deal
PM	Particulate Matter
ppm	Parts per million (1 mg/l $= 1$ ppm)
pS/m	Measure of electrical conductivity
1 pS/m	$1 \times 10^{-12}\ \Omega^{-1}\ M^{-1}$
PTL	Power-to-Liquid
RED	European Renewable Energy Directive
RON	Research Octane Number
SIP	Synthesized Iso-Paraffins
SPD	Synthesized Paraffinic Diesel
SPK	Synthesized Paraffinic Kerosine

SPK/A	Synthesized Paraffinic Kerosine plus Aromatics
TEU	Twenty-foot equivalent unit
ULO	Used Lubricating Oils
UN	United Nations
vol	Volume
wt	Weight
wsd	Wear scar diameter

1 Introduction

The members of the Organization for the Economic Cooperation and Development (OECD) are trying to limit the CO_2 emissions and to reach net zero emissions by 2050. Attempting to meet this challenge they "baptize" their programs' "Green Deal".
Examples are:

- the resolution in the US called "Green New Deal—GND" [1],
- the program in Canada "Pact for a Green New Deal—PGND" [2] and,
- in Europe "European Green Deal—EGD" [3].

The Global GNDs, while attempting to achieve their targets, they are pushing for the increment of the use of alternative fuels. The introduction of any fuel can be efficient only through the standardization process.

The energy challenges are not simply political challenges. Policy must be in line with technological change in order to be efficient and effective [4]. This is not always self-evident. The fact that European Renewable Energy Directive (RED) II has to be amended before even it could be transposed into National law shows that it has several shortcomings although it was adopted less than three years ago. Despite that the renewable fuels sector to date has realized the majority (>95%) of Green House Gas (GHG) emission savings in the transport sector, the lack of support and unfocused policy measures has resulted in that the fuel, the biofuel and alternative fuel industry has not been mobilized to its full capacity [5].

As the transportation platforms get more complicated, the equipment manufacturers face the challenge of adopting and constructing complex systems for their propulsion. The powertrains options have expanded beyond conventional fuels. Alternative powertrains may rely on alternative fuels from different primary sources or feedstocks from many different production paths with different degrees of efficiency.

This chapter will focus on the fuels which have been standardized to be commercially available. It will try to present the current state of the art of the fuel properties, which are responsible for the powertrain's continuous operability. In this attempt, the guide will be the international available standardization documents which are necessary first to establish the market and second to resolve any market failures.

The fuel technology influences the machines and powertrains design development and their environmental performance. Fuel is one of the pillars that helps powertrains be efficiently productive. Fuel characteristics depend on the specifications of the engine operating at different loads, in order for this cooperation to be efficient. It must always be borne in mind that the working environment of each powertrain determines its performance. On the other hand, the fuels must have sustainable characteristics. As a result of the above approach, the fuels used in the three domains (land, sea, and air) will be examined.

2   Land Domain

The fuel quality affects vehicle and engine operation, durability, and emissions. Fuel quality harmonization allows the introduction of common powertrain design worldwide. All the discussed advanced technologies will perform better when using a standard fuel of high quality [6]. Powertrain technology and certain fuel quality parameters influence not only the vehicle's fuel consumption but also its overall performance. The land-based powertrains, vehicle mainly, consumes liquid fuels which they have high energy density [7].

This chapter intends to cover the fuels for the transportation used in the land domain. Additionally, the housekeeping practices are essential to maintain the desired end use fuel quality. Chemical additives are very often used which aim to improve some quality characteristics of the fuel without having a holistic overview to the fuel-engine cooperation. The addition of additives must not affect the fuel standardized quality characteristics. The environmental requirements under the legislation in force must not reduce fuel efficiency and performance. The GHG emissions from the transportation sector are greater than the 90 s in the European Union [8, 9]. Unleaded Petrol and Diesel will be examined in this section as long with their bio replacements fuel Bioethanol and Fatty Acid Methyl Esters (FAME) well known as biodiesel, respectively.

2.1   Petrol

The current European quality requirements for the unleaded petrol specified mainly two types of unleaded petrol depending on the maximum ethanol content 5 or 10% vol. [10]. In Table 1 the main characteristics of unleaded petrol with 5–10% vol. ethanol are presented. The minimum requirements of the octane rating remain unchanged for at least the last two and a half decades. The octane number has impact to the engine performance, fuel efficiency, emissions and is associated with the resistance to auto ignition.

From 2008 until today the following changes have been made [10–12]:

- The introduction of bioethanol into the market [13]. This leads to the separation of the specifications into two categories with different bioethanol content. The

Table 1 European petrol specification based on [10]

Property	Units	Limits	5% ethanol	10% ethanol
Research Octane Number (RON)		Min	95	
Motor Octane Number (MON)		Min	85	
Density (at 15 °C)	kg/m^3	Max	775	
		Min	720	
Sulfur content	ppm	Max	10	
Manganese content	ppm	Max	2	
Oxidation stability	minutes		360	
Final boiling point (FBP)	°C	Max	210	
Distillation residue	% vol	Max	2	
Hydrocarbon type content	% vol			
Olefins		Max	18	
Aromatics		Max	35	
Oxygen content	% wt	Max	2.7	3.7

introduction of a higher percentage by volume of ethanol in the gasoline (10% vol.) has an impact on the refining and blending processes of the fuel as it influences the performance of the vehicle as the oxygenated organic compound increases the octane number. The first category contains bioethanol 5% vol. with a maximum oxygen content by weight of 2.7% wt., and the second 10% vol. with a maximum oxygen content of 3.7% wt.

- The addition of bioethanol changes the maximum volume of oxygenates in the petrol. This limit is important because the addition of ethanol (mainly bioethanol) change the volatility of the fuel and its distillation characteristics. The effect of this change is to affect the tailpipe and evaporative emissions. Car manufactures have introduced vehicles on the market that can use ethanol in greater amounts than 10% vol. [14, 15]. The current European specification for bioethanol is presented in Table 2.

- The change in the permittable manganese content is limited to 2 ppm. Manganese is the replacement of the lead in the petrol, and it was used as octane additive (octane booster). Its addition may affect the emission control systems and increase the low-speed pre-ignition in turbocharged engines. Car manufactures are strongly opposed the use of manganese and other metal additives such as Ferrocene which contains iron. The reason is that these metals poisoned the vehicle's catalyst and generating deposits in the spark plug.

- Sulfur reduced from 50 parts per million (ppm) to 10 ppm as result of stricter environmental legislation to reduce the sulfur oxides emissions [16–18]. Additionally, the sulfur in the fuel affects the lifetime of the catalyst and the sensors in the modern vehicles.

Table 2 Specification of bioethanol based on [14]

Property	Unit	Limits	Value
Ethanol plus higher saturated alcohols content	% wt	Min	98.7
Higher saturated (C3–C5) mono-alcohols content	% wt	Max	2
Methanol content	% wt	Max	1
Water content	% wt	Max	0.3
Total acidity (expressed as acetic acid)	% wt	Max	0.007
Electrical conductivity	pS/m	Max	2.5×10^6
Inorganic chloride content	ppm	Max	1.5
Sulfate content	ppm	Max	3
Copper content	ppm	Max	0.1
Phosphorus content	ppm	Max	0.15
Involatile material content	mg/100 ml	Max	10
Sulfur content	ppm	Max	10

It must be noticed that the unleaded petrol has a small amount of lead concentration (max. 5 ppm). Lead is a physical substance in the petrol. It has good lubrication performance but in amounts greater than 5 ppm contaminates the emission control system reducing the catalyst efficiency. The result of this is the increase of Hydrocarbons (HC) and Nitrogen Oxides (NO_x) emissions. The concentration of inorganic chlorine content in bioethanol is most of importance because chlorine forms highly corrosive acids during the combustion process which can damage the engine and destroy the fuel injection system.

2.2 Diesel

This section is trying to present the diesel fuel quality requirements reflecting the changes of standards in combination with the powertrains needs. The current diesel fuel standardization characteristics are presented in Table 3 [19]. The climate requirements are playing important role in the formulation of diesel characteristics because they influence the behavior of the hydrocarbon composition (paraffins, naphthene, aromatics). The property of the Cold Filter Plugging Point (CFPP) characterizes the cold flow performance of the fuel. The paraffinic hydrocarbons will come out of the fuel as waxes in low temperatures. Wax in the tank prevent the starting of the ignition process and is a source of operating problems. FAME is another parameter which can influence the cold flow performance of the finished fuel. It is important that FAME will be fully compliant with its standardized quality requirements.

Additionally, the climate regions are divided in two main categories, one is the temperate climate, and the other the arctic and severe climates.

Table 3 European diesel fuel specification based on [19]

Property	Unit	Limit	Value
Cetane number		Min	51
Cetane index		Min	46
Density at 15 °C	kg/m³	Max	845
		Min	820
Viscosity at 40 °C	mm²/s	Max	4.5
		Min	2
Distillation parameters			
% vol. recovered at 250 °C	% vol	Max	<65
% vol. recovered at 350 °C		Min	85
95% vol. recovered at	°C	Max	360
Polycyclic Aromatic Hydrocarbons (PAH)	% wt	max	8
Sulfur content	ppm	Max	10
Manganese content	ppm	Max	2
Carbon residue (on 10% distillation residue)	% wt	Max	0.3
Ash content	% wt	Max	0.01
Water content	ppm	Max	200
Total contamination	ppm	Max	24
FAME content	% vol	Max	7
Oxidation stability	Hours	Min	20
Lubricity, corrected wear scar diameter (wsd 1.4) at 60 °C	μm	Max	460

Arctic climates have been divided into four classes. Fuel flow and distillation properties are also different. The main characteristics of these classes are the lowest minimum value for density (800 kg/m³) and the minimum viscosity (1.5 mm²/s) to ensure the flow rate requirements of the powertrains.

The differences between the 2009 specifications and the current are the reduction of Polycyclic Aromatic Hydrocarbons (PAH) content from 11 to 8% wt. and the manganese content which is limited to 2 ppm. The PAH content is defined as the total aromatic hydrocarbon content minus the mono-aromatic-hydrocarbon content. The aromatic content of the diesel fuel is responsible for the formulation of the PAH and nitrogen oxide emissions because they influence the flame temperature during the combustion process into the chamber. The aromatic content is influencing the particulate matter emissions from the diesel engine. This reduction is in combination with the lower sulfur content of the fuel from 50 ppm, until the end of 2008, to 10 ppm.

Additionally, the new vehicle technology uses Diesel Particulate Filters (DPFs) in order to minimize the Particulate Matter (PM) emissions from the combustion of diesel fuel. The higher aromatic content of the fuel is shortening the regeneration interval of the filter and the result of this process is the plugging of the filter. Plugging of the filter not only increases the fuel consumption but also causes a backpressure in the exhaust system [6].

2.3 Fatty Acid Methyl Esters (FAME)

The current European specification for the FAME widely known as biodiesel is presenting in Table 4. The standard reference number is EN 14214 [20–22]. From 2009 until today the major changes are:

- the withdrawn of the carbon residue specification and
- the decrement of the maximum level of the finished fuel monoglycerides content.

The quality of the finished fuel has been improved throughout the years. The requirement to have minimum cetane number of 51 in combination with the limiting value of the carbon residue achieve the goal to reduce the excess amount of ignition improvers additives into the finished fuel.

The monoglyceride content along with the di- and tri- glyceride content is affecting the cold flow properties of the biodiesel. Car manufactures through the last decade

Table 4 FAME standardized characteristics based on [21]

Property	Unit	Limit	Value
FAME content	% wt	Min	96.5
Density at 15 °C	kg/m³	Max	900
		Min	860
Viscosity at 40 °C	mm²/s	Max	5
		Min	3.5
Cetane number		Min	51
Acid value	mg KOH/g	Max	0.5
Linolenic acid methyl ester	% wt	Max	12
Methanol content	% wt	Max	0.2
Monoglyceride content	% wt	Max	0.7
Diglyceride content	% wt	Max	0.2
Triglyceride content	% wt	Max	0.02
Water content	% wt	Max	0.05
Total contamination	ppm	Max	24
Sulfated ash content	% wt	Max	0.02
Sulfur content	ppm	Max	10
Sodium (Na) plus Potassium (K) content	ppm	Max	5
Calcium (Ca) plus Magnesium (Mg) content	ppm	Max	5
Phosphorus content	ppm	Max	4

introduced dedicated sophisticated vehicles for using 100% FAME fuel. If pure FAME diesel fuel is distributed in the market, then the fuel must have the same cold flow performance as the conventional diesel fuel. A computational guide has been introduced in order to help the producers to predict the performance of the fuel in cold weather conditions.

FAME is a good lubricity additive for the diesel fuel to avoid operation failures of the powertrains. It is important to have a good oxidation stability and for this reason additives were used to secure it, before the storage of the fuel. The amount of Sodium (Na), Potassium (K), Calcium (Ca) and Magnesium (Mg) must be between the referred limits because otherwise the fuel system of the vehicles will be poisoned.

2.4 Paraffinic Diesel

The paraffinic diesel fuels have been introduced into the European market from 2009. Through the last decade they have developed rapidly, and the standardization process has followed this development conducting a Common Workshop Agreement (CWA) on 2009 [23], a technical study on 2012 [24] and finally publishing the first European standard in 2016 [25]. The characteristics of the paraffinic diesel are illustrated in Table 5. The word "paraffinic" includes all the processes (everything to liquid) under the acronym XtL such as:

Table 5 Paraffinic fuel standardized characteristics based on [25]

Property	Unit	Limit	Fuel class A	Fuel class B
Cetane number		Min	70	51
Density at 15 °C	kg/m^3	Max	800	810
		Min	765	780
Flash point	°C	Min	>55	
Viscosity at 40 °C	mm^2/s	Max	4.5	
		Min	2	
Distillation parameters:				
Recovered at 250 °C	% vol	Max	<65	
Recovered at 350 °C	% vol	Min	85	
95% vol. recovered at	°C	Max	360	
Lubricity, wsd at 60 °C	μm	Max	460	
FAME content	% vol	Max	7	
Manganese content	ppm	Max	2	
Total aromatics content	% wt	Max	1.1	
Sulfur content	ppm	Max	5	
Total contamination	ppm	Max	24	

- BTL: Biomass-to-Liquid,
- GTL: Gas-to-Liquid,
- PTL: Power-to-Liquid and
- CTL: Coal-to-Liquid.

The Hydrotreatment can be used for the treatment of various vegetable oil feedstocks and produces the Hydrotreatment Vegetable Oils (HVO) which are covered by the same specifications. The distillation curve of the paraffinic diesel is different from the conventional as described above and covered by the EN 590 standard. Paraffinic fuels are divided into two (2) classes which are basically the same with only two differences:

- the cetane number and
- the range of the density.

These parameters are interconnected. The density is lower than the diesel fuel and this characteristic may affect the fuel economy of the powertrains.

This category of the fuel has good oxidation stability similar with the conventional diesel fuel, but they have, especially the class A fuels, high cetane number which can help the conventional fuel to have a clean burn. Paraffinic fuels have low concentrations of sulfur and aromatics. The cold flow properties are similar with the conventional replacement. They can be used securely as drop-in fuels.

On the other hand, the paraffinic fuels have low lubricity performance. Lubricity is one of the important parameters which always must be taken into consideration because it affects the performance of the engines and the powertrains. Not appropriate lubricity can cause:

- excessive pump wear leading to a catastrophic failure of the fuel pump,
- a fatal breakdown, and
- increased tailpipe emissions.

As mentioned above the FAME fuel has a very good lubricity and one of the reasons that this specification includes its addition is to improve the lubrication performance of the paraffinic fuels covered by this specification.

2.5 Liquefied Petroleum Gas (LPG)

The LPG is one of the products of the distillation process of the crude oil. LPG defined as low pressure liquefied gas composed of one or more light hydrocarbons. The LPG which consists mainly of propane, propene, butane, butane isomers, butenes with traces of other hydrocarbon gases and it is normally stored under pressure [26]. If the pressure is released, large volumes of gas will be produced which form flammable mixtures with air over the range of approximately 2–10% vol. The current standardization of the LPG is presented in Table 6. During the 2020 this specification is under amendment debating about the propane content of the fuel. The concentration

Table 6 LPG standardized characteristics based on [26]

Property	Unit	Limit	Value
MON		Min	89
1,3 butadiene	% wt	Max	<0.1
Propane content (debating)			
Until 30-04-2022	% wt	Min	20
From 01-05-2022	% wt	Min	30
Total sulfur content (after odorization)	ppm	Max	30
Evaporation residue	ppm	Max	60
Vapor pressure, gauge, at 40 °C	kPa	Max	1550
Vapor pressure, gauge, min. 150 kPa (or 200 kPa debating), for different climate requirements:			
Grade A	°C	Max	−10
Grade B			−5
Grade C			0
Grade D			10
Grade E			20

of propane is critical because is helping the cleaner combustion of the fuel and ensuring the ignition at very low temperatures in winter period. High vapor pressure is important for clean burning and for lean operation of the modern turbocharged engines.

The developments during the last decade in the LPG standardization are:

- the inclusion in the standard of the propane content of the fuel which as explained is important for the operational behavior of the gas engines.
- lower the limit for sulfur concentration of the fuel in order to be align with the environmental regulation and
- define separate limit for the 1,3 butadiene concentration because of the chemical safety legislation. The proposed concentration for 1,3 butadiene for the upcoming years will be 0.09% wt.

Table 7 Biomethane standardized characteristics based on [27]

Property	Unit	Limit	Value
Hydrogen	% wt	Max	2
Oxygen	% wt	Max	1
Sulfur	ppm	Max	30
$H_2S + COS$	mg/m^3	Max	5
Lower heating value (LHV)	MJ/kg	Min	39
		Max	48

2.6 Natural Gas - Biomethane

The use of natural gas and biomethane in the transport sector standardized in Europe with the technical standard document EN 16723-2 [27].

The concentrations of impurities are of importance of the fuel quality because they can harm the fuel system of the powertrains. Additionally, Biomethane contains silicon (Si) in amounts greater than 0.5 ppm. The current production methods cannot produce biomethane with concentration in silicon lower than this concentration. Great efforts of research are ongoing because higher concentrations than 0.1 ppm can harm many of the lambda oxygen sensors which are being available on the market today [28]. The biomethane must contain (Table 7):

- low concentrations of lubricating oil because otherwise they can harm the injection system.
- limited concentration of hydrogen because it can corrode the steel fuel tanks.

The hydrogen sulfide and the carbonyl sulfide (COS) used for safety purposes (odor mask) and contribute to the total sulfur of the fuel.

3 Maritime Domain

The world trade is depending on maritime transportation. According to United Nations (UN) in January 2020 [29] the world fleet reached the total carrying capacity of 2.1 billion dead-weight tons (dwt). 11.1 billion tons of cargo shipped globally in 2019, while the 7.9 billion tons of them were dry cargo [30].

Most of these ship vessels burns liquid fuels. In 2019 the total annual global marine fuel demand is more than 400 million tons. The containerships with more than 4,000 Twenty-foot Equivalent Unit (TEU) are responsible for the consumption of approximately 20% of all marine fuel. The fuel represents more than the 50% of the ship's total running costs [31].

The marine fuels classification is standardized by International Organization for Standardization (ISO) under the ISO 8216-99 standard document, in two categories. The one category contains the distillate marine fuels which are having the letter "D" in their definition and the other contains the residual marine fuels which are having the letter "R". For both categories, the letter "M" from the marine environment is used. The residual marine fuels accompanied by a number which defined the maximum kinematic viscosity of the fuel in square millimeters per second (mm^2/s) or centistoke (cSt) at 50 °C [32, 33]. The new era in marine fuel composition established in 2017. In the present state, the fuel shall contain hydrocarbons from synthetic or renewable sources such as HVO, GTL or BTL fuel and co-processed renewable feedstock which can be blended at the refineries with petroleum derived hydrocarbons.

3.1 Distillate Marine Fuels

The family of the distillate marine fuels currently includes seven categories of fuels as depicted in Table 8 based to the current standardization documents ISO 8216:2017 and ISO 8217:2017 [33, 34].

The characteristics of the DMX fuel remain unchanged during the last 12 years. As emergency fuel has lower flash point (>43 °C), has greater cetane index (>45) and its kinematic viscosity is between the range of 1.4–5.5 mm^2/s. Although, DMX is fuel for emergency purposes, in 2010 lubricity characteristic (max 520 μm at 60 °C) was added in order to provide the same lubrication performance with the other distillate marine fuels. The marine fuels for general purposes are mainly the DMA (DFA) and DMZ (DFZ). The difference between DFA and DMA is that the DFA marine fuel contains FAME (biodiesel). The same difference exists between DFZ and DMZ marine fuels. The characteristics of DMA are presented in Table 9. Through the last 12 years the major changes are the introduction of lubricity characteristic during 2010

Table 8 Classification of distillate marine fuels based on [33]

Category						
DMX	DMA	DFA	DMZ	DFZ	DMB	DFB
Fuel for emergency purposes	Fuel for general purposes. DFA can contain FAME up to 7% vol		Fuel for general purposes. DFZ can contain FAME up to 7% vol		Fuel for general purposes which may contain traces of residual fuel. DFB can contain FAME up to 7% vol	

Table 9 DMA fuel characteristics based on [34]

	Unit	Limit	ISO 8217:2017
Density at 15 °C	kg/m^3	Max	890
Kinematic viscosity at 40 °C	mm^2/s	Min	2
		Max	6
Flash point	°C	Min	60
Sulfur	% wt	Max	1
Cetane index		Min	40
Ash	% wt	Max	0.01
Hydrogen sulfide	ppm	Max	2
Acid number	mgKOH/gr	Max	0.5
Oxidation stability	gr/m^3	Max	25
Lubricity, corrected wear scar diameter (wsd 1,4) at 60 °C	μm	Max	520
Fatty acid methyl ester (FAME)	% vol	Max	7 (DFA)

to prevent the failure of the rotary mechanics and the reduction of sulfur concentration in 2017 due to lower the emissions of sulfur oxide in the marine sector. DMA (DFA) has a kinematic viscosity gradient of 4 mm²/s while the DMZ (DFZ) has 3 mm²/s.

The DMB (DFB) grades can include residual fuel, and this is the reason that the viscosity is limited between 2.0 and 11.0 mm²/s. The other characteristics that are necessary and depend on the nature of the fuel are: the concentration of water that has a maximum of 0.3% vol. and the total sediment having a maximum limit of 0.1% wt. In addition, the cetane index has a minimum value of five points lower than other marine distillation fuels, which indicates that this category does not have the same combustion efficiency as DMA and DMZ marine fuels.

For distillate marine fuels the amount of energy which the fuel can contribute during the combustion process can be expressed by the Gross Specific Energy (Q_{Dgv}) and the Net Specific Energy (Q_{Dnp}), both expressed in megajoules per kilogram (MJ/kg). These standardized equations must be used in the marine domain instead of Net Heating Value (NHV) and the Higher Heating Value (HHV) of the fuel.

The above-mentioned parameters can be calculated through the Eqs. 1 and 2 respectively [34]:

$$Q_{Dnp} = \left(46.423 - 8.792\rho_{15}^2 \cdot 10^{-6} + 3.170\rho_{15} \cdot 10^{-3}\right) \cdot [1 - 0.01(w_w + w_a + w_s)]$$
$$+ 0.094\,2w_s - 0.024\,49w_w \tag{1}$$

$$Q_{Dgv} = \left(51.916 - 8.792\rho_{15}^2 \cdot 10^{-6}\right) \cdot [1 - 0.01(w_w + w_a + w_s)] + 0.094\,2w_s \tag{2}$$

where:

- ρ_{15} = the density at 15 °C, kg/m³.
- w_w = the water content, %wt.
- w_a = the ash content, %wt. and
- w_s = the sulfur content, %wt.

Fleets around the world have issued their minimum quality requirements for Naval Distillate Fuel. The specifications change through the last decade are depicted in Table 10 [35–37]. The requirements are stricter than the distillate marine fuel according to ISO 8217.

Lubricity requirement of 460 μm, established in order to predict the wear on engine components, particularly the fuel pump in order to prevent engine breakdown. With the advent of low Sulfur marine diesel, lubricity became a requirement in most fuels for predictive maintenance purposes. In Fig. 1 a measurement of Lubricity parameter using the High Frequency Reciprocating Rig (HFRR) method for a distillate marine diesel fuel with sulfur concentration 0.092% wt. is illustrated [38]. From 2014 a minimum requirement for Aromatics content is added. This need is because of the introduction of Synthesized Paraffinic Diesel (SPD). Aromatics improve the

Table 10 Naval distillate fuel properties based on [37]

	Unit	Limit		Distillate marine fuel	Synthesized marine diesel
Density, at 15 °C	kg/m^3	Max		876	805
		Min		800	770
Distillation parameters:					
10% vol	°C			Record	191–290
50% vol				Record	Record
90% vol				Max. 357	Max. 357 Min. 290
Distillation end point				Max. 385	Max. 385 Min. 300
Residue + Loss	% vol	Max		3	3
cloud point	°C	Max		−1	−1
Flash point	°C	Min		60	60
Particulate contamination	ppm	Max		10	1
Viscosity, at 40 °C	mm^2/s	Max		4.3	4.3
		Min		1.7	1.7
Acid number	mg KOH/g	Max		0.3	0.08
Ash	% wt	Max		0.005	–
Aromatics	% wt			Min. 8.1	Max. 0.5
Carbon residue on 10% bottoms	% wt	Max		0.2	–
Hydrogen content	% wt	Min		12.5	14.5
Derived cetane number				Min. 42	Max. 80 Min. 42
Storage stability	mg/100 ml	Max		3	–
Sulfur content	% wt	Max		0.0015	0.0015
Trace metals:					
Calcium	ppm	Max		1	0.1
Lead	ppm	Max		0.5	0.1
Sodium plus Potassium	ppm	Max		1	0.2
Vanadium	ppm	Max		0.5	0.1
Lubricity, at 60 °C	μm	Max		460	–

stability of the fuel, but they have negative impact on ignition properties of the diesel fuel.

From 2014 the naval distillate marine fuel can be blended at a maximum of 50% vol. with SPD which is derived from Hydroprocessed Renewable Diesel (HRD) or Fischer–Tropsch (FT) produced SPD. The remaining 50% vol. of the blending

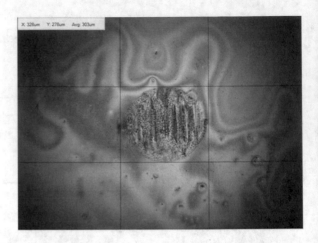

fuel must be from crude petroleum source derived hydrocarbons. The FT Hydroprocessed Synthetic Paraffinic Diesel (FT-SPD) and HRD derived blend components
shall conform to the requirements presented in Table 10. It is defined that FT blending
components shall be wholly derived from synthesis gas via the FT process using
iron or cobalt catalyst and HRD blend components shall be comprised of hydrocarbon fuel obtained from hydrogenation and deoxygenation of fatty acid esters and
free fatty acids. The processing of the product shall include hydrotreating, hydrocracking, or hydro-isomerization operational refinery processes such as polymeriza-
tion, isomerization, and fractionation. On the contrary FAME are limited to 0.1%
vol.

The synthesized paraffinic marine diesel has the limitation of 0.1 ppm per metal,
in the content of various metals. This is because due to the current processes various
metals can be included into the fuel. More specifically, this limitation applies to
Aluminum (Al), Calcium (Ca), Cobalt (Co), Chromium (Cr), Copper (Cu), Iron (Fe),
Magnesium (Mg), Manganese (Mn), Molybdenum (Mo), Nickel (Ni), Phosphorus
(P), Lead (Pb), Palladium (Pd), Platinum (Pt), Tin (Sn), Strontium (Sr), Titanium
(Ti), Vanadium (V) and Zinc (Zn). Additionally, the total concentration of alkali
metals and metalloids which are Boron (B), Sodium (Na), Potassium (K), Silicon
(Si) and Lithium (Li) must not exceed the 1 ppm. All the above limitation is because
the metals in the fuel can:

- damage the fuel systems,
- create corrosive environment in the chamber and
- increase the wear in the moving parts of every marine engine.

3.2 Residual Marine Fuels

The family of the residual marine fuels currently includes eleven categories of Fuels as depicted in Table 11 based to the current standardization documents ISO 8216:2017 and ISO 8217:2017 [39]. In order to predict the combustion characteristics of the residual fuel an indicator for the ignition performance was developed. This is the Calculated Carbon Aromaticity Index (CCAI), an index which is associated with the viscosity and the density of the fuel as illustrated in Eq. 3 [39].

$$\text{CCAI} = \rho_{15} - 81 - 141 \cdot \log_{10}\left[\log_{10}(v + 0.85)\right] - 483 \cdot \log_{10}\frac{T + 273}{323} \quad (3)$$

where:

- $\rho_{15} = $ is the density at 15 °C, kg/m^3,
- $v = $ is the kinematic viscosity at temperature T, mm^2/s (cSt) and
- $T = $ is the temperature, at which the kinematic viscosity is determined.

The CCAI is an important index because fuel with long ignition delays and bad combustion properties may contribute to an excessively high rate of pressure rise and thermal overload in the combustion chamber. This causes what is known as hard, knocking, or noisy engine running, especially at low load operation. The effects which are likely to follow, are poor fuel economy, loss of power, buildup of carbonaceous deposits and a possible engine damage from fatigue failure or metal-to-metal contact of shock loaded components. Examples of CCAI calculation, using Eq. 3, are illustrated in Fig. 2.

The residual marine fuels could be a potential place of discharging Used Lubricating Oils (ULO). This is inappropriate environmental behavior which can damage the heavy-duty marine engines. Thus, the residual fuels shall be free of ULO. In

Table 11 Classification of general-purpose residual marine fuels based on [33]

Category	Maximum kinematic viscosity at 50 °C (mm^2/s)
RMA	10
RMB	30
RMD	80
RME	180
RMG	180
RMG	380
RMG	500
RMG	700
RMK	380
RMK	500
RMK	700

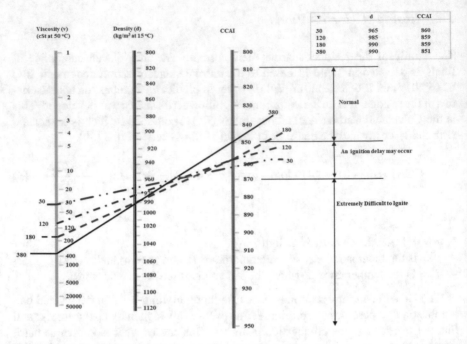

Fig. 2 Examples of CCAI calculation

order to identify the existence of ULO in the marine fuel the following Eqs. 4 and 5 must be considered.

$$K > 30\text{ppm} + Zn > 15\text{ppm} \qquad (4)$$

or

$$K > 30\text{ppm} + P > 15\text{ppm} \qquad (5)$$

The lubricating oils contain significant amounts of detergent and anti-wear additives. Calcium as well as calcium compounds are the base of the detergent additives of lubricating oils. Zinc and Phosphorus compounds are in high concentrations in the anti-wear additives. If the Eqs. 4 and 5 are not satisfied, then the residual marine fuel assumed to be free of ULO. The current characteristics of the RMA 10, RMB 30 and RMD 80 residual fuels are illustrated in the Table 12.

The limit of Vanadium and Sodium in the fuel is of great importance. Vanadium exists in the fuel oils as part of complex hydrocarbon molecules which also contain other metals and elements. The Sodium can be entered either into fuel through its contamination with sea water or into the combustion zone in a form of sea water aerosol entrained in the charging air. The limit in the water is a key parameter in order to avoid the entrance of sea water into the fuel. The concentration of these two

Table 12 Characteristics of residual fuel for the categories of RMA 10, RMB 30 and RMD 80 based on [34]

	Unit	Limit	RMA 10	RMB 30	RMD 80
Density at 15 °C	kg/m³	Max	920	960	975
Viscosity at 50 °C	mm²/s	Min	10	30	80
Flash point	°C	Min	60	60	60
Carbon residue	% wt	Max	2.5	10	14
Ash	% wt	Max	0.04	0.07	0.07
Water	% vol	Max	0.3	0.5	0.5
Vanadium	ppm	Max	50	150	150
Aluminum plus silicon	ppm	Max	25	40	40
Sodium	ppm	Max	50	100	100
CCAI		Max	850	860	860
Hydrogen sulfide	ppm	Max	2	2	2
Pour point (upper)					
Winter quality	°C	Max	0	0	30
Summer quality	°C	Max	6	6	30
Acid number	mgKOH/g	Max	2.5	2.5	2.5

elements is very important because the formulation of the extremely corrosive, for the cylinder metals, Na_2SO_4/V_2O_5 complex into the ash, can cause severe damage to the engine. The empirical knowledge has established the ratio of $Na/V = 1/3$ as very important in order to avoid a Vanadium rich ash.

The concentration of all the metals is important because they can formulate ash deposits which can cause severe damage to components (piston crowns, exhaust valves and turbocharger blade surfaces) [40].

For residual fuels, net specific energy, Q_{Rnp}, and gross specific energy, Q_{Rgv}, both expressed in megajoules per kilogram, can be calculated with a degree of accuracy acceptable for normal purposes from Eqs. 6 and 7, respectively. There is a difference in the calculation of the energy content of the fuel using the Lower Heating Value instead of the Net Specific Energy The examination of this difference for a gas turbine saws that it does not affect the calculations of the operating stages of the gas turbine except for the estimated specific fuel consumption which is greater [41].

$$Q_{Rnp} = \left(46.704 - 8.802\rho_{15}^2 \cdot 10^{-6} + 3.167\rho_{15} \cdot 10^{-3}\right) \cdot [1 - 0.01(w_w + w_a + w_s)] + 0.094\,2w_s - 0.024\,49w_w \tag{6}$$

$$Q_{Rgv} = \left(52,190 - 8,802\rho_{15}^2 \cdot 10^{-6}\right) \cdot [1 - 0,01(w_w + w_a + w_s)] + 0,094\,2w_s \tag{7}$$

where:

- ρ_{15} = the density at 15 °C, kg/m^3,
- w_w = the water content, %wt.,
- w_a = the ash content, %wt. and
- w_s = the sulfur content, %wt.

3.3 Liquefied Natural Gas (LNG) and Methanol

The use of Liquefied Natural Gas (LNG) increased during the last two decades. The ISO undertake considerable effort to issue a standard for this fuel. As a result, during 2020 the ISO 23306 specification of LNG was issued [42]. The limited standard-ized characteristics will be the standard quality of the fuel. Basically, the Nitrogen concentration must not exceed the 1% mol. The theoretical mixture of the LNG has the composition of 99% mol methane and 1% mol nitrogen. The knock resistance of the fuel depends on its composition and for prediction the method of Propane Knock Index developed by Det Norske Veritas (DNV) was presented. The Net Calorific Value (NCV) of the fuel must be at least 33.6 MJ/m^3. The energy output of the fuel when it is not used in reciprocating powertrains must be calculated using the Wobbe index which is the division of the calorific value with the square root of relative density of the fuel. Existence of the same impurities (siloxanes and ammonia) as mentioned in biomethane above can be found in LNG.

Also, the use of methanol as bunkering fuel has been increased. The use of methanol as a fuel introduced as a potential solution for the shipping industry to reduce the GHG emissions by 2050. Methanol powertrains are already available in the market. During the 2020 the first step in standardization of the fuel became available with the publication of a CWA 17540 from CEN [43]. This is the first approach for standardization of the use of methanol as marine fuel incorporating the guidelines for the safety of ships using methanol as fuel from International Maritime Organization (IMO).

4 Aviation Domain

The aviation fuel is synonym with the power. The combustion quality in association with the energy content are the dominant parameters which must ensure stable power production. Even though, the consumption of the global aviation fuel was decreased during the last eight years, while the need for high quality fuel increased [44].

The fuel has important roles in the engine control and in the cooling of engine's fuel system components. Other key parameters are the stability, lubricity, cleanliness, non-corrosivity and fluidity [45]. The years before 2020, the aviation sector was steadily growing. The forecast for the global aircraft fleet was 39,000 aircrafts [46, 47]. Jet fuel is the main cost for the global airline industry [48, 49]. Traditionally, the global specifications of the civil aviation turbine fuels are issued and maintained by the American Society for Testing and Materials (ASTM). The compositions of jet fuel are main hydrocarbons from 8 to 16 carbon atoms, in various chemical compounds such as alkanes, iso-alkanes, naphthenics or naphthenic derivatives and aromatic compounds [50, 51]. Napthenes and Isoparaffines with less than 12 carbon atoms, help to reduce the freezing point which is critical in high altitude flying [49]. Additionally, the jet engines rely on the aviation turbine fuel to lubricate the moving parts in fuel pumps and in the flow control units. The production process influences the fuel's lubricity behavior. The "straight-run" fuels have good lubricity behavior because they contain trace amounts of certain oxygen, nitrogen, and sulfur compounds. Hydrotreatment of the fuel reduces sulfur and aromatic content and removes the natural provide lubricity.

No change has been observed for the properties of the Jet A-1 during the last 5 years [52–55]. During 2018, a blend of conventionally produced kerosene with synthetic iso-paraffinic kerosene produced by SASOL, has been acceptable for use in the civil aviation sector. Only the Co-processed mono-, di-, and triglycerides, free fatty acids, and fatty acid esters producing co-hydroprocessed hydrocarbon synthetic kerosene is recognized as being acceptable for manufacturing jet fuel. All other feedstocks are excluded from the jet fuel processing [56]. There is also the restriction that the refinery units, where the processed streams are used for aviation fuel production, shall not exceed 5% vol. of mono-, di-, and triglycerides, free fatty acids, and fatty acid esters in feedstock.

The main changes in the extended requirements of aviation turbine fuels which are containing co-hydroprocessed esters and fatty-acids is the limit on the viscosity at -40 °C. The maximum value must be 12 mm^2/s and the maximum content of unconverted esters and fatty acids will not exceed the 15 ppm. The fuel performance enhancing additives and the fuel handling and maintenance additives maintained their dosage ratings.

One important safety parameter in all the aviation turbine fuels is the electrical conductivity. The generation of static electric charge when fuel moves through a pipe, hose, valves, or filters, must be avoided. The aviation turbine fuel can be a conductor because it contains trace amount of ionizable compounds like water, phenols and napthenic acids. The potential initiation of a spark must be avoided, and this is the reason why the use of electrical conductivity additive is a requirement. If electrical conductivity additive is used, the conductivity shall not exceed 600 pS/m at the point of use of the fuel. When electrical conductivity additive is specified by the user, the conductivity shall be between 50 and 600 pS/m under the conditions at point of delivery.

Simultaneously, the evolution of the standard specification for the Aviation Turbine Fuel Containing Synthesized Hydrocarbons has undergone [57–60]. The term Synthetic Paraffinic Kerosene (SPK) usually used for describing all the synthesized aviation fuels. The basic requirements of the aviation kerosene continue to be active but additional batch requirements for each of the production process have been established. These processes are the:

- Fischer–Tropsch (FT) hydroprocessed synthesized paraffinic,
- Synthesized Paraffinic Kerosine produced from Hydroprocessed Esters and Fatty Acids (HEFA),
- Synthesized Iso-Paraffins (SIP) produced from Hydroprocessed Fermented Sugars,
- FT Synthesized Paraffinic Kerosine plus Aromatics (SPK/A),
- Alcohol-To-Jet Synthetic Paraffinic Kerosene (ATJ-SPK)
- Catalytic Hydrothermolysis Jet (CHJ) and
- Bio-derived Hydroprocessed Hydrocarbons (BHH), esters and fatty acids.

The comparison of the detailed requirements for the finished fuel from various synthesized hydrocarbon processes are illustrated in Tables 13 and 14. From 2016 no alteration has been observed for the FT hydroprocessed SPK. For SPK produced by the HEFA process, from 2016 the only change was the slight increment of the upper limit of density to 772 kg/m^3 the rest of the properties remained unchanged.

FAME is considered as incidental contaminant for the aviation fuels which can be picked up during fuel conveyance. FAME like water can be the source of microbial growth. The organism can be either aerobic or anaerobic or both. They use the fuel as "food", but jet fuel can provide also other element nutrients, except for phosphorus content. The best approach to microbial contamination is prevention.

The fuel specifications trying to control the properties which are having dependency with each other. If the total aromatics content increases, density, final boiling point temperature, and freezing point increases, on the other hand the smoke point decreases. The properties of the fuel influence the thermal stability of the fuel. In all engines the fuel is also used to remove heat. In aircrafts the fuel removes heat from the engine oil, the hydraulic fluid, and the air conditioning system. The uncontrolled thermal increment of the fuel can accelerate the reactions which can that lead to the formation of gum and particulates matters. These two unwanted products may clog fuel filters and deposit on the surfaces of aircraft fuel systems or/and restricting flow in small diameter passageways causing faults in fuel flow. These deposits may lead to operational problems and increased maintenance.

For all synthesized paraffinic fuels there is a limitation of 0.1 ppm in the content of various metals. More specifically, this limitation applied to Aluminum (Al), Calcium (Ca), Cobalt (Co), Chromium (Cr), Copper (Cu), Iron (Fe), Potassium (K), Lithium (Li), Magnesium (Mg), Manganese (Mn), Molybdenum (Mo), Sodium (Na), Nickel (Ni), Phosphorus (P), Lead (Pb), Palladium (Pd), Platinum (Pt), Tin (Sn), Strontium (Sr), Titanium (Ti), Vanadium (V) and Zinc (Zn).

Almost in all the synthetic kerosene production routes the maximum Halogen presence, which is the sum of the Fluorine (F), Chlorine (Cl), Bromine (Br), Iodine (I), and Astatine (At) content, will not exceed 1 ppm. On the contrary, if the fuel produced following the Synthesized Iso-Paraffins (SIP) process from hydroprocessed fermented sugars, each element of the halogen group will not exceed 1 ppm.

These limits are important because the fuel can corrode any of the materials which were used in the construction of aircrafts fuel systems including sealants, coatings, and elastomers. Additionally, these elements can form solid particulates which can

Table 13 Detailed requirements for synthetic kerosene produced by different synthesized hydrocarbons processes based on [55, 60]

Property	Unit	Limit	SPKA	HEFA	SIP	SPK/A	ATJ-SPK
Distillation parameters:							
10% vol	°C	Max	205	205	205	205	205
50% vol		Max	Report	Report	Report	Report	Report
90% vol		Max	Report	Report	Report	Report	Report
Distillation end point		Max	300	300	255	300	300
T90–T10		Min	22	22	5	22	21
Distillation residue	% vol	Max	1.5	1.5	1.5	1.5	1.5
Distillation loss	% vol	Max	1.5	1.5	1.5	1.5	1.5
Flash point	°C	Min	38	38	100	38	38
Density at 15 °C	kg/m³	Max	770	772	780	800	770
		Min	730	730	765	755	730
Freezing point	°C	Max	−40	−40	−60	−40	−40
Cycloparaffins	% wt	Max	15	15	Not specified	15	15
Saturated hydrocarbons	% wt	Min	Not specified	Not specified	98	Not specified	Not specified
Farnesane	% wt	Min	Not specified	Not specified	97	Not specified	Not specified
2,6,10-trimethyldodecane	% wt	Max	Not specified	Not specified	1.5	Not specified	Not specified
Aromatics	% wt	Max	0.5	0.5	0.5	20	0.5
Paraffins	% wt		Report	Report	Not specified	Report	Report
Carbon and Hydrogen	% wt	Min	99.5	99.5	99.5	99.5	99.5
Nitrogen	% wt	Max	2	2	2	2	2
Water	ppm	Max	75	75	75	75	75
Sulfur	ppm	Max	15	15	2	15	15
FAME	ppm	Max	Not specified	5	Not specified	Not specified	Not specified

cause plug of the fuel filters and wear increment in all the moving parts, especially fuel pumps. Contamination from trace amounts of sodium, potassium, and other alkali metals in the fuel can cause corrosion in the turbine section of the engine [45].

The routes of the Catalytic Hydrothermolysis Jet (CHJ) and the production of synthesized paraffinic kerosene produced from bio-derived hydroprocessed hydrocarbons, esters and fatty acids are added as standardized procedures in 2020. These processes include combination of hydrotreating, hydrocracking, or hydroisomerization, and other conventional refinery processes, and shall include fractionation [60].

Table 14 Detailed requirements for synthetic kerosene produced by catalytic hydrothermolysis jet (CHJ) and bio-derived hydroprocessed hydrocarbons (BHH), esters and fatty acids based on [55, 60]

Property	Unit	Limit	CHJ	BHH
Distillation parameters:				
10% vol	°C	Max	205	205
50% vol		Max	Report	Report
90% vol		Max	Report	Report
Distillation end point		Max	300	300
T90–T10		Min	40	22
Distillation residue	% vol	Max	1.5	1.5
Distillation loss	% vol	Max	1.5	1.5
Flash point	°C	Min	38	38
Density at 15 °C	kg/m^3	Max	840	800
		Min	775	730
Freezing point	°C	Max	40	40
Cycloparaffins	% wt	Max	Report	50
Aromatics	% wt	Max	21.2	0.5
		Min	8.4	–
Paraffins	% wt		Report	Report
Carbon and Hydrogen	% wt	Min	99.5	99.5
Nitrogen	% wt	Max	2	2
Water	ppm	Max	75	75
Sulfur	ppm	Max	15	15
FAME	ppm	Max	5	5

- The Kerosene type aviation turbine fuel (Jet Propulsion-JP) which is used from fighter jets categorized as follows [61–65]:
- JP-8: Kerosene type turbine fuel which will contain a static dissipater additive, corrosion inhibitor/lubricity improver, and fuel system icing inhibitor, and may contain antioxidant and metal deactivator.
- JP-8 + 100: JP-8 type kerosene turbine fuel which contains thermal stability improver additive.
- JP-8 + static dissipater additive: Kerosene type turbine fuel which will contain a static dissipater additive, may contain antioxidant, corrosion inhibitor/lubricity improver, and metal deactivator but will not contain fuel system icing inhibitor.
- JP-5: High flash point (>60 °C), kerosene type turbine fuel.
- Jet A-1 and JP-8 having high similarity because most of the properties are identical. There are two differences as shown in Table 15. The one is the sulfur mercaptan content higher in the Jet-A1 and the other is the smoke point level which is higher in the JP-8.

Table 15 Existing differences between Jet A1 and JP-8 based on [55, 63]

Property	Unit	Limit	Jet A1	JP-8
Sulfur, mercaptan	% wt	Max	0.003	0.002
Smoke point	mm	Min	18	19

The maximum height of the flame that can be achieved without smoking is the smoke point definition. The difference in smoke point unveils the use (commercial fuel or fighter jets fuel). Another difference is that not all the synthetic routes, which used to produce synthetic kerosene, have been standardized and approved as drop-in fuels of JP-8. The routes that were standardized is the FT Process and the Hydroprocessed Esters and Fatty Acids (HEFA) synthetic kerosene.

The JP-5 is a high flashpoint kerosine-type fuel which is used for navy application (basically for air powertrains used from ships) because of safety considerations. The fuel has the same flash point (>60 °C) like the marine diesel fuels which were used for ships. In order to achieve this safety property, its composition is different than the JP-8 or/and Jet-A1. These differences are illustrated in the Table 16.

When the JP-5 contains synthesized hydrocarbons the additional requirements which are illustrated in Table 17 must be followed. The Derived Cetane Number, the

Table 16 Existing differences between Jet A1 and JP-5 based on [55, 65]

Property	Unit	Limit	Jet A1	JP-5
Sulfur, mercaptan	% wt	Max	0.003	0.002
Sulfur, total	% wt	Max	0.3	0.2 from 2013
Flash point	°C	Min	38	60
Density at 15 °C	kg/m^3	Max	840	845
		Min	775	788
Viscosity −20 °C	mm^2/s	Max	8	7 (until 2013 was 8.5)
Freezing point	°C	Max	−47	−46
Net heat of combustion	MJ/kg	Min	42.8	42.6
Hydrogen content	% wt	Min	Not specified	13.4
Smoke point	mm	Min	18	25 (until 2016 was 19)

Table 17 Additional requirements of JP-5 containing synthesized hydrocarbons based on [65]

Property	Unit	Limit	Value
Aromatics	% vol	Min	8
Distillation parameters:			
T50–T10	°C	Min	15
T90–T10	°C	Min	40
Derived cetane number		Min	40
Viscosity at −40 °C	mm^2/s	Max	12

viscosity and the distillation gradient were established from 2016. The net heat of combustion is usually predicted by fuel density, which is also a function of composition. The flow rate of the fuel in the fuel systems has close relationship with density [66].

A fuel for fighter jets applications, must have a high volumetric energy content because it is a way to maximizes the energy that can be stored in a fixed volume (fuel tanks) and thus provides the longest flight range. A fuel for commercial applications must have high gravimetric energy content giving the advantage to the aircraft to have lower fuel weight [67].

5 Conclusion

The range of the fuels used by engines with different "missions" for the road transport, shipping and the aviation transport sector was examined and presented in this chapter. There is no "silver bullet" fuel but fuels produced from different technologies and raw materials.

The different properties that influence their performance examined and illustrated. The future progress of the energy transition from the three transport domains will not be feasible without the use of conventional and alternative gaseous and liquid fuels. Fuels with greater energy densities and lower carbon footprint are likely to be required to achieve the transportation goals of every load in each domain.

References

1. Galvin, R. Healy, N.: The green new deal in the United States: what it is and how to pay for it. Energy Res. Soc. Sci. **67** (2020). https://doi.org/10.1016/j.erss.2020.101529
2. MacArthur, L.J., Hoicka, E.C., Castleden, H., Das R., Lieu, J.: Canada's green new deal: forging the socio-political foundations of climate resilient infrastructure? Energy Res. Soc. Sci. **65** (2020). https://doi.org/10.1016/j.erss.2020.101442
3. European Commission: A European green deal: striving to be the first climate-neutral continent (2019). https://ec.europa.eu/info/strategy/priorities-2019-2024/european-green-deal_en (Accessed 3 Jan 2020)
4. Sovacool, B.K.: How long will it take? Conceptualizing the temporal dynamics of energy transitions. Energy Res. Soc. Sci. **13**, 202–215 (2016). https://doi.org/10.1016/J.ERSS.2015.12.020
5. Maniatis, K., Landälv, I., Waldheim, L., van den Heuvel, E., Kalligeros, S.: REDII-revision: an opportunity to build on SGAB's 2017-recommendations and significantly increase the Renewable Fuels targets in EU transport (2020)
6. ACEA, Auto Alliance, EMA, JAMA: World Wide Fuel Charter 6th Edition. European Automobile Manufacturers Association (ACEA) Avenue des Nerviens 85, B-1040 Brussels, Belgium (2019)
7. Paoli, L. Teter, J. Tattini, J. Raghavan, S.: Fuel consumption of cars and vans. international energy agency (IEA) (2020). https://www.iea.org/reports/fuel-consumption-of-cars-and-vans. (Accessed 28 March 2021)

8. Gray, N. McDonagh, S. O'Shea, R. Smyth, B. Murphy, D. J.: Decarbonising ships, planes and trucks: an analysis of suitable low-carbon fuels for the maritime, aviation and haulage sectors. Adv. Appl. Energy **1** (2021). https://doi.org/10.1016/j.adapen.2021.100008

9. Transport & Environment. How to decarbonise European transport by 2050. Square de Meeûs, 18, 2nd floor, B-1050, Brussels, Belgium (2018)

10. EN 228:2012+A1:2017: Automotive fuels—unleaded petrol—requirements and test methods. European committee for standardization. Rue de la Science 23, B-1040 Brussels, Belgium (2017)

11. EN 228:2008: Automotive fuels—unleaded petrol—requirements and test methods. European committee for standardization. Rue de la Science 23, B-1040 Brussels, Belgium (2008)

12. EN 228:2012: Automotive fuels—unleaded petrol—requirements and test methods. European committee for standardization. Rue de la Science 23, B-1040 Brussels, Belgium (2012)

13. EN 15376:2011: Automotive fuels—ethanol as a blending component for petrol—requirements and test methods. European committee for standardization. Rue de la Science 23, B-1040 Brussels, Belgium (2011)

14. EN 15376:2014: Automotive fuels—ethanol as a blending component for petrol—requirements and test methods. European committee for standardization. Rue de la Science 23, B-1040 Brussels, Belgium (2014)

15. ACEA: E10 compatibility list. European automobile manufacturers association (ACEA) Avenue des Nerviens 85, B-1040 Brussels, Belgium (2018).

16. Directive 98/70/EC of the European parliament and of the council of 13 October 1998 relating to the quality of petrol and diesel fuels and amending council directive 93/12/EEC, OJ L 350, 28 Dec 1998

17. Directive 2003/17/EC of the European parliament and of the council of 3 March 2003 amending Directive 98/70/EC relating to the quality of petrol and diesel fuels, OJ L 76, 22 March 2003

18. Directive 2009/30/EC of the European parliament and of the council of 23 April 2009 amending directive 98/70/EC as regards the specification of petrol, diesel and gasoil and introducing a mechanism to monitor and reduce greenhouse gas emissions and amending council directive 1999/32/EC as regards the specification of fuel used by inland waterway vessels and repealing Directive 93/12/EEC, OJ L140, 5 June 2009

19. EN 590:2013+A1:2017: Automotive fuels—diesel—requirements and test methods. European committee for standardization. Rue de la Science 23, B-1040 Brussels, Belgium (2017)

20. EN 590:2009: Automotive fuels—diesel—requirements and test methods. European committee for standardization. Rue de la Science 23. B-1040 Brussels, Belgium (2009)

21. EN 14214:2012+A2:2019: Liquid petroleum products—fatty acid methyl esters (FAME) for use in diesel engines and heating applications—requirements and test methods. European Committee for Standardization. Rue de la Science 23, B-1040 Brussels, Belgium (2019)

22. EN 14214:2012: Liquid petroleum products—fatty acid methyl esters (FAME) for use in diesel engines and heating applications - Requirements and test methods. European Committee for Standardization. Rue de la Science 23. B-1040 Brussels, Belgium (2012).

23. CEN/CWA 15940:2009: Automotive fuels—paraffinic diesel fuel from synthesis or hydrotreatment—requirements and test methods. European committee for standardization. Rue de la Science 23. B-1040 Brussels, Belgium (2009)

24. CEN/TS 15940:2012: Automotive fuels—paraffinic diesel fuel from synthesis or hydrotreatment—requirements and test methods. European committee for standardization. Rue de la Science 23. B-1040 Brussels, Belgium (2012)

25. EN 15940:2016+A1:2018+AC:2019: Automotive fuels – Paraffinic diesel fuel from synthesis or hydrotreatment – Requirements and test methods. European Committee for Standardization. Rue de la Science 23. B - 1040 Brussels, Belgium (2019).

26. EN 589:2018: Automotive fuels—LPG—requirements and test methods. European committee for standardization. Rue de la Science 23, B-1040 Brussels, Belgium (2018)

27. EN 16723–2:2017: Natural gas and biomethane for use in transport and biomethane for injection in the natural gas network—part 2: automotive fuels specification. European committee for standardization. Rue de la Science 23. B-1040 Brussels, Belgium (2017)

28. ACEA, Auto Alliance, EMA, JAMA: World wide fuel charter, methane—based transportation fuels, 1st Edn. European automobile manufacturers association (ACEA) Avenue des Nerviens 85, B-1040 Brussels, Belgium (2019)
29. UNCTAD Handbook of Statistics 2020—maritime transport: fact sheet no. 14: Merchant fleet. United Nations development statistics and information branch, Palais des Nations, 1211 Geneva 10, Switzerland (2020)
30. UNCTAD Handbook of statistics 2020—Maritime transport: fact sheet no. 13: World seaborne trade. United Nations development statistics and information branch, Palais des Nations, 1211 Geneva 10, Switzerland (2020)
31. IHS Markit Ltd.: IMO 2020: what every shipper needs to know. (2019)
32. ISO 8216-99:2002: Petroleum products—fuels (class F)—classification—part 99: general. International organization for standardization (ISO), Chemin de Blandonnet 8. CP 401-1214, Vernier, Geneva, Switzerland (2002)
33. ISO 8216-1:2017: Petroleum products—fuels (class F) classification—part 1: categories of marine fuels. International organization for standardization (ISO), Chemin de Blandonnet 8. CP 401-1214 Vernier, Geneva, Switzerland (2017)
34. ISO 8217:2017: Petroleum products—fuels (class F)—specifications of marine fuels. International organization for standardization (ISO), Chemin de Blandonnet 8. CP 401-1214 Vernier, Geneva, Switzerland (2017)
35. MIL-DTL-16884L: detail specification fuel naval distillate. Naval sea systems command, 1333 Isaac Hull Avenue, SE, Stop 5160, Washington Navy Yard, DC, 20376–5160 (2006)
36. MIL-DTL-16884M: detail specification fuel naval distillate. Naval sea systems command, 1333 Isaac Hull Avenue, SE, Stop 5160, Washington Navy Yard, DC, 20376–5160 (2012)
37. MIL-DTL-16884N: detail specification fuel naval distillate. Naval sea systems command, 1333 Isaac Hull Avenue, SE, Stop 5160, Washington Navy Yard, DC, 20376–5160 (2014)
38. Krivokapic, M.: The effect of using aviation fuel JP8 in mixtures with F76 on the injection profile of MTU 538, MTU 232 engine burners. Diplomatic Thesis, Hellenic Naval Academy, Piraeus, (2019)
39. The International Council on Combustion Engines (CIMAC): fuel quality guide—ignition and combustion. CIMAC e. V. Lyoner Strasse 18, 60528 Frankfurt, Germany (2011)
40. The International Council on Combustion Engines (CIMAC): Recommendations regarding fuel quality for diesel engines. CIMAC e. V. Lyoner Strasse 18, 60528 Frankfurt, Germany (2003).
41. Savvakis, A.C., Venardos, T., Roumeliotis, I., Kalligeros, S., Aretakis, N.: The influence of the synthetic fuels use in the operation of naval gas turbine. Hellenic Institute of Marine Technology, Book of Marine Technology, pp. 35–45, Athens, Greece (2017)
42. ISO 23306:2020: Specification of liquefied natural gas as a fuel for marine applications. International organization for standardization (ISO), Chemin de Blandonnet 8, CP 401-1214 Vernier, Geneva, Switzerland (2020)
43. CWA 17540:2020: Ships and marine technology—specification for bunkering of methanol fueled vessels. European committee for standardization. Rue de la Science 23. B-1040 Brussels, Belgium (2020)
44. IEA, Global aviation fuel consumption, 2013–2021, IEA, Paris. https://www.iea.org/data-and-statistics/charts/global-aviation-fuel-consumption-2013-2021. Last Accessed 09 March 2021
45. Chevron Corporation: Aviation Fuels Technical Review. Chevron Global Aviation, Chevron Products Company, 1500 Louisiana Street, Houston, TX 77002 (2006)
46. Cooper, T., Reagan, I., Porter, C., Precourt, C.: Global fleet & MRO market forecast commentary (2019)
47. Gray, N., McDonagh, S., O'Shea, R., Smyth, B., Murphy, J. D.: Decarbonising ships, planes and trucks: an analysis of suitable low-carbon fuels for the maritime, aviation and haulage sectors. Adv. Appl. Energy 1 (2021). https://doi.org/10.1016/j.adapen.2021.100008
48. International Air Transport Association: Annual Review 2020. 76th Annual General Meeting, Amsterdam (2020)

49. Weia, H., Liua, W., Chen, X., Yanga, Q., Li, J., Chen, H.: Renewable bio-jet fuel production for aviation: a review. Fuel. **254** (2019). https://doi.org/10.1016/j.fuel.2019.06.007
50. Hileman, J.I., Stratton, R.W.: Alternative jet fuel feasibility. Transport. Policy **34**, 52–62 (2014). https://doi.org/10.1016/j.tranpol.2014.02.018
51. ASTM: Physical constants of hydrocarbon and non-hydrocarbon compounds. ASTM International, West Conshohocken, PA (1991). https://doi.org/10.1520/DS4B-EB
52. ASTM D1655-16c: Standard specification for aviation turbine Fuels. ASTM International, 100 Barr Harbor Drive, PO Box C700, West Conshohocken, PA 19428–2959. United States (2016)
53. ASTM D1655-17: Standard specification for aviation turbine fuels. ASTM International, 100 Barr Harbor Drive, PO Box C700, West Conshohocken, PA 19428–2959. United States (2017)
54. ASTM D1655-18a: Standard specification for aviation turbine fuels. ASTM International, 100 Barr Harbor Drive, PO Box C700, West Conshohocken, PA 19428–2959. United States (2018)
55. ASTM D1655-20d: Standard specification for aviation turbine fuels. ASTM International, 100 Barr Harbor Drive, PO Box C700, West Conshohocken, PA 19428–2959. United States (2020)
56. Defence Standard 91-091: Turbine fuel, Kerosine type, Jet A-1; NATO code: F-35; Joint Service Designation: AVTUR. UK Defence Standardization, Kentigern House, 65 Brown Street. Glasgow, G2 8EX (2018)
57. ASTM D7566-16b: Standard specification for aviation turbine fuel containing synthesized hydrocarbons. ASTM International, 100 Barr Harbor Drive, PO Box C700, West Conshohocken, PA 19428–2959. United States (2016)
58. ASTM D7566-17a: Standard specification for aviation turbine fuel containing synthesized hydrocarbons. ASTM International, 100 Barr Harbor Drive, PO Box C700, West Conshohocken, PA 19428–2959. United States (2017)
59. ASTM D7566-18: Standard specification for aviation turbine fuel containing synthesized hydrocarbons. ASTM International, 100 Barr Harbor Drive, PO Box C700, West Conshohocken, PA 19428–2959. United States (2018)
60. ASTM D7566-20c: Standard specification for aviation turbine fuel containing synthesized hydrocarbons. ASTM International, 100 Barr Harbor Drive, PO Box C700, West Conshohocken, PA 19428–2959. United States (2020)
61. MIL-DTL-83133G: Detail specification turbine fuel, aviation, kerosene type, JP-8, and JP-8+100. US Department of Defense, AFPET/PTPS, 2430 C Street, Building 70, Area B, Wright-Patterson AFB OH 45433–7631 (2010)
62. MIL-DTL-83133H (AMENDMENT 2): Detail specification turbine fuel, aviation, kerosene type, JP-8, and JP-8+100. US Department of Defense, AFPET/PTPS, 2430 C Street, Building 70, Area B, Wright-Patterson AFB OH 45433–7631 (2013)
63. MIL-DTL-83133J: Detail specification turbine fuel, aviation, kerosene type, JP-8, and JP-8+100. US Department of Defense, AFPET/PTPS, 2430 C Street, Building 70, Area B, Wright-Patterson AFB OH 45433–7631 (2015)
64. MIL-DTL-5624V: Detail specification turbine fuel, aviation, grades JP-4 AND JP-5. Naval air warfare center, Aircraft division Lakehurst, Code 4.1.2, Mail Stop 120–3, Route 547, Joint Base MDL, NJ 08733–5100 (2013)
65. MIL-DTL-5624W: Detail Specification Turbine Fuel, Aviation, Grades JP-4 AND JP-5. Naval Air Warfare Center, Aircraft Division Lakehurst, Code 4.1.2, Mail Stop 120-3, Route 547, Joint Base MDL, NJ 08733–5100 (2016).
66. Wang, X., Jia, T.,·Lun Pan, L., Liu, Q., Fang, Y., Zou, J-J., Zhang, X.: Review on the relationship between liquid aerospace fuel composition and their physicochemical properties. Trans. Tianjin Univ. **27**, 87–109. https://doi.org/10.1007/s12209-020-00273-5
67. Chevron Corporation: Aviation fuels technical review, 1500 Louisiana Street, Houston, TX 77002, (2006)

Hybrid and Electrified Powertrains

Methodology for TurboGenerator Systems Optimization in Electrified Powertrains

Charbel Mansour⬤, **Wissam Bou Nader**⬤, **and Maroun Nemer**

Abstract Due to the continuous tightening of the Corporate Average Fuel Economy standards on passenger vehicles, several options are being explored in the automotive industry to reduce the consumption further in electrified powertrains, with one of the options being replacing internal combustion engines with alternative energy converters. Turbogenerator systems are among potential energy converters as they present fundamental benefits to powertrain applications, such as high efficiency, multi-fuel use, cogeneration capability, reduced vibration and noise, fewer components and compactness, as well as reduced weight compared to the engine and other energy converters. These systems are typically suitable for extended-range electric vehicles with a series-hybrid powertrain configuration, given their capacity to recharge the batteries at high efficiency and consequently extend the vehicle electric range beyond the typical urban driving limits. This chapter presents a methodology to design turbogenerators and optimize their system configuration to replace the engine in the auxiliary-power-unit of a series-hybrid powertrain. It consists of conducting first an exergo-technological analysis to identify the optimal turbogenerator system configuration, and second, assessing the resulting energy consumption while accounting for the additional consumption from thermal comfort and other auxiliaries on the Worldwide harmonized Light vehicle Test Cycle. For reference, the proposed methodology is applied to several turbogenerator systems, namely a simple gas-turbine, an external combustion gas-turbine, and a combined-cycle gas-turbine. The optimal configuration and design parameters for each of the three systems are identified while respecting automotive technological constraints. Consumption results show fuel savings up to 25% as compared to a reference extended-range electric vehicle equipped with an engine, depending on the battery size, the trip length,

C. Mansour (✉) · M. Nemer
Center for Energy Efficiency of Systems, Mines ParisTech, Palaiseau, France
e-mail: charbel.mansour@mines-paristech.fr

C. Mansour
Industrial and Mechanical Engineering Department, Lebanese American University, Byblos, Lebanon

W. Bou Nader
PSA Group, Technical Center of Vélizy, Vélizy, France

T. Parikyan (ed.), *Advances in Engine and Powertrain Research and Technology*,
Mechanisms and Machine Science 114,
https://doi.org/10.1007/978-3-030-91869-9_10

239

and the maximum turbine inlet temperature. Consequently, the proposed methodology helps design optimal turbogenerator systems presenting a serious alternative to engines on series-hybrid electrified vehicles, should the cost-effectiveness of these systems be proven.

Keywords Turbogenerator · Series-hybrid · Powertrain ·
Extended-range-electric-vehicle · Genetic algorithm · Dynamic programming

1 Introduction

In order to meet the continuously evolving regulations to reduce the fuel consumption from passenger cars, vehicle manufacturers are continually investing efforts to electrify vehicle powertrains and also introduce alternative technologies to the internal combustion engines (ICE). The introduction of new alternatives is observed through the development of innovative energy converters such as fuel cell technologies [1] and turbogenerator systems.

Several manufacturers investigated in past years the integration of gas turbines (GT) in conventional powertrains as the main energy converter instead of conventional engines. Early GT vehicle models in the 60s and 70s showed poor acceleration response and higher fuel consumption compared to internal combustion engine vehicles (ICEV) [2]. These drawbacks were mainly due to operating the turbine at high speed even at idle conditions, in addition to mechanically coupling the turbine to the vehicle driving load, which resulted in a low efficiency operating range of the GT system. Despite the numerous technological advancements made later on GT such as the variable turbine geometry, the injection of water, and the increase of turbine inlet temperature; the acceleration lag and the poor fuel efficiency of these systems remained the main reasons hindering their deployment in conventional powertrains.

A review of recent research and development programs of automotive manufacturers revealed renewed interests in GT for automotive applications, demonstrated in several vehicle concept cars in the form of turbogenerators, where the turbine is coupled to an electric generator to produce electricity. Moreover, the review of the recent literature showed interesting insights on turbogenerator consumption and emissions reductions. A study on turbogenerators for automotive applications at the Chalmers University of Technology presented an interest in operating GT at the optimal efficiency point compared to the ICE [3]. A complementary study at the University of Rome showed that turbogenerator emissions at optimal efficiency operation meet the Euro 6 emissions levels of CO, NOx, and soot even without the use of after-treatment systems [4]. Also, turbogenerator systems offer other intrinsic benefits for vehicle powertrains such as the reduced number of moving parts, vibration-free operation, low maintenance cost, high durability, and the absence of a cooling system [5].

Based on the aforementioned findings, turbogenerator systems present a forthcoming potential for improving modern vehicle efficiency and emissions, with the benefit of fuel-use flexibility when compared to ICEVs; particularly, in series hybrid electric powertrain configurations. These powertrains combine a thermal and an electric powertrain in a series energy-flow arrangement. The thermal powertrain is constituted of an energy converter and an electric generator and is referred to as Auxiliary Power Unit (APU). The APU is mainly used to recharge the battery once depleted, and the electric powertrain provides the necessary power to overcome the driving load. Consequently, the APU operating speed is cinematically decoupled from the vehicle velocity, and the energy converter operating point is easily controllable to meet its best efficiency.

In this context, several turbogenerator system options could be considered for integration in series hybrid vehicles, combining a basic GT to regenerative systems and single or multi-stage compressions and expansions. There have been numerous studies published over the past decade in the academic literature covering a multitude of GT-system configurations and performance analysis in different applications, such as power generation [6–12] and aeronautics [13, 14]. The survey of these studies confirms that most turbogenerator systems are designed based on efficiency optimization, power density optimization, or a compromise between the two criteria. For instance, industrial turbogenerator system studies focused on finding an acceptable compromise between maximizing the system efficiency and the power density, with less concern on reducing the weight. Along these lines, aeronautical GT-system studies focused on maximizing the power density due to the high weight-reduction priority constraint for such applications. In both applications, a combination of several measures was required such as the need to increase the turbine inlet temperature, to add intercooler, regenerator and reheat systems to achieve the needed optimization [8, 13, 14]. However, there have only been a few recent papers on turbogenerator systems suitable for automotive applications [15–17] due to the lack of competitiveness of GT compared to ICE in conventional powertrains. Other papers date for more than 20 years [18–21].

The review of these recent and old studies underlines few gaps such as there is no specific methodology was developed for the design of the optimal turbogenerator system for automotive applications. The studies' focus is on the performance investigations of some pre-defined GT architectures, without taking into consideration any optimization requirement or technological constraints. Another gap found in the literature concerns the inexistence of assessment of vehicle consumption under driving conditions and consequently the lack of any benchmark of these turbogenerator technologies against ICEV.

Therefore, based on the above synthesis of the insights and gaps in the literature for re-adopting turbogenerators in automotive applications, this chapter proposes a comprehensive methodology to identify the potential system options and select the optimal system configuration dedicated for a series hybrid extended-range electric vehicle (EREV).

2   Methodological Framework

The proposed framework, illustrated in Fig. 1, consists of two consecutive modeling steps. It starts by modeling the thermodynamic performance of turbogenerator systems to identify the potentially suitable system architectures to replace the ICE in an EREV, and then it is followed by system-level modeling of the EREV powertrain including the identified turbogenerator systems to select the best suitable configuration according to specific criteria related to minimizing consumption and reducing the integration complexity of the system in the powertrain.

The first modeling step consists of performing a thermodynamic exergy and energy assessment of a simple turbogenerator configuration and of investigating possible optimization options through combining the simple configuration to reheat gas turbine systems and single or multiple-stage compressions and expansions, as delineated in Sect. 3. Several system configurations are then derived, and only one or

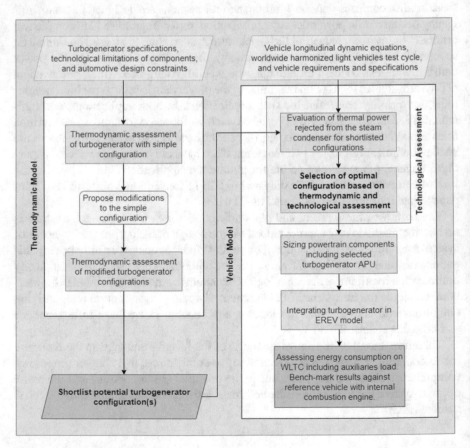

Fig. 1 A methodological framework for the design and optimization of turbogenerator systems, suitable to replace the ICE in a series-hybrid EREV application

a few are shortlisted, those which present a convenient compromise between reducing the exergy destruction while considering state-of-the-art component specifications, and meeting automotive design constraints.

Then, the shortlisted systems derived from the first modeling step are integrated into a vehicle model, and results are used for the prioritization and the selection of the optimal turbogenerator system configuration. To that end, an EREV model is presented in Sect. 4, along with a methodology for sizing the powertrain components, namely the auxiliary power unit, according to speed, acceleration, and gradeability performance specifications required from typical EREV. The criteria considered in the selection are optimizing the system efficiency and reducing the vehicle integration complexity, mainly the thermal heat rejected through the frontal vehicle surface from the turbogenerators. The system efficiency and vehicle consumption are calculated on the WLTC while accounting for the vehicle thermal cabin and auxiliary power needs.

Finally, the overall vehicle consumption of the EREV model with the optimized turbogenerator configuration is benchmarked against a reference EREV model, equipped with an ICE auxiliary power unit, and using the Dynamic Programming (DP) as Energy Management Strategy to provide the optimal global strategy for powering ON and OFF the APU [22, 23].

3 Thermodynamic Model

The thermal model aims at assessing the thermodynamic performance of turbogenerator systems to identify the optimal system configuration for integration as an APU into an EREV. It consists of a two-steps assessment, as illustrated in Fig. 2.

The first step consists of explicit energy and exergy assessments conducted on a simple turbogenerator system configuration, where the overall efficiency and net specific work are calculated using Refprop to identify the significant sources of exergy destruction in the system. The exergy destruction is calculated for each component, and based on the observed results, modifications to the studied turbogenerator configuration are proposed such as expansion reheat, compression intercooling, internal regeneration, and increasing the turbine inlet temperature. The effect of each of these modifications on the total exergy destruction in the system is then investigated.

Accordingly, in the second step of the thermodynamic assessment, energy and exergy analyses are conducted for each of the proposed modified configurations, as illustrated in Fig. 3. State-of-the-art components' specifications and technological constraints are first used to perform the calculations of the overall efficiency, net specific work, and exergy destruction for each configuration. Then, a multi-objective genetic optimization algorithm, the Non-dominated Sorting Genetic Algorithm (NSGA) is used to generate optimal design parameters, namely the expansion ratios, compression ratios, and maximum pressure in the cycle, while taking the overall efficiency and net specific work as objective functions. Consequently, Pareto curves are obtained for the studied configurations. These curves are used to detect

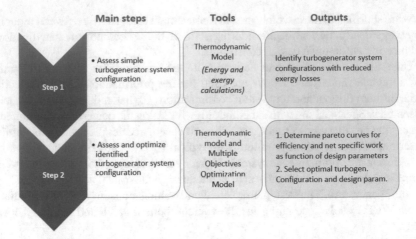

Fig. 2 Assessment steps for the identification of the highest efficiency GT-system and its optimal design parameters

the optimal operating point for each of the configurations and select the most optimal topology.

Finally, one or more of the modified-turbogenerator configurations resulting from this thermodynamic assessment are shortlisted and will be further assessed from a technological point of view to check their suitability for automotive applications.

In the following subsections, the proposed methodology is detailed following the proposed two-steps framework for the case of three turbogenerator systems presenting high potential to replacing the ICE in EREV: (1) Brayton gas turbine system (GT), (2) external combustion gas turbine system (ECGT), and (3) combined-cycle gas turbine system (CCGT). Section 3.1 presents the thermodynamic assessment of the basic configuration of each of the three turbogenerator systems to determine the exergy losses and identify improved configurations for each of them. Section 3.2 demonstrates the assessment and optimization of the identified configurations to select the optimal one for integration in the EREV.

3.1 Energy and Exergy Analysis of the Simple Turbogenerator System Configuration

The simple system configurations of the studied turbogenerators are illustrated in Fig. 4a–c. The basic Brayton GT cycle consists of a compressor, a combustion chamber, and a turbine (Fig. 4a), with air as the working fluid. The ECGT builds on the basic Brayton GT and consists of an additional air-loop with a combustion chamber and a combustion chamber blower (CCB), sharing a common air heat

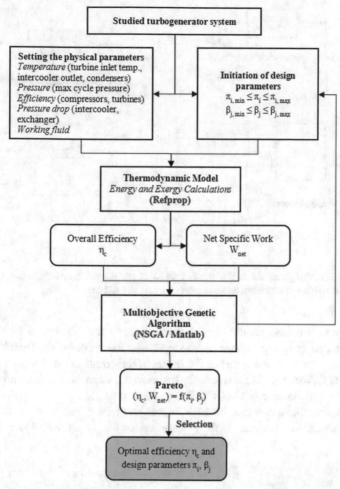

Fig. 3 Assessment methodology for the identification of the highest efficiency GT-system and its optimal design parameters

exchanger through which heat is exchanged between the two loops (Fig. 4b). Similarly, the CCGT consists of two-loops, a basic Brayton GT and a Steam Rankine Cycle (SRC), with water as the working fluid, sharing heat through a common steam evaporator (Fig. 4c). The Rankine-loop includes a water pump, steam condenser, and steam turbine.

The first law of thermodynamics is applied to each component to deduce each system's efficiency and trace the exergy losses in the components, their types, and quantities, to better inform on the possible options to reduce the inefficiencies. Energy and exergy model equations for each component are explicitly developed in [24–26].

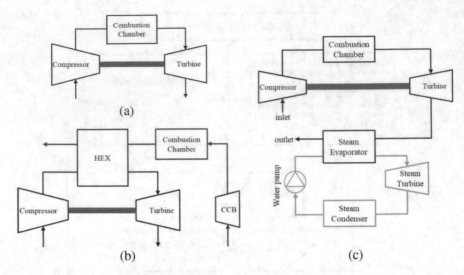

Fig. 4 Simple configurations of the studied turbogenerator systems. **a** Brayton gas-turbine, **b** external combustion gas turbine, and **c** combined cycle gas turbine

Exergy destruction results for the three systems are illustrated in Fig. 5. The highest shares of exergy losses are observed in the combustion chamber and the exhaust gas at the turbine outlet or the outlet of the combustion chamber in the case of the ECGT. Also, considerable exergy losses in the steam condenser and evaporator are observed in the case of the CCGT.

Several modifications can be investigated to reduce exergy destruction in the components of the system. Table 1 summarises the observed sources of exergy

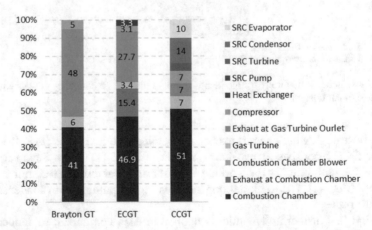

Fig. 5 Distribution of exergy destruction in the turbogenerator systems with a turbine inlet temperature of 1250 °C and maximum cycle pressure of 1.2 MPa

Table 1 Sources of exergy destruction in simple turbogenerator systems and proposed modifications to reduce losses

Exergy destruction source	Modifications to reduce exergy destruction
Combustion chamber	• Increase the average temperature in the combustion chamber through expansion reheat • Increase the combustion chamber outlet temperature (or the turbine inlet temperature) while respecting the metallurgical and technological constraints
Gas turbine outlet	• Implement an external waste heat recovery system (such as Rankine cycle, case of the CCGT) • Implement an internal heat recovery system using a regenerator
Steam condenser	• Implement external waste heat recovery systems such as bottoming cycles, which is disregarded in the case of automotive applications given the high level of integration complexity • Implement internal regenerators to recover heat from the turbine outlet
Steam evaporator	• Decrease the minimum temperature difference between the hot and cold streams (pinch), which is disregarded in automotive applications since it requires a large heat exchange surface • Use a supercritical Rankine cycle, where water is pumped to a pressure higher than its critical value (disregarded due to technological constraints)
Pump, compressor, and Turbines	• Consider isothermal compression and expansion (disregarded given the technical challenges in its implementation) • Consider intercooling during compression • Consider reheating during expansion *Note* Exergy destruction in the pump is negligible compared to the other components. It is recommended to disregard improving the efficiency of pumps

destruction in the three systems along with the corresponding modifications to reduce these losses.

Based on the solutions presented in Table 1, a list of improved system configurations for the three investigated turbogenerators is proposed. These systems are classified according to the combination of the suggested techniques for exergy loss reduction, such as the intercooled compression, the reheat expansion, and the use of internal regenerators in the Rankine-loop. For convenience purposes, an example is illustrated in Fig. 6 for the case of the Brayton GT, however, the names of the derived configurations for the three Turbogenerators are summarized in Table 2.

The corresponding system architectures of each of the modified turbogenerator systems are detailed in [24–26]. These architectures are considered in the rest of the chapter for further assessment to determine the most suitable turbogenerator system configuration for an EREV application with a series-hybrid powertrain.

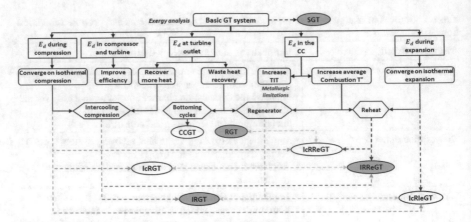

Fig. 6 Exergy assessment methodology for the identification of the Brayton GT system configurations with reduced exergy losses

Table 2 List of the different system configurations derived from the exergy assessment methodology

Brayton gas turbine system	
RGT	Regenerative GT
IRGT	Intercooled regenerative GT
IRReGT	Intercooled regenerative reheat GT
External combustion gas turbine system	
S-ECGT	Simple ECGT
R-ECGT	Regenerative ECGT
DS-ECGT	Downstream simple ECGT
I-ECGT	Intercooled ECGT
IR-ECGT	Intercooled regenerative ECGT
DI-ECGT	Downstream intercooled ECGT
IRe-ECGT	Intercooled reheat ECGT
IRRe-ECGT	Intercooled regenerative reheat ECGT
DIRe-ECGT	Downstream intercooled reheat ECGT
Combined cycle gas turbine	
GT-SRC	Simple combined cycle gas turbine
IGT-SRC	Intercooled gas turbine coupled to steam Rankine cycle
IReGT-SRC	Intercooled reheat gas turbine coupled to steam Rankine cycle
IReGT-TReSRC	Intercooled reheat gas turbine coupled to a turbine reheat SRC
IReGT-CRTReSRC	Intercooled reheat GT with turbine-reheat and condenser-regenerative SRC
ReGT-SRC	Reheat GT coupled to steam Rankine cycle
ReGT-TReSRC	Reheat GT coupled to turbine reheat steam Rankine cycle
ReGT-CRTReSRC	Reheat GT with turbine-reheat and condenser-regenerative SRC

3.2 Energy and Exergy Analysis of the Modified Turbogenerator System Configurations

The identified GT system configurations of Table 2 are assessed here to prioritize these options based on their respective efficiency and net specific work, as detailed in the assessment methodology illustrated in Fig. 3. The energy and exergy calculations are performed first with Refprop software, using the set of physical parameters such as the turbine inlet temperature, the cycle pressure, the efficiencies of the components, among others; as summarized in Table 3. These parameters correspond to the state-of-the-art specifications and limitations of turbogenerator component technologies and automotive design constraints.

The energy and exergy calculations are made as a function of the different design parameters: the compression ratio (π_i) and the expansion ratio (β_i), with i and j

Table 3 Thermodynamic simulation parameters based on state-of-the-art component specifications, component technologies limitations, and automotive design constraints

Parameter	Unit	Value
GT working fluid	–	Air
Compressor technology	–	Radial
Maximum number of compression stages	–	2
Compressor maximum pressure ratio	–	3.5
Compressors efficiency	%	80
Compressor inlet pressure drop	%	0.5
Combustion chamber pressure drop	%	4
GT maximum cycle pressure	MPa	1.2
Intercoolers pressure drop	%	5
Intercoolers outlet temperature	°C	60
Regenerator efficiency	%	85
Regenerator pressure drop cold side	%	4
Regenerator pressure drop hot side	%	3
HEX pinches	°C	100
Maximum number of expansion stages	–	2
Turbines technology	–	Radial
Turbine inlet temperature	°C	1250
Turbines isentropic efficiency	%	85
Turbine expansion ratio	–	4
SRC pump isentropic efficiency	%	70
SRC turbine isentropic efficiency	%	85
SRC working fluid	–	Water
Steam condensing temperature	°C	100
Steam condenser sub-cooling	K	3
Steam maximum pressure	MPa	10

referring to the number of compressors and turbines in the GT cycle, respectively. Also, other design parameters are considered such as the maximum steam pressure ($P_{max, SRC}$), the maximum steam temperature ($T_{max, SCR}$), the steam turbine reheat maximum temperature ($T_{max, TRe-SRC}$), and the high-pressure steam turbine expansion ratio (β_{HP-SRC}), whenever a Rankine cycle is included in the system architecture. Therefore, the second step of the thermodynamic simulation is an optimization utilizing the multi-objective Non-dominated Sorting Genetic Algorithm (NSGA) [27] to determine the Pareto optimal efficiency and net specific work solutions for the optimal π_i, β_j, $P_{max, SRC}$, $T_{max, SRC}$, $T_{max, TRe-SRC}$, and β_{HP-SRC}. These design parameters were considered in the optimization among others given their significant impact on the system efficiency and net specific work, as demonstrated in [24].

Figure 7 illustrates the resulting Pareto optimal solution curves for the different

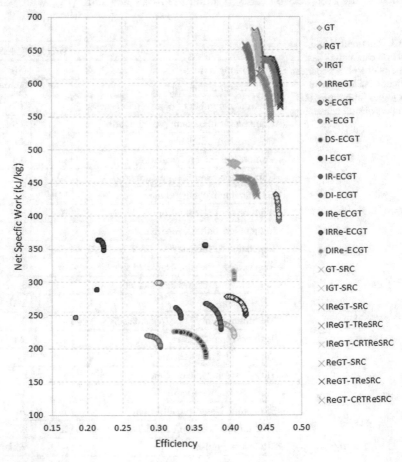

Fig. 7 Pareto optimal efficiency and net specific work solutions of the different turbogenerator systems

assessed systems. For the rest of the study, the comparison between the different GT-systems will be based on the best efficiency points on the Pareto curves. Therefore, the selection of the best-suited turbogenerator system for our EREV will be made according to the highest efficiency among the compared systems.

The figure shows that CCGT systems present the highest energy efficiency, in particular the ReGT-TReSRC and ReGT-CRTReSRC configurations, with 47.2% and 46.8% respectively. The second best-ranked configuration in terms of efficiency is the IRReGT configuration, which achieves an efficiency of 47%, however, for lower net specific work compared to the CCGT systems. Given the slight difference in efficiency and the higher number of components in the CCGT systems, the IRReGT configuration is prioritized and considered in the rest of this chapter for demonstration purposes. The CCGT systems will be disregarded for the sake of conciseness, however, the integration performance of these systems in the EREV model could be consulted in [26]. Section 4 presents the vehicle modeling of an EREV with the prioritized IRReGT serving as a turbogenerator substitute to the ICE in the APU.

4 Vehicle Model

The vehicle model aims at assessing the consumption performance of an EREV to evaluate the benefits of turbogenerator systems and compute their fuel-saving potential compared to ICE. In this study, the EREV powertrain has a series-hybrid architecture, which consists of the prioritized IRReGT-APU (referred to in the text as GT-APU) and an electric traction unit, as illustrated in Fig. 8.

Series-hybrid configuration presents the advantage of tackling the two main deficiencies preventing the integration of GT systems in conventional powertrains as revealed in the literature and OEM interrupted development programs: (1) the poor system efficiency when the turbine is operated to follow the dynamic load of the car, and (2) the acceleration lag. On one hand, the turbine is mechanically decoupled from the vehicle load in a series-hybrid setup, and therefore, the vehicle is propelled by an electric machine only, powered by a battery and/or the APU, and properly sized to ensure the vehicle's acceleration and velocity performance without deficiency. On the other hand, the turbine operates at steady power corresponding to the GT-APU optimum efficiency, which is higher than the maximum efficiency of the ICE.

Typically, the assessment of energy consumption is conducted on the Worldwide harmonized Light vehicles Test Cycle (WLTC), which consists of evaluating the energy consumed by the powertrain for driving the vehicle on a defined velocity profile, emulating different driving conditions from urban to the highway. In the proposed methodological framework of Fig. 2, the authors suggest accounting for the additional energy consumption related to the vehicle auxiliaries, such as the cabin thermal comfort equipment, under specific weather conditions, given their considerable impact on the vehicle consumption and range. Therefore, the steps and the required data for conducting this comprehensive energy modeling for the EREV with a GT-APU and an ICE-APU are illustrated in Fig. 9.

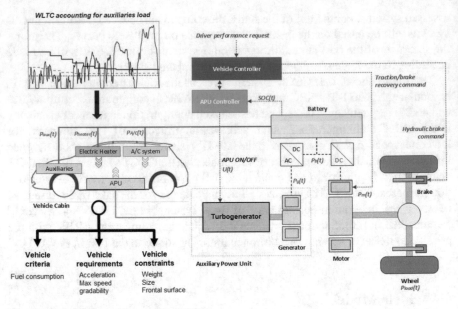

Fig. 8 Configuration of the modeled EREV

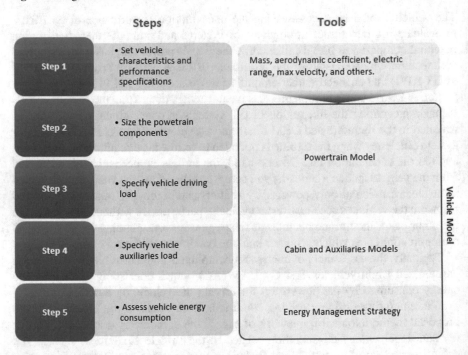

Fig. 9 Methodology for the energy modeling of EREV

The first step consists of collecting the required vehicle performance in terms of maximum velocity and gradeability, minimum electric driving range, among others, and the vehicle's basic characteristics such as the aerodynamic and rolling resistance coefficients, chassis mass, and others, to properly size the powertrain components. Consequently, the APU, the electric machine, and the battery are properly sized in step 2 to meet the performance requirements. This requires the development of a powertrain model.

The following steps 3 and 4 complete the vehicle model by adding the cabin and other thermal comfort and electric auxiliaries models, to account for the consumption of the different energy systems in the vehicle. This helps capture the additional consumption observed by these systems, particularly for thermal comfort when operating under severe cold and hot weather conditions. Several recommendations highlight the importance of accounting for these additional conditional consumption, particularly when sizing batteries, to effectively meet the requested electric driving range [28, 29].

Finally, an energy management strategy is added to the vehicle model to manage the power flow within the powertrain and adequately meet the load demand while optimizing consumption.

In the following subsections, the proposed methodology is detailed for the case of an EREV with a GT-APU and another reference vehicle with an ICE-APU. Section 4.1 introduces the sizing of the components. Sections 4.2 and 4.3 summarizes the powertrain and the auxiliaries model, and finally, Sect. 4.4 presents the energy management strategy.

4.1 Components Sizing

The adopted methodology to size the powertrain components is illustrated in Fig. 10. The motor, APU, and battery are sized to meet the vehicle performance requirements summarized in Table 4.

Sizing the Electric Machine. The electric motor is sized to ensure the required performance of a midsize vehicle, in terms of velocity, acceleration, and gradeability. The maximum velocity performance consists of reaching a top speed of 160 km/h and maintaining a continuous speed of 120 km/h without depleting the battery. The acceleration performance requires achieving a 0–100 km/h in 9.6 s, and the gradeability criteria are maintaining a constant velocity of 90 km/h at a 10% slope and 110 km/h at a 5% slope. Thus, the motor power must overcome the vehicle load for each of these performance requirements. The maximum velocity and gradeability vehicle load are determined using Eq. (1), and the vehicle acceleration load is calculated using Eq. (2).

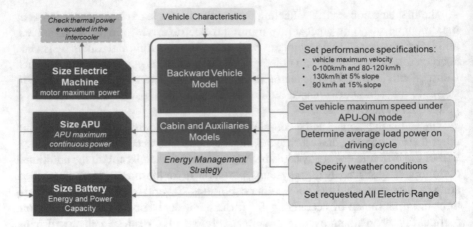

Fig. 10 Methodology for the powertrain component sizing

Table 4 Vehicle performance requirements

Performance criteria	Value
Maximum continuous velocity	120 km/h
Maximum velocity	160 km/h
Acceleration time from 0 to 100 km/h	9.6 s
Gradeability criteria 1	110 km/h at a 5% slope
Gradeability criteria 2	90 km/h at a 10% slope
Urban electric driving range	50 km

$$P_{\text{load}}(t) = \left(M_t.g.\sin\theta.v(t) + \frac{1}{2}\rho SC_x v(t)^2 + M_t.g.f_r.v(t) + \delta.M_t.\frac{dv(t)}{dt} \right) \times v(t) \tag{1}$$

$$P_{m_{\text{acc}}} = \frac{M_t}{2t_a}\left(V_f^2 + V_b^2\right) + \frac{2}{3}M_t g f_r V_f + \frac{1}{5}\rho C_x S V_f^3 \tag{2}$$

With,

M_t Vehicle total mass (kg)
δ Ratio of the equivalent mass of rotational components
θ Slope angle
f_r Friction rolling coefficient
g Gravity (m/s²)
ρ Air density (kg/m³)
S Vehicle frontal area (m²)
Cx Vehicle drag coefficient
V Vehicle velocity (m/s)

$P_{m_{acc}}$ Traction power required to accelerate the vehicle from zero to V_f in t_a seconds (W)
V_f Vehicle final velocity (m/s)
V_b Vehicle velocity corresponding to the motor base speed (m/s)
t_a Time to accelerate from 0 to V_f (s).

Results show that 80 kW is needed for the acceleration test, whereas 40 kW is for cruising at the maximum velocity of 160 km/h, 38 kW for driving at 110 km/h on a 5% slope, and 47 kW at 90 km/h on a 10% slope. Therefore, the considered motor size is 80 kW.

Sizing the Auxiliary Power Unit. The main role of the APU in an EREV is to sustain the battery energy under all driving conditions. Hence, the APU is sized according to the different driving conditions of a vehicle, namely the stop-and-go patterns and the high-speed cruising patterns. The stop-and-go patterns are represented in this study by the WLTC, and the average vehicle load power is computed using Eq. (3). The high-speed cruising pattern is emulated by driving for a long distance at the maximum continuous velocity of 120 km/h without the need for battery support, using Eq. (4). Note that in both cases, the vehicle auxiliary power, such as the electric auxiliaries and the cabin thermal need under extreme hot and cold weather conditions, were considered as presented in Sect. 4.3.

$$P_{APU} = \frac{1}{\eta_g}\left(\frac{1}{t_c}\int_0^{t_c}\left(M_t g f_r + \frac{1}{2}\rho_a C_x S v(t)^2\right)V\,dt + \frac{1}{t_c}\int_0^{t_c} M_t \frac{dv(t)}{dt}dt\right) + P_{aux}$$

(3)

$$P_{APU} = \frac{1}{\eta_g}\left(\frac{v(t)}{\eta_t \eta_m}\cdot\left(M_v g f_r + \frac{1}{2}\rho_a C_x S v(t)^2\right)\right) + P_{aux}$$ (4)

With,

t_c Driving cycle time length (s)
η_t Efficiency of the transmission
η_m Efficiency of the motor
η_g Efficiency of the generator
P_{aux} Electric auxiliaries and cabin thermal comfort needs (3250 W in the case of this study)
P_{APU} Power of the auxiliary power unit (W).

Results show that 19 kW is required to propel the vehicle at 120 km/h on the highway compared to 10.5 kW for the WLTP stop-and-go pattern. Hence, accounting for the generator efficiency losses in sizing the APU, a 25 kW IRReGT system is considered in the rest of the study to power the APU.

At this stage, it is essential to evaluate the thermal power rejected from the 25 kW APU to validate the feasibility of the system integration in the vehicle. In typical ICE vehicles, the engine radiator, typically mounted at the vehicle front end to receive airflow from the vehicle movement, can evacuate to the ambient up to 100 kW of heat from the engine coolant. Therefore, the heat evacuated from the steam condenser must not exceed 100 kW to fit in the allocated space for heat exchangers at the vehicle's front end. Calculations show that the observed thermal power rejected from the steam condenser of the prioritized IRReGT system is less than 30 kW, and therefore, meets the 100 kW constraint. Therefore, the 25 kW GT-APU could be considered for integration in the EREV.

Sizing the Battery. As for battery sizing, power and capacity have to be considered. The battery is expected to deliver the needed traction power under any driving condition, while the APU intervenes under extreme power demand. Consequently, the battery's maximum power is sized with respect to the electric motor maximum power and the APU power, using Eq. (5).

$$P_b \geq \frac{P_{m_{max}}}{\eta_m} - P_{CCGT}\eta_g \qquad (5)$$

With

P_b Battery power (W)
$P_{m_{max}}$ Motor's maximum power (W)
P_{GT} GT system power (W).

Results show that a 60 kW battery is needed. As for the energy capacity, it is determined based on ensuring 50 km of electric driving range in urban areas in moderate weather conditions. The urban driving cycle is considered to be the low-speed portion of the WLTC, where the vehicle load is 2.75 kW, and the average velocity is 25 km/h. Assuming the battery's state of energy ranges between 80 and 30%, and considering the auxiliary power observed under extreme weather conditions, the needed battery energy capacity is 12 kWh. A battery mass of 280 kg is considered as part of the total vehicle's mass; this value was retrieved from a survey of commercialized battery specifications illustrated in Fig. 11.

4.2 Powertrain Model

Based on the above, Table 5 summarizes the vehicle parameters considered in this study for modeling the GT-APU-EREV and the ICE-APU-EREV, respectively. The latter is considered as the reference EREV model used as a fuel consumption benchmark.

Equations (1) and (6)–(14) present the powertrain backward model of the GT-APU-EREV and the ICE-APU-EREV. The equations are used to calculate the power needs for traction, electric consumption, and cabin thermal comfort to derive the final energy consumption of both vehicles.

Equation (6) provides the traction and braking power of the electric motor using the efficiencies of the motor and the transmission.

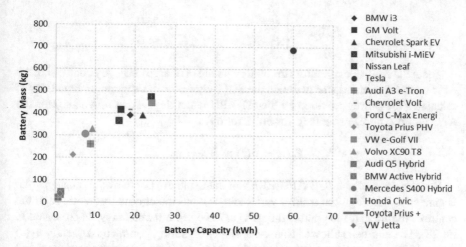

Fig. 11 Battery mass as a function of capacity retrieved from commercialized electrified vehicles

Table 5 Vehicle and component specifications

Vehicle specifications	Symbol	Unit	Value
Vehicle and driver mass (excl. the APU)	M_v	kg	1210
Rotational components equivalent mass	δ	–	1.05
Frontal area	S	m^2	2.17
Drag coefficient	C_x	–	0.29
Wheel friction coefficient	f_r	–	0.0106
Air density	ρ	kg/m^3	1.205
Wheel radius	R_w	m	0.307
Battery maximum power	$P_{b\,max}$	kW	60
Battery capacity	C_b	kWh	12
Battery mass	M_b	kg	280
GT system power	P_{GT}	kW	25
GT efficiency	η_{GT}	%	47.2
GT-APU weight	M_{GT}	kg	75
ICE efficiency	η_{ICE}	%	37
ICE-APU weight	M_{ICE}	kg	150
Generator maximum power	P_g	kW	45
Generator maximum efficiency	η_g	%	95
Motor maximum power	P_m	kW	80
Transmission ratio	i	–	5.4
Transmission efficiency	η_t	%	97
Vehicle total mass	M_t	kg	$M_v + M_{GT} + M_b$
Fuel heating value (Gasoline)	H_v	MJ/kg	42.5

$$P_m(t) = \begin{cases} \frac{P_{\text{load}}(t)}{\eta_t \times \eta_m}, & \frac{dv}{dt} \geq 0 \\ P_{\text{load}}(t) \times \eta_t \times \eta_m, & \frac{dv}{dt} < 0 \end{cases} \tag{6}$$

The APU controller monitors the battery's state of charge (SOC) by controlling the APU operations in order to maintain the SOC within the desired range. Equation (7) calculates the power provided by the GT-APU when the GT is on, where (t) is the APU on/off control variable (0 for off, 1 for on).

$$P_g(t) = u(t) \times P_{\text{GT}} \times \eta_g \tag{7}$$

For the case of the ICE-APU-EREV model, the APU controller manages the engine speed through the control variable $u_1(t)$ and the engine torque through the control variable $u_2(t)$ to operate the APU at the optimum efficiency, as presented in Eq. (8). The engine is allowed to operate at any operating point, consequently $u_1(t)$ ranges between 0 and the maximum engine speed, while $u_2(t)$ ranges between 0 and the maximum engine torque.

$$P_{\text{ICE}}(t) = u_1(t) \times u_2(t) \tag{8}$$

Equations (9) and (10) express respectively the generator power and the battery power consumption, which depends on the power of the electric motor, the auxiliaries, the electric heater, the air-conditioning, and the APU when turned on.

$$P_g(t) = u_1(t) \times u_2(t) \times \eta_g \tag{9}$$

$$P_b(t) = P_m(t) + P_{\text{aux}}(t) + P_{\text{heater}}(t) + P_{A/C}(t) - P_g(t) \tag{10}$$

Then, the battery's current and state of charge (SOC) are computed using Eqs. (11) and (12).

$$I_b(t) = \frac{V_{\text{oc}}(\text{SOC}(t)) - \sqrt{V_{\text{oc}}^2(\text{SOC}(t)) - 4P_b(t)R_i(\text{SOC}(t))}}{2R_i(\text{SOC}(t))} \tag{11}$$

$$\text{SOC}(t) = \text{SOC}_i(t) + \frac{1}{C_b} \int_{t_0}^{t} I_b(t)dt \tag{12}$$

Finally, the fuel consumption is computed using Eqs. (13) and (14) for the GT-APU-EREV and the ICE-APU-EREV, respectively.

$$\dot{m}_f(t) = \begin{cases} \frac{P_{\text{GT}}(t)}{\eta_{\text{GT}} \times H_v}, & \text{APU:ON} \\ 0, & \text{APU:OFF} \end{cases} \tag{13}$$

$$\dot{m}_f(t) = \begin{cases} \frac{P_{ICE}(t)}{\eta_{ICE} \times H_v}, & \text{APU:ON} \\ 0, & \text{APU:OFF} \end{cases} \tag{14}$$

4.3 Auxiliaries and Cabin Thermal Comfort Models

The impact of thermal comfort of the EREV was included in the fuel assessment, due to the electric range sensitivity of these vehicles to auxiliary consumption and in particular the auxiliaries used for thermal comfort. The authors presented in [30] a framework for modifying the WLTP to account for the vehicle thermal and electrical energy needs. These power profiles would be used in combination with the WLTC velocity profile for a more representative fuel assessment.

The modified WLTP consists of combining the WLTC velocity profile with heating and cooling thermal power profiles, to consider both the mechanical power on wheels required to drive the vehicle at a given speed, and the heating and cooling thermal needs required to ensure the vehicle comfort. These thermal power profiles were developed for three different climate conditions: cold, moderate, and hot. Note also that constant power consumption is also considered to account for the average load of typical electric auxiliaries on vehicles. For reference, the modified WLTP under cold climate conditions is shown in Fig. 12. Additional information on the developed framework and the results for moderate and hot climates can be retrieved from [30].

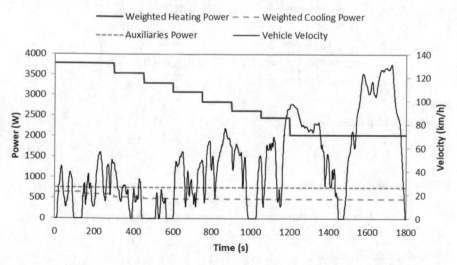

Fig. 12 Modified WLTP accounting for thermal and electrical auxiliaries load under cold climate

4.4 Energy Management Strategy

Two distinct controllers are considered in the model as illustrated in Fig. 8: the vehicle controller and the APU controller. The vehicle controller is in charge of meeting the driver's request in terms of performance. Hence, its main objective is to control the electric motor power to meet the traction and brake energy recovery demand, as presented in Eq. (6). The APU controller monitors the battery SOC; thus, it controls the APU operations to maintain the SOC in the desired range. Therefore, an engine on/off variable $u(t)$ is considered in Eq. (7) to control the APU start operations. $u(t)$ takes the value of 0 for APU-off and 1 for APU-on.

Dynamic programming (DP) is used to provide the global optimal strategy to control the APU operations [23]. The choice of DP over traditional rule-based energy management strategies is to ensure that only the impact of the cabin thermal and auxiliary needs are assessed, thereby excluding the potential bias of rule-based strategies on fuel consumption [22]. It decides on the optimal strategy $U_{opt} = \{u(1), \ldots, u(N)\}_{opt}$ for the scheduled route at each instant t while minimizing the fuel cost function J presented in Eqs. (15) and (16). Consequently, DP computes backward in time from the final desired battery state of charge SOC_f to the initial state SOC_i the optimal fuel mass flow rate $\dot{m}_f(SOC(t), u(t))$ in the discretized state time–space as per Eqs. (17)–(19). The generic DP function presented in [31] is considered in this study, with the battery SOC as a state variable $x(t)$ and the APU on/off as a control variable $u(t)$.

$$J_{GT} = \min\left\{\sum_{t=1}^{N} \dot{m}_f(SOC(t), u(t)) \times dt_s\right\} \tag{15}$$

$$J_{ICE} = \min\left\{\sum_{t=1}^{N} \dot{m}_f(SOC(t), u_1(t), u_2(t)) \times dt_s\right\} \tag{16}$$

With,
Discrete step time:

$$dt_s = 1 \tag{17}$$

Number of time instances:

$$N = \frac{n}{dt_s} \quad \text{(with } n \text{ the time length of the driving cycle)} \tag{18}$$

State variable equation:

$$SOC(t + 1) = f(SOC(t), u(t)) + SOC(1) \tag{19}$$

Initial SOC:

$$SOC(1) = SOC_i \qquad (20)$$

Final SOC:

$$SOC(N) = SOC_f \qquad (21)$$

SOC constraint:

$$SOC(t) \in [0.2, 0.9] \qquad (22)$$

Battery power constraint:

$$P_{b_{\min}} \le P_b(t) \le P_{b_{\max}} \qquad (23)$$

Motor torque constraint:

$$P_{m_{\min}}(\omega_m(t)) \le P_m(t) \le P_{m_{\max}}(\omega_m(t)) \qquad (24)$$

Motor speed constraint:

$$0 \le \omega_m(t) \le \omega_{m_{\max}}(t) \qquad (25)$$

Generator power constraint:

$$P_{g_{\min}}(\omega_m(t)) \le P_g(t) \le P_{g_{\max}}(\omega_m(t)) \qquad (26)$$

Generator speed constraint:

$$0 \le \omega_g(t) \le \omega_{g_{\max}}(t) \qquad (27)$$

During APU operations, the GT-APU is designed to operate at its optimal operating point and delivers 25 kW of mechanical power, whereas the ICE is allowed to operate at any point of its torque-speed map; however, the energy management strategy tends to maximize the powertrain efficiency by operating the engine on its optimal operating line.

Note that the resulting optimal APU on/off strategy U_{opt} must not cause the components to violate their relevant physical boundary constraints in terms of speed, power or SOC, to ensure their proper functioning within the normal operating range. These constraints are included in the DP model and summarized in Eqs. (20)–(27).

5 Results

Two midsize EREV models are compared in this section: the prioritized GT-APU-EREV with the IRReGT turbogenerator and a reference ICE-APU-EREV. The GT-APU is designed to deliver 25 kW of mechanical power, operating at its optimal point. As for the ICE-APU, it uses a 1-L spark ignition engine with a maximum efficiency of 37% and operates at points of its torque-speed map falling between 1000 and 4000 RPM and between 25% full load and 80% full load, delivering up to 34 kW of mechanical power. The operating zone has been limited as such to avoid high load at low engine speed, which causes irritating vibrations, and to limit thermal stress and fatigue at high load. The operating point of the ICE is decided by the DP energy management strategy, presented in Sect. 4.4.

Both vehicle models use the same battery size of 12 kWh, and simulations are performed on a sequence of one to ten-repeated WLTC, with an initial battery SOC of 80% and a final SOC of 30%, reached at the end of each trip. Gasoline is the fuel used in the two models.

Simulations are performed under the cold climate scenario, where heating is ensured through an electric heater and cooling in summer through an air conditioning system. A 750 W constant power is imposed to emulate other electric auxiliaries' need for midsize vehicles. Note also that transient thermal load is only considered during the first WLTC and then the steady-state load is imposed in the following repeated cycles. Results are illustrated in Fig. 13, showing the fuel consumption of both models.

Although the ICE was not constrained to operate at one operating point, results showed that the ICE operated at power values between 19 and 24 kW, and on its optimal operating line where the efficiency is very close to its maximum value of 37%. This implies the superiority of the GT-APU compared to the ICE-APU in terms of fuel savings, as confirmed by the results, with fuel savings up to 25%.

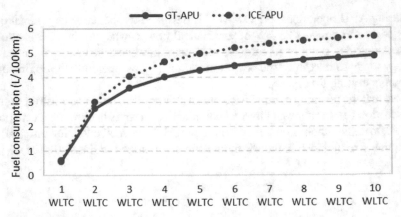

Fig. 13 Energy consumption of the GT-APU-EREV and the ICE-APU-EREV models on one to ten-repeated WLTC, accounting for thermal and electrical auxiliaries load under cold climate

In addition to benchmarking the consumption performance of the proposed GT-APU against the ICE-APU, it is also interesting to compare this performance to the other turbogenerator systems also suitable to replace the ICE in EREV, such as the "Downstream intercooled reheat ECGT" (DIRe-ECGT) and the "Reheat GT coupled to turbine reheat steam Rankine cycle" (ReGT-TReSRC-CCGT). Figure 14 illustrates the efficiency and the net specific work of these systems and compares them to the ICE.

Turbogenerator systems present superior efficiency to the ICE, and the ReGT-TReSRC (CCGT) configuration offers the highest efficiency value of 47.2% and net specific work of 564 kJ/kg among the compared turbogenerators. This gives the CCGT a slight advantage over the IRReGT system; however, when accounting for the system weight, the CCGT is found to be heavier and consequently more costly, given the higher number of system components; which brings the consumption saving advantage back to the IRReGT. These consumption savings are illustrated in Fig. 15 as compared to the ICE.

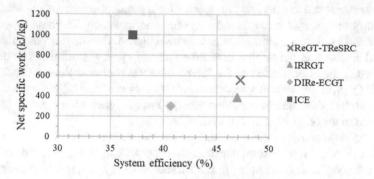

Fig. 14 Efficiency and net specific work of typical conventional engines and turbogenerator systems

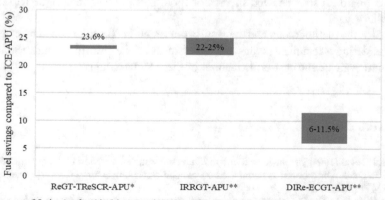

* Fuel savings for 12 kWh battery and 3750 W auxiliary power consumption
** Fuel savings range function of battery capacity (2kWh-20kWh) and 750 W auxiliary power consumption

Fig. 15 Fuel savings of the evaluated GT systems in comparison to the ICE-APU on 5 WLTC

Therefore, turbogenerator systems in general, and the IRReGT system in particular, present a high potential to replace the ICE in series hybrid powertrains in EREV, should the cost-effectiveness of these technologies in replacing the ICE be proven. Thus, an explicit technical–economic study should be conducted to quantify the advantages and drawbacks of using a turbogenerator-APU in terms of fuel savings compared to the added cost of manufacturing, integration, and maintenance of these systems.

6 Concluding Remarks

This chapter presents a methodology for the prioritization and optimization of turbogenerator systems to replace the internal combustion engine in series-hybrid extended-range electric vehicles.

An exergo-technological explicit method considering energy and exergy analysis as well as components and automotive technological constraints is introduced to identify the suitable turbogenerator system configuration. The case of the most relevant three turbogenerators systems was investigated: the Brayton gas turbine, the external combustion gas turbine, and the combined cycle gas turbine. The methodology led to the prioritization of the Intercooled Regenerative Reheat Gas Turbine (IRReGT), given the system's best compromise between high efficiency and power density on one side, and on the other side the limited integration complexity when compared to the other investigated systems.

An extended-range electric vehicle, based on a series-hybrid powertrain was modeled to assess the consumption reduction benefits of the prioritized turbogenerator when replacing the engine in the vehicle auxiliary power unit. Results were compared to those of a reference ICE-APU and the consumption was computed on the WLTC while accounting for the thermal comfort load and other electric auxiliaries onboard. Results showed savings of up to 25%, depending on the trip length and battery size.

Finally, the methodology and the results presented in this chapter can be extended by considering a technical–economic assessment of the prioritized turbogenerator system to evaluate the cost-effectiveness of these technologies.

References

1. Lü, X., et al.: Energy management of hybrid electric vehicles: a review of energy optimization of fuel cell hybrid power system based on genetic algorithm. Energy Convers. Manag. **205**, 112474 (2020)
2. Huebner, G.J.: The Chrysler regenerative turbine-powered passenger car. SAE Technical Paper 640414. SAE World Congress and Exhibition
3. Cunha, H.E., Kyprianidis, K.G.: Investigation of the potential of gas turbines for vehicular applications. Proc. ASME Turbo Expo **3**, 51–64 (2012)

4. Capata, R., Coccia, A., Lora, M.: A proposal for CO_2 abatement in urban areas: the UDR1-Lethe© turbo-hybrid vehicle. Energies **4**(3), 368–388 (2011)
5. History of Chrysler Corporation Gas Turbine Vehicles. Chrysler Corporation (1979)
6. Briesch, M.S., Bannister, R.L., Diakunchak, I.S., Huber, D.J.: A combined cycle designed to achieve greater than 60 percent efficiency. J. Eng. Gas Turbines Power **117**(4), 734–741 (1995)
7. Nada, T.: Performance characterization of different configurations of gas turbine engines. Propuls. Power Res. **3**(3), 121–132 (2014)
8. Bhargava, R.K., Bianchi, M., De Pascale, A., di Montenegro, G.N., Peretto, A.: Gas turbine based power cycles—a state-of-the-art review. In: Challenges of Power Engineering and Environment, pp. 309–319. Springer (2007)
9. Dellenback, P.A.: Improved gas turbine efficiency through alternative regenerator configuration. J. Eng. Gas Turbines Power **124**(3), 441–446 (2002)
10. El-Masri, M.A.: Thermodynamics and performance projections for inter-cooled/reheat/recuperated gas turbine systems. In: Aircraft Engine Marine Microturbines and Small Turbomachinery, vol. 2 (1987)
11. El-Masri, M.A.: A modified, high-efficiency, recuperated gas turbine cycle. J. Eng. Gas Turbines Power **110**(2), 233–242 (1988)
12. Cheng, D.Y., Nelson, A.L.C.: The chronological development of the Cheng cycle steam injected gas turbine during the past 25 years. In: Turbo Expo 2002, Parts A and B, vol. 2 (2002)
13. Sirignano, W.A., Liu, F.: Performance increases for gas-turbine engines through combustion inside the turbine. J. Propuls. Power **15**(1), 111–118 (1999)
14. Andriani, R., Gamma, F., Ghezzi, U.: Main effects of intercooling and regeneration on aeronautical gas turbine engines. In: 46th AIAA/ASME/SAE/ASEE Joint Propulsion Conference and Exhibit (2010)
15. Christodoulou, F., Giannakakis, P., Kalfas, A.I.: Performance benefits of a portable hybrid micro-gas turbine power system for automotive applications. J. Eng. Gas Turbines Power **133**(2) (2010)
16. Capata, R., Sciubba, E., Toro, C.: The gas turbine hybrid vehicle LETHE at UDR1: the on-board innovative orc energy recovery system feasibility analysis. In: Energy, Parts A and B, vol. 6 (2012)
17. Shah, R.M.A., McGordon, A., Amor-Segan, M., Jennings, P.: Micro gas turbine range extender—validation techniques for automotive applications. In: Hybrid and Electric Vehicles Conference 2013 (HEVC 2013) (2013)
18. Mackay, R.: Development of a 24 kW gas turbine-driven generator set for hybrid vehicles. SAE Technical Paper Series (1994)
19. Leontopoulos, C., Etemad, M.R., Pullen, K.R., Lamperth, M.U.: Hybrid vehicle simulation for a turbogenerator-based power-train. Proc. Inst. Mech. Eng. Part D J. Automob. Eng. **212**(5), 357–368 (1998)
20. Lampérth, M.U., Pullen, K.R., Mueller, K.G.: Turbogenerator based hybrid versus dieselelectric hybrid—a parametric optimisation simulation study. SAE Technical Paper Series (2000)
21. Juhasz, A.J.: Automotive gas turbine power system—performance analysis code. SAE Technical Paper Series (1997)
22. Mansour, C.J.: Trip-based optimization methodology for a rule-based energy management strategy using a global optimization routine: the case of the prius plug-in hybrid electric vehicle. Proc. Inst. Mech. Eng. Part D J. Automob. Eng. **230**(11) (2016)
23. Mansour, C., Clodic, D.: Optimized energy management control for the Toyota hybrid system using dynamic programming on a predicted route with short computation time. Int. J. Automot. Technol. **13**(2), 309–324 (2012)
24. Bou Nader, W.S., Mansour, C.J., Nemer, M.G., Guezet, O.M.: Exergo-technological explicit methodology for gas-turbine system optimization of series hybrid electric vehicles. Proc. Inst. Mech. Eng. Part D J. Automob. Eng. (2017)
25. Bou Nader, W.S., Mansour, C.J., Nemer, M.G.: Optimization of a Brayton external combustion gas-turbine system for extended range electric vehicles. Energy **150**, 745–758 (2018)

26. Barakat, A.A., Diab, J.H., Badawi, N.S., Bou Nader, W.S., Mansour, C.J.: Combined cycle gas turbine system optimization for extended range electric vehicles. Energy Convers. Manag. **226** (2020)
27. Deb, K., Member, A., Pratap, A., Agarwal, S., Meyarivan, T.: A fast and elitist multiobjective genetic algorithm. IEEE Trans. Evol. Comput. **6**(2), 182–197 (2002)
28. Mansour, C., Haddad, M., Zgheib, E.: Assessing consumption, emissions and costs of electrified vehicles under real driving conditions in a developing country with an inadequate road transport system. Transp. Res. Part D Transp. Environ. (2018)
29. Basma, H., Mansour, C., Haddad, M., Nemer, M., Stabat, P.: Comprehensive energy modeling methodology for battery electric buses. Energy **207**, 118241 (2020)
30. Mansour, C., Bou, W., Breque, F., Haddad, M.: Assessing additional fuel consumption from cabin thermal comfort and auxiliary needs on the worldwide harmonized light vehicles test cycle. Transp. Res. Part D **62**, 139–151 (2018)
31. Sundström, O., Guzzella, L.: A generic dynamic programming Matlab function. Proc. IEEE Int. Conf. Control Appl. **7**, 1625–1630 (2009)

On the Road Towards Zero-Prototype Development of Electrified Powertrains via Modelling NVH and Mechanical Efficiency

S. Theodossiades⬤, N. Morris⬤, and M. Mohammadpour⬤

Abstract In this chapter we review, discuss and critically evaluate the progress made towards zero-prototype development of electrified powertrains. The chapter will focus on fully electric powertrains rather than any combination of electrical and internal combustion engine power units (hybridisation). Emphasis will be given on disciplines that have previously, in the first instance, been looked at in isolation e.g. dynamics without coupling to tribological effects and powertrain transient dynamics without coupling to motor electromagnetics. Computational, numerical, and analytical simulation methods will be evaluated for prediction of mechanical efficiency and wear, as well as Noise, Vibration and Harshness. Areas in the available literature that require further investigation will be highlighted, with suggestions for future developments and directions of future research.

Keywords Electrified powertrains · Mechanical efficiency · NVH · Zero-prototype

1 Introduction

The uptake of electric vehicles globally is accompanied by an ever-increasing demand and supply of new vehicle models. As an example, in 2017 and 2019 only 6 and 5 new battery electric vehicle models respectively were introduced to the European market; however, it is expected that by 2025 there will be at least 172 different models available to consumers [1]. The new models are required to overcome the barriers to the adoption of electric vehicle technology, with the most significant of these being: cost, autonomy range and battery charge time [2]. To develop such a variety of functional (and marketable) new vehicle models, in such a limited time frame and without the legacy of a long-term engineering refinement constitutes a significant opportunity but also an immense engineering challenge.

To meet such a challenge, multi-physics modelling and virtual prototype development must be integrated within each step of the design process in order to establish

S. Theodossiades (✉) · N. Morris · M. Mohammadpour
Loughborough University, Loughborough, Leicestershire L11 3TU, UK
e-mail: S.Theodossiades@lboro.ac.uk

© The Author(s), under exclusive license to Springer Nature Switzerland AG 2022   267
T. Parikyan (ed.), *Advances in Engine and Powertrain Research and Technology*,
Mechanisms and Machine Science 114,
https://doi.org/10.1007/978-3-030-91869-9_11

the integrity, functionality and robustness of the product well ahead of manufacturing. This complete assimilation of design and multiphysics simulations of (battery) electric vehicle powertrains (with very limited development and testing of physical prototypes, also known as zero-prototype development) gives the aspirational focus of this chapter.

In this chapter we focus on two specific aspects of mechanical powertrain design and development, namely Noise, Vibration and Harshness (NVH) and mechanical efficiency (tribology and wear). These two critical aspects of electric powertrain development were chosen to provide a case study of the zero-prototype concept. These subject areas often present contradictory design challenges and, therefore, only through informed and salient multiphysics simulations it is possible to arrive at highly refined solutions within short and restricted time spans. Furthermore, NVH and mechanical efficiency are of key importance to the success of any new electric vehicle model. The NVH refinement of a vehicle is the hallmark of its quality, as the removal of broadband noise masking created by the Internal Combustion Engine (ICE) power unit exposes and promotes NVH in electric vehicles. This is of particular importance due to the objectionable, tonal nature, of many common electric vehicle noise issues [3]. End-users (customers) are widely expecting electric vehicles to be "silent", without any powertrain noises. However, unwanted noises may surface in the late vehicle development phases and can only be suppressed with palliative measures. A more sophisticated approach calls for improved CAE-optimization and definition of design criteria (targets) for unacceptable noises. Electric motor *whistling* [3], transmission *whining* [4], *squeak rattle* and *buzz* [5] are typical NVH examples in electric powertrains.

The integration of electric drive systems that transmit significant amounts of power face challenges with mechanical oscillations caused by electromechanical interactions. The resulting severe dynamic loads accelerate component wear, potentially leading to unstable operation and failure [6–8]. On the other hand, powertrain mechanical efficiency and tribology are of paramount importance, as energy is stored at a premium in a battery electric vehicle (as a result of the batteries' specific energy density) and therefore energy consumed by frictional losses is highly undesirable. It is estimated that the e-motor friction and powertrain transmission account for approximately 1% and 3% of the energy consumption, respectively [9]. On the other hand, the reliability of a vehicle powertrain is frequently limited by the characteristics of its surfaces in relative motion and therefore consideration of tribological wear is also a key concern.

This chapter presents three sections, each focusing on the above mentioned design aspects, with the aim of reviewing the state of the art literature in battery electric vehicle powertrain simulations. The intensification of the race to bring products to market and the short amount of time that has so far been dedicated to electric vehicle investigations determine the research and development landscape to be in a formative state. Thus, the authors have noted only a limited number of publications available in the open literature, where the fundamentals of the underlying science are discussed.

2 Noise, Vibration and Harshness (NVH)

The past literature on electric powertrain modelling and virtual prototyping can be classified into various headings depending on the features of the simulation models and the methodologies employed. An attempt is made below to present some key categories:

- Component-level versus System-level modelling approaches.
- Structural dynamics (involving quasi-static and/or transient analysis) and/or Noise analysis.
- Simulation models involving multiple disciplines (Multidisciplinary), different scales of motion (Multiscale) and various physics (Multiphysics) in an integrated manner as coupled (interacting) analysis and simulations.
- Numerical versus Analytical methods, simulation strategies and techniques.

An example of detailed system modelling and simulation flowchart for electric powertrains containing several key features is depicted in Fig. 1. Variations can be found in the literature depending on the specific aims and objectives of each investigation [10–12].

The reader should bear in mind that it can be quite challenging (and perhaps meaningless) to allocate a piece of work under a single of the above mentioned categories, since the majority of literature can combine several key features. A generic description of the past literature will be provided thereafter, with references to the above headings where appropriate.

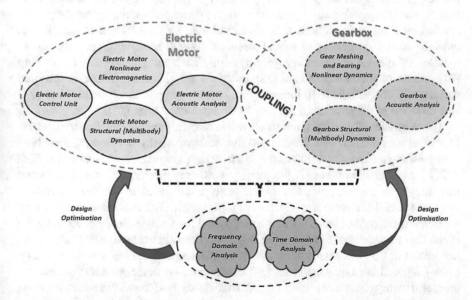

Fig. 1 Typical flowchart of electric powertrain system modelling and simulation

A typical approach in electric powertrain (simulation) investigations (as well as in internal combustion engine driven powertrain simulations) would be to consider and model specific components and/or subsystems in isolation, establishing some type of connection (using boundary conditions) with the rest of the powertrain. In this respect, several authors have focused on developing electromechanical models of the **electric motor system** without attaching this to the drivetrain [7, 8]. In other attempts, the drivetrain was represented by lumped parameter models connected to the electric motor. More specifically, detailed models of **gear system** oscillations have been studied in the context of electric drivetrains by [13–15]. The multibody dynamics of a wind turbine drivetrain, focusing on dynamic gear mesh forcing and transmission error were studied in [16]. An electric traction gearbox dynamic model was integrated with locomotive vertical dynamics to study the system's oscillatory response under combined excitation due to gear meshing and wheel rail motion [17]. The transient oscillations of a single-stage gearbox during electric motor start-up were studied [18] whereas a numerical tool to provide quick estimation of the optimal transmission layout for electric powertrains was presented [19]. In the latter, two-dimensional representations were used, supported by analytical calculations to estimate gear widths, bearing loads and their dimensions. A coupled gear–shaft–bearing–housing model was developed for an electric vehicle differential, allowing for component flexibility and elastic supports [20]. Gear tooth modification schemes were explored to highlight the effects on teeth contact pattern and dynamic transmission error as metrics for vibration and noise.

Lumped parameter models have been developed to simulate the dynamics of **planetary gear systems** (which are gradually becoming an attractive configuration for electric powertrains) by considering the generated oscillatory motions only (excluding rigid-body motions that are introduced by the rotation of the electric machine rotor). A number of model attempts contain sensitivity analysis of key features of interest in planetary gear systems, such as: gear teeth load sharing [21], modes of oscillatory motion [22], tooth wedging, bearing clearance and time varying stiffness effects [23], time varying pressure angles and contact ratios [24] and variation in the position of the gear meshing contact lines [25].

The **acoustics of gear systems** employed in electric powertrains have been investigated using commercial software and dynamic simulations. Parameters such as gear meshing stiffness, transmission error, bearing parameters and component flexibility have been considered [26] to predict oscillatory motions and sound radiation response in an electric powertrain. Gear teeth micro geometry parametric studies were conducted for noise and vibration optimization. The oscillatory response of the component surface led to sound pressure level far-field results by combining the Finite Element Method (FEM) and the Boundary Element Method (BEM). A numerical model of a helical gear system was developed to study the mechanism of *whine* noise [27], validated by modal testing. The structure-borne noise of the system was predicted using a multibody finite element model and the dynamic transmission error as excitation. The effects of gear parameters, such as pressure angle, helix angle and contact ratio on *whine* noise were examined (see Fig. 2 [27]).

Fig. 2 Gearbox radiation
noise during
macro-parameter study: **a**
Pressure angles
(15.5°—dashed line,
18.5°—solid line and
21.5°—dotted line), **b** Helix
angles (20°—dotted line,
24°—solid line and
28°—dashed line) and **c**
Overall contact ratios
(1.9—solid line, 2.0—dotted
line and 2.2—dashed line)
[27]

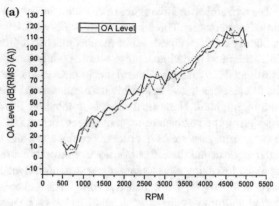

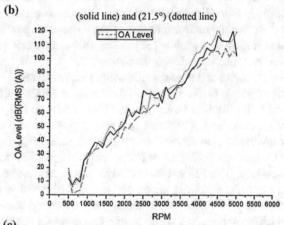

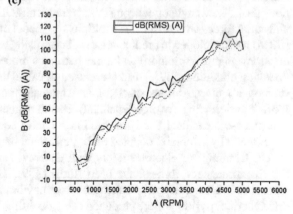

The **dynamics and acoustics of electric motors** have naturally attracted the attention of researchers. The effects of rotor eccentricity on the magnetically induced oscillatory motion of permanent-magnet electric motors were predicted analytically [28]. The sound radiation of pulse-width modulation (PWM) induction machines was investigated [29] using analytical mechanical and acoustics two-dimensional models. The interactions between audible harmonics in the electromagnetic noise spectrum were highlighted. Numerical models combining FEM and BEM were developed, considering the permanent-magnet direct current (PMDC) motor electromagnetics and structural mechanics in order to predict noise and vibration response [30]. Transient response and the effect of rotor eccentricity were investigated. A multiphysics approach involving electromagnetics, structural dynamics and acoustics (via FEM) was developed [31] to simulate dynamic loads and noise radiation in electric motors. The methodology assumes weak coupling between the various physics involved in the investigation. The stability of the nonlinear, coupled lateral/torsional response caused by electromagnetic excitation in *permanent magnet synchronous motors* (PMSM) was studied analytically [32]. A key aim of the work was to control rotor eccentricity and parametric studies were conducted.

The noise and vibration in PMSMs was analysed by examining the generated electromagnetic forces using two-dimensional FEM and an analytical model to predict the radial motion along the teeth of the stator [33]. Different PMSM topologies were examined. The noise generated by an electric motor has been calculated via FEM/BEM simulations, using the dynamic response of the structure due to the electromagnetic force [34]. The motor's modal response has been experimentally obtained. The PMSM rotor design was optimised to improve the generated noise using FEM and wave shape optimisation of the air-gap magnetic density [35]. The electric motor electromagnetics were predicted using two-dimensional FEM and the vibroacoustic response was investigated for different slot/pole number combinations [36]. Skew angle effects on NVH were highlighted, as well as the influence of tangential forces on acoustics. A Multiphysics model of a switched reluctance motor (SRM) was developed to predict noise radiation using two-dimensional FEM (for the electromagnetics excitation) and a mechatronics model to simulate the appropriate operating conditions [37]. BEM were used to calculate sound power. Multiphysics (numerical) investigations of SRM and mutually coupled SRM were presented [38]. The occurring (two-dimensional) electromagnetics, mechanical oscillations and (three-dimensional FEM) radiated noise, have been investigated and validated experimentally. Mutually coupled SRM performed better than SRM.

The dynamics of electric motors with different slot/pole configurations due to the electromagnetic forces have been studied [39]. The work aims to distinguish resonance and torque ripple from other sources of vibration and noise. Radial forces at different rotor positions were predicted using two-dimensional FEM simulations (for the electromagnetics and the structure) to identify vibration modes. A comparative study of the NVH performance between interior PMSM, induction and switched reluctance motors has been conducted using FEM electromagnetics [40]. The options of distributed and concentrated windings have been considered. NVH analysis of an e-truck electric motor has been presented [41], where four different motors of similar output power and torque have been discussed using FEM electromagnetics with respect to their slot and pole ratio characteristics. Design guidelines to reduce motor output perturbations (ripples) have been proposed. An analytical model was developed to investigate the root causes behind unbalanced magnetic pull (UMP) in permanent magnet motors due to magnetic asymmetry or static rotor eccentricity [42]. UMP often leads to high loads and potentially wear. The method was validated against FEM results. An analytical method to predict the unbalanced magnetic force (UMF) of permanent magnet electric motors with eccentricity has been developed [43]. The method has examined both rotating asymmetric and symmetric electric motors and was validated against FEM and experimental measurements.

Generally, the electric motor sound comprises aerodynamic, electromagnetic and mechanical components [44, 45]. The aerodynamic noise is dominant at high motor speeds, whereas electromagnetic noise is generated by the operation of the power electronic converter. The mechanical noise originates from the stator and is dominant at intermediate motor speeds. Popular (deterministic) methodologies to predict low frequency noise response are the FEM and BEM approaches [46]. On the other hand, Statistical Energy Analysis (SEA) is effective at higher frequency spectra [47]. The medium frequency spectrum is more difficult to predict due to the modal density not being sufficiently high (for employing SEA), neither too low for allowing the evolvement of separated modes (for BEM or FEM methods). This deficiency is dealt with by various methods [48–50] (see Fig. 3 [49]).

The **coupling** between the dynamics of an electric motor and the flexibility of the powertrain structure (casing) were considered to investigate electric powertrain NVH, using a two-dimensional electromagnetic model (electric motor) and a three-dimensional structural FEM (casing) [51]. The electric motor electromagnetics and the coupled (rotational and translational) motions of a multi-stage gearbox (electromechanical interactions) were modelled numerically in an electric drivetrain context [6]. The effect of electromagnetics on the system natural frequencies was investigated, as well as that of mechanical vibrations on the electric motor (see Fig. 4 [6]).

The electric motor *whistling* noise was modelled, calculating the generated electromagnetic forces and the occurring powertrain structural response [52]. A two-dimensional electromagnetic model was coupled to the three-dimensional structural FEM model for predictions up to 4.5 kHz. NVH predictions were conducted using an electric driveline with nonlinear electromagnetic (FEM) and mechanical (multi-body dynamics) excitation [53]. The effects of eccentricity and rotor skew angle were examined. The normal surface velocity of the structure was used for sound calculation purposes. In a later publication [12], NVH simulations and validation of a system equipped with an interior permanent magnet electric motor, a gearbox and inverter were conducted. The combination of electrical and mechanical excitations led to a linear system in the frequency domain.

Simulations of the radiated noise were conducted in a complete powertrain [54] by using a frequency domain methodology, whereas the dynamic response (NVH) of an electric powertrain was determined by coupling a two-dimensional electromagnetic

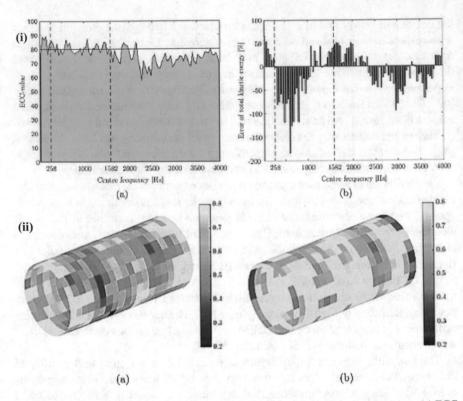

Fig. 3 **i** Correlation of cylindrical shell test-structure with interpolated dynamic response (a) ECC applied on the kinetic energies and (b) Kinetic energy error, **ii** (a) Spatial ECC without interpolated responses and (b) Spatial ECC with interpolated responses [49]

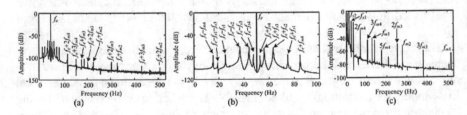

Fig. 4 Dynamic response of the electrical signals in the frequency domain at 1475 rpm: **a** stator phase current; **b** stator phase current in the frequency range of (0–100 Hz); **c** electromagnetic torque [6]

model with a three-dimensional structural FEM [55]. In a later work [10], the vibroacoustics of an electric powertrain comprising a PMSM and gearbox were determined. A two-dimensional electromagnetics transient solver combined with a mode superposition method contributed to the input conditions for powertrain noise predictions by BEM. Indeed the state-of-the-art to NVH simulations of electric powertrains is a coupled electromechanical analysis within a multiphysics environment. However, the computational burden of this approach has forced researchers to alternatives such as a combination of (faster) analytical and numerical approaches [56].

As a final point in this section, **bearings** constitute an important subsystem in every powertrain applications. Some representative literature on the subject area of bearing dynamics was deemed necessary in order to complete the picture, since the role of bearings in vibration (and structure-borne noise) transmissibility is crucial. The dynamics of bearings within the context of a gear system have been investigated considering component flexibility [57]. Vibration transmissibility has been predicted using lumped parameter models and FEM. A dynamic model of deep-groove ball bearings was developed including local nonlinearities (Hertzian and elastohydrodynamic contacts) of the bearing elements [58]. The effects of waviness and localized defects were also considered in this work.

3 Mechanical Efficiency and Tribology

In this section, the effect of electrification on the tribological aspects of powertrains, specifically on frictional power loss is discussed. The tribological behaviour of different conjunctions in components of electric powertrains is significantly altered in comparison to ICE driven powertrains due to reduction (or complete elimination) of reciprocating parts and higher pressure in the component contacts. It was expected to gain significant amount of frictional efficiency in electrified systems, but these gains were not quantified until comprehensive scientific research was devoted in recent years to characterise the nature of losses in these modern powertrains. A clear picture of the impact of electrification on frictional losses, comparing with equivalent ICE powered vehicle was provided in [9]. Their analysis, based on assumption of a light electric vehicle with 75 kW electric motor and Li-ion battery technology showed that frictional losses will account for 57% of all losses in that case study (Fig. 5). Out of this, 41% was attributed to the rolling resistance, 1% to motor friction, 3% to transmission losses, 19% to brakes and 17% to air drag. As it can be observed, the ratio contribution of frictional losses from the power source significantly reduces in electric powertrains. The losses associated with electric motor are only 6% in comparison with 16.5% for ICEs.

Despite the overall picture provided by Holmberg and Erdemir, specific components and subsystems such as gear pairs and bearings might be more vulnerable to frictional losses in electric vehicles, revealing higher losses [59, 60]. This is shown for a planetary gear set under ICE and electric motor powered modes [59], where the system shows higher inefficiency under the electric motor mode despite better NVH refinement as shown in Figs. 6 and 7. The figure shows the dynamic transmission error (DTE) which is the instantaneous approach between the engaged gear teeth. It is shown that the electric motor mode provides less teeth separation (DTE < 0) as a source of gear *whine* as well as lower DTE peak-peak amplitude. But, for the same speed sweep, the electric motor mode provides overall higher inefficiency

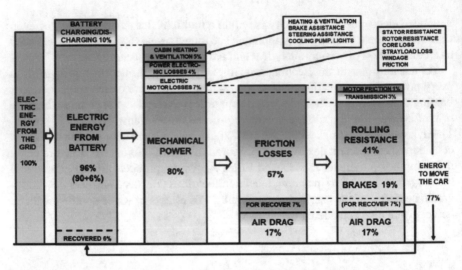

Fig. 5 Comparison of power loss sources in an electrified vehicle [9]

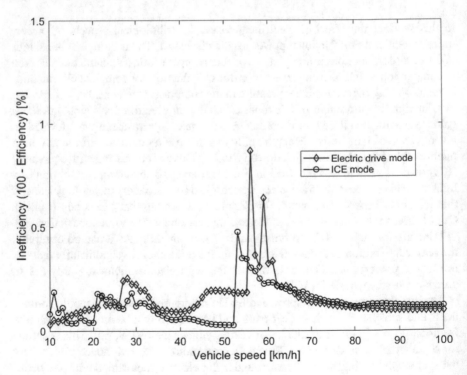

Fig. 6 Frictional inefficiency under IC and EM modes [59]

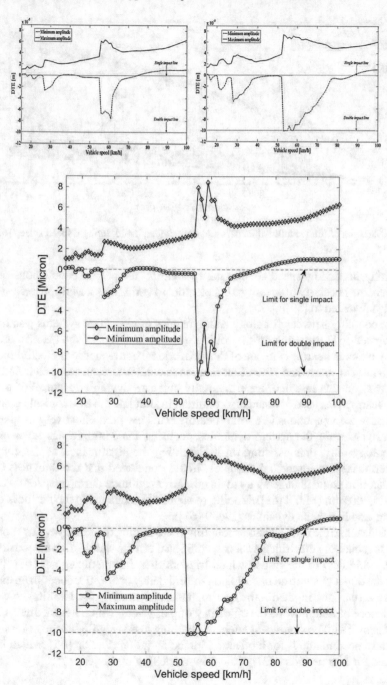

Fig. 7 DTE under EM (top) and IC (below) modes [59]

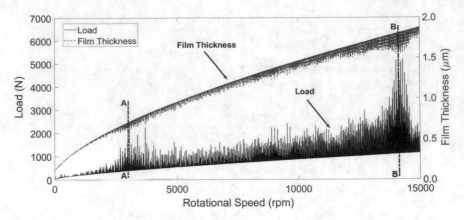

Fig. 8 Variation of lubricant film thickness and contact load based during a speed sweep [61]

especially under low-speed conditions. In addition to specific conclusions of this research, the necessity of investigating electric powertrain as a Multiphysics system was also revealed in [59].

The coupling between tribology and dynamics of electrified systems was further investigated [61], focusing on the rolling element bearings of the electric motors. Rolling element bearings are one of the most critical components of electric motors, responsible for more than 70% of lubrication related failures in some electric motor types [62]. In this research, experimentally obtained dynamic response of a high-speed bearing was used as boundary condition to simulate the tribological behaviour of lubricated conjunctions in a roller bearing. This research effectively used experimental values for the system dynamics, focusing the Multiphysics investigation to tribology only, thus assisting on highlighting the significance of the coupling between these two research disciplines. It was concluded that the thickness of the lubricant film could change by up to an order of magnitude during a speed sweep of up to 15,000 rpm (Fig. 8 [61]), leading to significant changes in the stiffness of the bearing, and hence the dynamics (Fig. 9 [61]).

In addition to the tribo-dynamic coupling necessitating the Multiphysics approach towards modelling electrified systems, additional effects should also be considered, such as fluid flow in the system which triggers churning inefficiencies [63]. In this work, the effect of sump oil churning on the efficiency of speed reducer in an electric powertrain was investigated both experimentally and numerically. Results showed the significance of fluid flow on the efficiency of the system, which in turn affects NVH refinement [59]. The significant effect of churning losses highlights the necessity for more advanced methods for lubrication including jet lubrication [64], which brings the need for 3-dimentional (3D) fluid flow modelling of the jet.

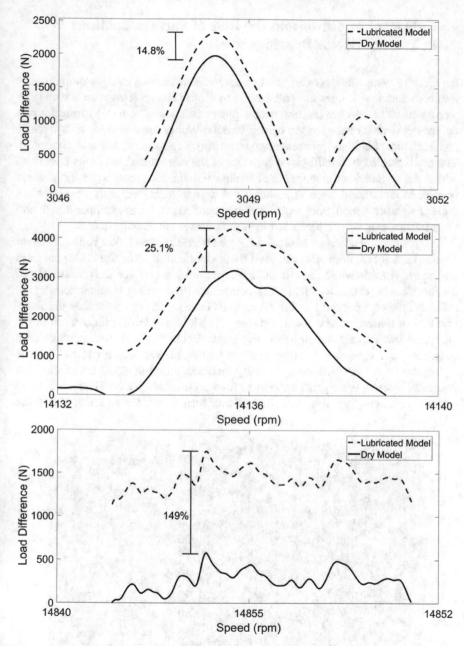

Fig. 9 Effect of inclusion of lubricant film on the dynamic load in contact [61]

4 Lubricant Additives and the Role of Surface Coatings in Traction Motor Bearings

An array of compact integrated power unit and transmission designs are currently available and new designs are still emerging. These designs comprise a variety of arrangements of: electric traction motor power unit, speed reduction transmission and power division between the driven wheels. Within these designs heat transfer and lubricating fluids and greases are engineering components of critical importance. The multi-faceted and multi-physics nature of the roles that these fluids undertake within the system determine that their inclusion in the zero-protype design process is of paramount importance to a successful design. Simultaneously, these factors present significant modelling and simulation challenges due to multiple length and temporal scales at which the multi-physics processes are coupled across.

In both the lubricating greases used for the electric traction motor rolling element bearings and for the fluid lubricant used for the lubrication of the transaxle gear train and bearings as shown in Fig. 10, additives are used to enhance the capabilities of the lubricant beyond that of its fluidic component. The rolling element bearing are commonly grease packed and therefore must have a life span to enable the functionality of the component for its service life. The transmission fluid services the gear train but also acts as an automatic transmission fluids in two speed electric transmission designs where it also services multi-plate wet clutch packs. The role of lubricant anti-wear additives, viscosity modifiers, friction modifiers, dispersant, foam inhibitors and detergents are critical for system durability, NVH behaviour and sustained efficiency performance. The additives form part of a complex fluid with the

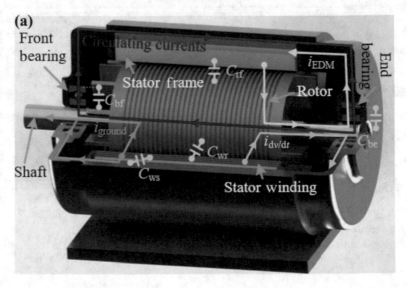

Fig. 10 Bearing currents, their paths and capacitive conjunctions of an induction motor taken from He et al. [73]

constituent molecules interacting synergistically and antagonistically to the desired constituents objective in isolation. The additives interact with each other, but also with the surfaces which they lubricate and are bounded by. The observed behaviour of the surface active additives changes with the thermal conditions and shear stress to which they are subjected. The wear behaviour of high loaded conjunctions can be ameliorated by thermally activated and stress enhanced growth of nanoscale anti-wear additive films [65]. The surface topography stabilizes when the simultaneous rate of removal through mechanical abrasion of contiguous surfaces and adsorption are equal. The lubricant additive pack also plays an important role in controlling the frictional behaviour of multi-plate wet clutch packs used to facilitate two speed transmission architectures. The design of various shift strategies for multispeed transmission have been demonstrated by Walker et al. [66] using a multi-body dynamic using constant friction coefficient and an electrical machine model. It has been shown by Lou et al. [67] that the magnitude of the friction, and friction coefficient variation with relative surface speed must be carefully selected to avoid stick–slip vibrational behaviour. In addition to the stick slip clutch vibrational behaviour other coupling instabilities resulting from frictional interactions of the slipping discs should also be accounted for using suitable friction models in dynamical simulations [68, 69].

The zero-prototype design including thermo-mechanical activation of surface active molecules, mutual additive interactions and mechanical and electrical wear phenomena present the electrical vehicle designer a significant challenge. The challenge derives from the intensity and duration of the current state the of art computational molecular dynamics simulations required to determine the behaviour of complex molecules across realistic time scales and geometries. To address the limited time and length scales which can be approximated using molecular dynamics simulations, advanced approaches have been designed to enable prediction of component scale engineering tribological challenges [70]. The challenge is addressed by using hierarchical or hybrid schemes to couple (where appropriate, at least), the physical mechanisms apparent at separate scales. The behaviour of additive tribo-films has been well described by Arrhenius type, shear stress enhanced, thermal activation models [65, 71]. The inclusion of abrasive wear through Archard type wear models has, for simplified systems at least, presented promising results [72]. For complex transmission systems in which boundary conditions are unknown or poorly defined determination of the various coefficients for complex and transient systems prior to experimentation is a challenging task. Multiple other factors related to the surface condition, formation of nascent surface sites, lubricant degradation and contamination further complicate the issue. As a result, this is an area where some level of component level experimental development will be necessary for the foreseeable future.

Critically, for an electrical power train the introduction of additives and impurities to lubricants has been shown to significantly affect the electrical properties of the lubricated conjunctions [74]. This can be of particular importance due to the presence of shaft voltages that result from "electromagnetic induction, electrostatic coupled from internal sources or electrostatic coupled from external sources" (Erdman et al. [75]). An illustrative diagram of the current flow paths and capacitance of an electric

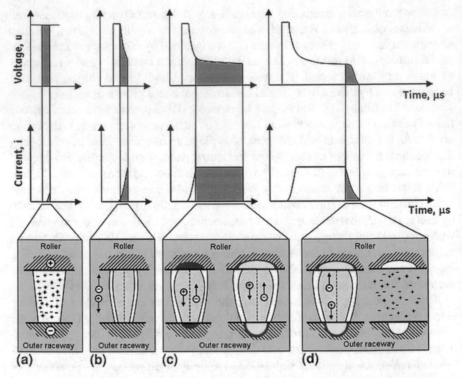

Fig. 11 Electrical discharge machining mechanism From [76] and originally edited from [77]

motor is provided He et al. [73] in Fig. 10. The dielectric strength and conductivity of the lubricant which intercedes the two surfaces is critical to controlling electrical pitting and fluting (electrical discharge machining) that can occur due to sufficient electrical current flow across the interface [75]. A mechanism for Electrical discharge machining is shown in Fig. 11.

Once initiated the wear can be propagated through mechanical adhesion. In addition, it has been shown that electrical wear can affect the composition of organic compounds found within greases [78]. Any change in grease electrical impedance alters the prevalent mechanism of rolling element bearing wear [79]. Clearly, the tribological and electrical phenomena present is, at least to some degree, coupled. More recently, the formation of micro and nano bubbles at electrically charged elastohydrodynamic point contacts have been observed. The rate of bubble formation like fluting behaviour previously has been shown to correlate with current density at the tribological bearing interface [80, 81]. These studies have shown that the introduction of an anti-wear additive to the lubricant increases peak current densities across the tribological conjunction with a corresponding increase in the rate of nano-bubble production. This again shows a coupling between the electrical contact behaviour and the tribological conjunction. An electrical model and reduced order model of the rolling element shaft bearings and electrical machine by Erdman et al. [75] showed an effective approach to capture discharge events across rolling element bearings in use with electrical machinery.

Other experimental research has clearly associated variations in additive chemistry to the fatigue life of tribological conjunctions exposed to electrical currents as shown in Fig. 12 [82]. Similar to the electrical properties of the intervening fluid film, surface adsorbed anti-wear films are known to tenaciously grow on suitable surfaces to such an extent that a measurable change in the surface roughness can be observed [83]. The surface roughness has been shown to affect the local charge density and field strength by Lazic and Persson [84] by solving the electrostatic rough surface

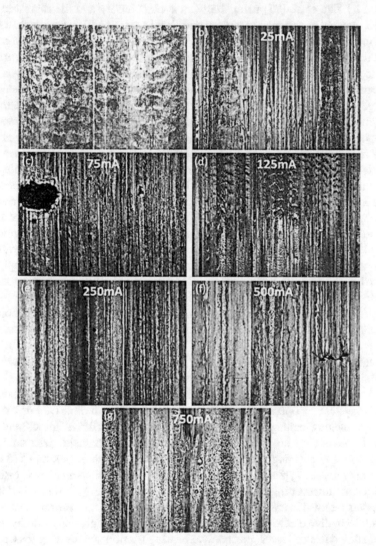

Fig. 12 Surface images showing wear and tribofilm adsorption after exposure to different currents. Figure taken from [82]

problem using a boundary element method. This shows a further coupling of surface active additive absorption and electrical properties of the electrical powertrains tribological conjunctions. Any zero-prototype development strategy should be capable of predicting the electrical contact properties at a variety of surface roughness, lubricant additive compositions and throughout the degradation of the components that may result during the bearings operation.

Aside from the lubricant itself other aspects of the tribological contacts of the contiguous bearing's surfaces contribute to the electrical properties of the conjunction. The capacitance of rolling element bearing electrical conjunction has been predicted by Zhu et al. [85] using Hertzian contact theory and the regressed elliptical EHL contact equations of Dowson and Higginson [86]. This approached linked capacitance shaft speed and load which control lubricant film thickness and the contact footprint. The current density is closely linked with fluting behaviour and is shown to vary with contact patch area [87–89]. Close attention must be given to the engineering design challenge posed by the objectives of reducing unwanted electrical discharge through low resistance discharge paths and the build-up of charge that can lead to the dielectric strength of the contact being overcome and sudden electrical discharge [90]. Zika et al. [91] demonstrated experimentally that bearings exposed to high frequency pulsed currents experienced earlier wear behaviour than those exposed to direct or lower frequency alternating current. While compatibility with exposed material of the electrical system is critical, electric vehicles engineers can use a variety of surface coatings at their disposal to refine the frictional, wear and electrical properties of powertrain bearings. The use of non-conductive ceramic rolling components in bearings or electrical isolation of the bearing can prevent unwanted electrical discharge across the bearing conjunctions [92]. Such steps however need to be considered in context of the system as the discharge currents can simply relocated to the next path of least resistance [93]. Busse et al. [89] presented an reduced order electrical model capable of determining the role of system parameters such as stator/frame and stator rotor capacitance. Clearly, the details of the tribological conjunction should be considered in the simulations of the electrical system for an effective zero prototype methodology.

Air, lubricant and bespoke heat transfer fluids are frequently employed within the power unit and transmission/transaxle unit to maintain suitable thermal conditions for both the mechanical, magnetic and electrical components. Excessive temperatures cause the system to operate inefficiently and can affect durability. For example: permanent magnet demagnetisation, increase bearing frictional losses and Joule heating. Lubricant is a key component for heat transfer of parasitic frictional losses sourced from the meshing gear teeth, shaft bearings, clutch engagement (2 speed) and churning losses. Typically, the Battery electric vehicle transmission is estimated 95% efficient, however this a sizable proportion of input energy is dissipated as heat and requires removal to maintain a stable temperature. There are various heat sources such as: dissipative electrical resistive losses in the windings, magnetic hysteresis, induced eddy currents losses, mechanical bearing friction and air drag losses.

The electric vehicle traction motor can either be indirectly cooled through a series of coolant galleries in the housing or coolant jacket and/or rotor. In addition, direct cooling of the motor surfaces can enhance heat transfer and limit maximum temperature excursions [94]. Rotor cooling is of particular importance in brushless motors, where the permanent magnet is at risk of demagnetisation if excessive temperatures are reached.

A common heat transfer fluid for the e-motor cooling and lubrication of the transmission system is a potentially advantageous but challenging possibility due to the requirement of surface-active additives for the transmission systems and their electrical behaviour and material compatibility with the electrical components of the traction motor. Highly effective simulation techniques are available to predict detail electro-magnetic coupling to the motor thermal management using FEA [95] and thermal resistant networks [96–98]. Thermal analyses have also be shown to be critical to predict Wheel hub mounted motor cooling [98]. Thermal network models have been shown to be time efficient and a reasonably accurate tool for analysis of electric vehicle transmissions [99].

5 Suggestions for Future Developments

The authors hope that this (review) chapter has clearly outlined the diverse fields of physical sciences and engineering that must be considered, and in many cases coupled, to design electric vehicle powertrains to the expected quality. The inclusion of essential system physics such as dynamics, tribology, electromagnetics and fluid flow under the same platform is therefore a critical step towards the zero-prototype design objective. The challenges posed by the physical behaviour that takes place from nano-seconds to years of operation (in the time domain) and from nanometres to meters (in the space domain) create significant challenges in terms of developing time efficient and accurate solutions of these coupled problems. Model reduction techniques or methods such as Artificial Intelligence (AI) to enhance computational efficiency of such complex systems (at system level) are emerging promising research fields (techniques) to capture the key physical behaviour in a time efficient manner. For the moment at least while these coupled Multiphysics models are still yet to fully mature and are under development there is still room for experimental validation and testing of the available models for the simulation and design of the electric vehicles.

To conclude, further developments in the below aspects seem inevitable:

- Inclusion of required physics such as dynamics, tribology, electromagnetics and fluid flow under the same modelling (simulation) platform.
- Inclusion of the system's local nonlinearities.
- Development of suitable and accurate metrics for NVH and mechanical efficiency.
- Model reduction techniques or other methods such as Artificial Intelligence (AI) to enhance computational efficiency of electric drivetrains at system level.

References

1. Bannon, E.: Electric surge: carmakers' electric car plans across Europe 2019–2025. In: European Federation for Transport and Environment. https://www.transportenvironment.org/public ations/electric-surge-carmakers-electric-car-plans-across-europe-2019-2025. Last accessed 1 June 2020
2. BP Castrol: Accelerating the evolution: the tipping points to mainstream electric vehicle adoption (2020). https://bit.ly/3ifaAPS. Last accessed 1st Oct 2020
3. Wellmann, T., Tousignant, T., Govindswamy, K., Tomazic, D., Steffens, C., Janssen, P. (2019) NVH aspects of electric drive unit development and vehicle integration (No. 2019-01-1454). SAE Technical Paper
4. Lepoittevin, G., Horak, J., Caprioli, D.: The new challenges of NVH package for BEVs. SAE Technical Paper 2019-01-1452 (2019). https://doi.org/10.4271/2019-01-1452
5. Gavric, L.: NVH refinement issues for BEV. In: Siebenpfeiffer W. (eds.) Automotive Acoustics Conference 2019. Proceedings. Springer, Wiesbaden (2020)
6. Yi, Y., Qin, D., Liu, C.: Investigation of electromechanical coupling vibration characteristics of an electric drive multistage gear system. Mech. Mach. Theory **121**, 446–459 (2018)
7. Gustavsson, R.K., Aidanpää, J.: The influence of nonlinear magnetic pull on hydropower generator rotors. J. Sound Vib. **297**(3–5), 551–562 (2006)
8. Chen, X., Yuan, S., Peng, Z.: Nonlinear vibration for PMSM used in HEV considering mechanical and magnetic coupling effects. Nonlinear Dyn. **80**(1–2), 541–552 (2015)
9. Holmberg, K., Erdemir, A.: The impact of tribology on energy use and CO_2 emission globally and in combustion engine and electric cars. Tribol. Int. **135**, 389–396 (2019)
10. Fang, Y., Zhang, T.: Vibroacoustic characterization of a permanent magnet synchronous motor powertrain for electric vehicles. IEEE Trans. Energy Convers. **33**(1), 272–280 (2018). https://doi.org/10.1109/TEC.2017.2737483
11. Meier, C., Lieske, D., Bikker, S.: NVH-development of electric powertrains—CAE-methods and NVH-criteria. SAE Technical Paper 2014-01-2072 (2014). https://doi.org/10.4271/2014-01-2072
12. Mehrgou, M., Garcia de Madinabeitia, I., Graf, B., Zieher, F., et al.: NVH aspects of electric drives-integration of electric machine, gearbox and inverter. SAE Technical Paper 2018-01-1556 (2018). https://doi.org/10.4271/2018-01-1556
13. Ramtharan, G., Jenkins, N., Anaya-Lara, O.: Influence of rotor structural dynamics representations on the electrical transient performance of FSIG and DFIG wind turbines. Wind Energy **10**(4), 293–301 (2007)
14. Mandic, G., Ghotbi, E., Nasiri, A.: Mechanical stress reduction in variable speed wind turbine drivetrains. IEEE Energy Convers. Congr. Exposit. **27**(8), 306–312 (2011)
15. Chen, K., Hu, J., Peng, Z.: Analysis of torsional vibration in an electromechanical transmission system. Ad. Mech. Eng. **8**(6), 1–9 (2016)
16. Qin, D., Wang, J., Lim, T.C.: Flexible multibody dynamic modeling of a horizontal wind turbine drivetrain system. J. Mech. Des. **131**(11), 114501 (2009)
17. Chen, Z., Zhai, W., Wang, K.: A locomotive-track coupled vertical dynamics model with gear transmissions. Veh. Syst. Dyn. **55**(2), 244–267 (2017)
18. Khabou, M.T., Bouchaala, N., Chaari, F.: Study of a spur gear dynamic behavior in transient regime. Mech. Syst. Sig. Process. **25**(8), 3089–3101 (2011)
19. A. Barreng, Optimization tool for EV-transmissions, 2018. M.Sc. Thesis 2018:96. Department of Mechanics and Maritime Sciences Division of Combustion and Propulsion Systems, Chalmers University of Technology, Göteborg, Sweden
20. Xiang, C., Song, C., Zhu, C., et al.: Effects of tooth modifications on the mesh and dynamic characteristics of differential gearbox used in electric vehicle. Iran J. Sci. Technol. Trans. Mech. Eng. **43**, 537–549 (2019). https://doi.org/10.1007/s40997-018-0176-7
21. Kahraman, A.: Load sharing characteristics of planetary transmissions. Mech. Mach. Theory **29**(8), 1151–1165 (1994)

22. Lin, J., Parker, R.G.: Analytical characterization of the unique properties of planetary gear free vibration, J. Vib. Acoust. Trans. ASME **121**(3), 316–321 (1999)
23. Guo, Y., Parker, R.G.: Dynamic modeling and analysis of a spur planetary gear involving tooth wedging and bearing clearance nonlinearity. Eur. J. Mech. A. Solids **29**(6), 1022–1033 (2010)
24. Kim, W., Lee, J.Y., Chung, J.: Dynamic analysis for a planetary gear with time-varying pressure angles and contact ratios. J. Sound Vib. **331**(4), 883–901 (2012)
25. Liu, C., Qin, D., Lim, T.C., et al.: Dynamic characteristics of the herringbone planetary gear set during the variable speed process. J. Sound Vib. **333**(24), 6498–6515 (2014)
26. Wang, P., Yu, P., Zhang, T.: Gear tooth modification of EV powertrain for vibration and noise reduction. SAE Technical Paper 2018-01-1289 (2018). https://doi.org/10.4271/2018-01-1289
27. Zhang, S.: Parameter study and improvement of gearbox whine noise in electric vehicle. Automot. Innov. **1**, 272–280 (2018). https://doi.org/10.1007/s42154-018-0029-5
28. Kim, U., Lieu, D.K.: Effects of magnetically induced vibration force in brushless permanent-magnet motors. IEEE Trans. Magn. **41**(6), 2164–2172 (2005)
29. Besnerais, J.L., Lanfranchi, V., Hecquet, M., Brochet, P., Friedrich, G.: Acoustic noise of electromagnetic origin in a fractional-slot induction machine. COMPEL: Int. J. Comput. Math. Electr. Electron. Eng. Emerald (2008). hal-01730150. https://hal.archives-ouvertes.fr/hal-01730150
30. He, G., Huang, Z., Qin, R., Chen, D.: Numerical prediction of electromagnetic vibration and noise of permanent-magnet direct current commutator motors with rotor eccentricities and glue effects. IEEE Trans. Magn. **48**(5), 1924–1931 (2012). https://doi.org/10.1109/TMAG.2011.2178100
31. Dupont, J., Aydoun, R., Bouvet, P.: Simulation of the noise radiated by an automotive electric motor: influence of the motor defects. SAE Int. J. Alt. Power. **3**(2), 310–320 (2014). https://doi.org/10.4271/2014-01-2070
32. Chen, X., Chen, R., Deng, T.: An investigation on lateral and torsional coupled vibrations of high power density PMSM rotor caused by electromagnetic excitation. Nonlinear Dyn. **99**, 1975–1988 (2020). https://doi.org/10.1007/s11071-019-05436-1
33. Islam, R., Husain, I.: Analytical model for predicting noise and vibration in permanent-magnet synchronous motors. IEEE Trans. Ind. Appl. **46**(6), 2346–2354 (2010). https://doi.org/10.1109/TIA.2010.2070473
34. Arabi, S., Steyer, G., Sun, Z., Nyquist, J.: Vibro-acoustic response analysis of electric motor. SAE Technical Paper 2017-01-1850 (2017). https://doi.org/10.4271/2017-01-1850
35. Zheng, H., Wei, C., Sun, D., Ren, Y., Zeng, L., Pei, R.: Rotor design of permanent magnet synchronous reluctance motors in rail traffic vehicle. In: 2018 21st International Conference on Electrical Machines and Systems (ICEMS), pp. 2762–2765. Jeju, Korea (South) (2018). https://doi.org/10.23919/ICEMS.2018.8549125
36. Du, H.Y.I., Hao, L., Lin, H.: Modeling and analysis of electromagnetic vibrations in fractional slot PM machines for electric propulsion. In: 2013 IEEE Energy Conversion Congress and Exposition, pp. 5077–5084. Denver, CO, USA (2013). https://doi.org/10.1109/ECCE.2013.6647386
37. dos Santos, F.L.M., Anthonis, J., Naclerio, F., Gyselinck, J.J.C., Van der Auweraer, H., Góes, L.C.S.: Multiphysics NVH modeling: simulation of a switched reluctance motor for an electric vehicle. IEEE Trans. Industr. Electron. **61**(1), 469–476 (2014). https://doi.org/10.1109/TIE.2013.2247012
38. Liang, X., Li, G., Ojeda, J., Gabsi, M., Ren, Z.: Comparative study of classical and mutually coupled switched reluctance motors using multiphysics finite-element modeling. IEEE Trans. Industr. Electron. **61**(9), 5066–5074 (2014). https://doi.org/10.1109/TIE.2013.2282907
39. Islam, M., Islam, R., Sebastian, T.: Noise and vibration characteristics of permanent magnet synchronous motors using electromagnetic and structural analyses. In: 2011 IEEE Energy Conversion Congress and Exposition, pp. 3399–3405. Phoenix, AZ, USA (2011). https://doi.org/10.1109/ECCE.2011.6064228
40. Yang, Z., Shang, F., Brown, I.P., Krishnamurthy, M.: Comparative study of interior permanent magnet, induction, and switched reluctance motor drives for EV and HEV applications. IEEE Trans. Transp. Electrif. **1**(3), 245–254 (2015). https://doi.org/10.1109/TTE.2015.2470092

41. Lan, I.W., Ho, H.: Slot and pole ratio of permanent magnet synchronous motor for cogging torque and torque ripple performance. In: 2018 International Conference of Electrical and Electronic Technologies for Automotive, Milan, Italy, 2018, pp. 1–5. https://doi.org/10.23919/EETA.2018.8493159
42. Dorrell, D.G., Popescu, M., Ionel, D.M.: Unbalanced magnetic pull due to asymmetry and low-level static rotor eccentricity in fractional-slot brushless permanent-magnet motors with surface-magnet and consequent-pole rotors. IEEE Trans. Magn. 46(7), 2675–2685 (2010). https://doi.org/10.1109/TMAG.2010.2044582
43. Li, Y., Lu, Q., Zhu, Z.Q.: Unbalanced magnetic force prediction in permanent magnet machines with rotor eccentricity by improved superposition method. IET Electr. Power Appl. 11(6), 1095–1104 (2017)
44. Dupont, J.B., Bouvet, P., Wojtowicki, J.L., Simulation of the airborne and structure-Borne noise of electric powertrain: validation of the simulation methodology. SAE Technical Paper No. 2013-01-2005 (2013)
45. Gurav, R., Udawant, K.D., Rajamanickam, R., Karanth, N.V., Marathe, S.R.: Mechanical and aerodynamic noise prediction for electric vehicle traction motor and its validation. SAE Technical Paper No. 2017-26-0270 (2017)
46. Fuchs, A., Nijman, E., Priebsch, H.H. (eds.) Automotive NVH Technology. Springer International Publishing. ISBN: 978-3-319-24053-4 (2016)
47. Y. Gur, J. Pan, D. Wagner, Sound package development for lightweight vehicle design using statistical energy analysis (SEA). SAE Technical Paper No. 2015-01-2302 (2015)
48. Schaefer, N., Bergen, B., Keppens, T., Desmet, W.: A design space exploration framework for automotive sound packages in the mid-frequency range. SAE Technical Paper No. 2017-01-1751 (2017)
49. Biedermann, J., Winter, R., Wandel, M., Boswald, M.: Energy based correlation criteria in the mid-frequency range. J. Sound Vib. 400, 457–480 (2017)
50. Yin, S., Yu, D., Yin, H., Lu, H., Xia, B.: Hybrid evidence theory-based finite element/statistical energy analysis method for mid-frequency analysis of built-up systems with epistemic uncertainties. Mech. Syst. Signal Process. 93, 204–224 (2017)
51. Humbert, L., Pellerey, P., Cristaudo, S.: Electromagnetic and structural coupled simulation to investigate NVH behavior of an electrical automotive powertrain. SAE Int. J. Alt. Power. 1(2) (2012). https://doi.org/10.4271/2012-01-1523
52. Wang, S., Jouvray, J.-L., Kalos, T.: NVH technologies and challenges on electric powertrain. SAE Technical Paper 2018-01-1551 (2018). https://doi.org/10.4271/2018-01-1551
53. Mehrgou, M., Zieher, F., Priestner, C.: NVH and acoustics analysis solutions for electric drives. SAE Technical Paper 2016-01-1802 (2016). https://doi.org/10.4271/2016-01-1802
54. James, B., Hofmann, A.: Simulating and reducing noise excited in an EV powertrain by a switched reluctance machine. SAE Technical Paper 2014-01-2069 (2014). https://doi.org/10.4271/2014-01-2069
55. Fang, Y., Zhang, T.: Modeling and analysis of electric powertrain NVH under multi-source dynamic excitation. SAE Technical Paper 2014-01-2870 (2014). https://doi.org/10.4271/2014-01-2870
56. Holehouse, R., Shahaj, A., Michon, M., James, B.: Integrated approach to NVH analysis in electric vehicle drivetrains. In: The 9th International Conference on Power Electronics, Machines and Drives (PEMD 2018). J. Eng. 2019(17), 3842–3847 (2019)
57. Lim, T.C., Singh, R.: Vibration transmission through rolling element bearings. Part III: geared rotor system studies. J. Sound Vib. 151(1), 31–54 (1991). https://doi.org/10.1016/0022-460X(91)90650-9
58. Sopanen, J, Mikkola, A.: Dynamic model of a deep-groove ball bearing including localized and distributed defects. Part 1: theory. Proc.. Inst. Mech. Eng. Part K J. Multi-Body Dyn. 217(3), 201–211 (2003). https://doi.org/10.1243/14644190360713551
59. Mohammadpour, M., Theodossiades, S., Rahnejat, H.: Dynamics and efficiency of planetary gear sets for hybrid powertrains. Proc. Inst. Mech. Eng. C J. Mech. Eng. Sci. 230(7–8), 1359–1368 (2016)

60. Becker, E.P.: Lubrication and electric vehicles. Tribol. Lubr. Technol. **75**(2), 60–60 (2019)
61. Questa, H., Mohammadpour, M., Theodossiades, S., Garner, C.P., Bewsher, S.R., Offner, G.: Tribo-dynamic analysis of high-speed roller bearings for electrified vehicle powertrains. Tribol. Int. **154**, 106675 (2021)
62. Sheng, S.: Report on wind turbine subsystem reliability—a survey of various databases. National Renewable Energy Laboratory. NREL/PR-5000-59111 (2013)
63. Jia, F., Lei, Y., Liu, X., Fu, Y., Hu, J.: Simulation and experimental study on lubrication of high-speed reducer of electric vehicle. Ind. Lubr. Tribol. (2021)
64. Fatourehchi, E., Shahmohamadi, H., Mohammadpour, M., Rahmani, R., Theodossiades, S., Rahnejat, H.: Thermal analysis of an oil jet-dry sump transmission gear under mixed-elastohydrodynamic conditions. J. Tribol. **140**(5) (2018)
65. Spikes, H.: Stress-augmented thermal activation: Tribology feels the force. Friction **6**(1), 1–31 (2018)
66. Walker, P., Zhu, B., Zhang, N.: Powertrain dynamics and control of a two speed dual clutch transmission for electric vehicles. Mech. Syst. Signal Process. **85**, 1–15 (2017)
67. Lou, Z., Duan, Y., Zhang, Y.: Dynamics and Control of Gearshifts in Wet-Type Dual Clutch Transmission for BEVs (No. 2020-01-0767). SAE Technical Paper (2020)
68. Li, M., Khonsari, M., Yang, R.: Dynamics analysis of torsional vibration induced by clutch and gear set in automatic transmission. Int. J. Automot. Technol. **19**(3), 473–488 (2018)
69. Minas, I., Morris, N., Theodossiades, S., O'Mahony, M.: Noise, vibration and harshness during dry clutch engagement oscillations. Proc. Inst. Mech. Eng. C J. Mech. Eng. Sci. **234**(23), 4572–4588 (2020)
70. Ewen, J.P., Heyes, D.M., Dini, D.: Advances in nonequilibrium molecular dynamics simulations of lubricants and additives. Friction **6**(4), 349–386 (2018)
71. Gosvami, N.N., Bares, J.A., Mangolini, F., Konicek, A.R., Yablon, D.G., Carpick, R.W.: Mechanisms of antiwear tribofilm growth revealed in situ by single-asperity sliding contacts. Science **348**(6230), 102–106 (2015)
72. Akchurin, A., Bosman, R.: A deterministic stress-activated model for tribo-film growth and wear simulation. Tribol. Lett. **65**(2), 59 (2017)
73. He, F., Xie, G., Luo, J.: Electrical bearing failures in electric vehicles. Friction **8**(1), 4–28 (2020)
74. Prashad, H.: Investigation of damaged rolling-element bearings and deterioration of lubricants under the influence of electric fields. Wear **176**(2), 151–161 (1995)
75. Erdman, J.M., Kerkman, R.J., Schlegel, D.W., Skibinski, G.L.: Effect of PWM inverters on AC motor bearing currents and shaft voltages. IEEE Trans. Ind. Appl. **32**(2), 250–259 (1996)
76. Didenko, T., Pridemore, W.D.: Electrical fluting failure of a Tri-lobe roller bearing. J. Fail. Anal. Prev. **12**(5), 575–580 (2012)
77. König, W., Klocke, F.: Fertigungsverfahren 3: Abtragen Und Generieren, vol. 3. Springer, Berlin (1997)
78. Romanenko, A., Ahola, J., Muetze, A.: Influence of electric discharge activity on bearing lubricating grease degradation. In: 2015 IEEE Energy Conversion Congress and Exposition (ECCE), pp. 4851–4852. IEEE (2015)
79. Prashad, H.: Variation and recovery of resistivity of industrial greases—an experimental investigation. Lubr. Sci. **11**(1), 73–103 (1998)
80. Xie, G.X., Li, G., Luo, J.B., Liu, S.H.: Effects of electric field on characteristics of nano-confined lubricant films with ZDDP additive. Tribol. Int. **43**(5–6), 975–980 (2010)
81. Xie, G., Luo, J., Liu, S., Guo, D., Zhang, C.: Bubble generation in a nanoconfined liquid film between dielectric-coated electrodes under alternating current electric fields. Appl. Phys. Lett. **96**(22), 223104 (2010)
82. Gould, B., Demas, N., Erck, R., Lorenzo-Martin, M.C., Ajayi, O., Greco, A.: The effect of electrical current on premature failures and microstructural degradation in bearing steel. Int. J. Fatigue **145**, 106078 (2021)
83. Topolovec-Miklozic, K., Forbus, T.R., Spikes, H.A.: Film thickness and roughness of ZDDP antiwear films. Tribol. Lett. **26**(2), 161–171 (2007)

84. Lazić, P., Persson, B.N.J.: Surface-roughness–induced electric-field enhancement and tribolu-minescence. EPL (Europhys. Lett.) **91**(4), 46003 (2010)
85. Zhu, X., Shi, N., Li, Y., Yang, Y., Wang, X.: Bearing capacitance estimation of electric vehicle driving motor. IOP Conf. Ser. Ear. Environ. Sci. **199**(3), 032065 (2018)
86. Dowson, D., Higginson, G.R.: A numerical solution to the elasto-hydrodynamic problem. J. Mech. Eng. Sci. **1**(1), 6–15 (1959)
87. Busse, D.F., Erdman, J.M., Kerkman, R.J., Schlegel, D.W., Skibinski, G.L.: An evaluation of the electrostatic shielded induction motor: a solution for rotor shaft voltage buildup and bearing current. IEEE Trans. Ind. Appl. **33**(6), 1563–1570 (1997)
88. Busse, D.F., Erdman, J.M., Kerkman, R.J., Schlegel, D.W., Skibinski, G.L.: The effects of PWM voltage source inverters on the mechanical performance of rolling bearings. IEEE Trans. Ind. Appl. **33**(2), 567–576 (1997)
89. Busse, D., Erdman, J., Kerkman, R.J., Schlegel, D., Skibinski, G.: System electrical parameters and their effects on bearing currents. IEEE Trans. Ind. Appl. **33**(2), 577–584 (1997)
90. Beyer, M., Brown, G., Gahagan, M., Higuchi, T., Hunt, G., Huston, M., Jayne, D., McFadden, C., Newcomb, T., Patterson, S., Prengaman, C.: Lubricant concepts for electrified vehicle transmissions and axles. Tribol. Online **14**(5), 428–437 (2019)
91. Zika, T., Gebeshuber, I.C., Buschbeck, F., Preisinger, G., Gröschl, M.: Surface analysis on rolling bearings after exposure to defined electric stress. Proc. Inst. Mech. Eng. Part J J. Eng. Tribol. **223**(5), 787–797 (2009)
92. Hadden, T., Jiang, J.W., Bilgin, B., Yang, Y., Sathyan, A., Dadkhah, H., Emadi, A.: A review of shaft voltages and bearing currents in EV and HEV motors. In: IECON 2016–42nd Annual Conference of the IEEE Industrial Electronics Society, pp. 1578–1583. IEEE (2016)
93. Willwerth, A., Roman, M.: Electrical bearing damage—a lurking problem in inverter-driven traction motors. In: 2013 IEEE Transportation Electrification Conference and Expo (ITEC), pp. 1–4. IEEE (2013)
94. Guo, F., Zhang, C.: Oil-cooling method of the permanent magnet synchronous motor for electric vehicle. Energies **12**(15), 2984 (2019)
95. Tikadar, A., Johnston, D., Kumar, N., Joshi, Y., Kumar, S.: Comparison of electro-thermal performance of advanced cooling techniques for electric vehicle motors. Appl. Therm. Eng. **183**, 116182 (2020)
96. Zhang, B., Qu, R., Xu, W., Wang, J., Chen, Y.: Thermal model of totally enclosed water-cooled permanent magnet synchronous machines for electric vehicle applications. In: 2014 International Conference on Electrical Machines (ICEM), pp. 2205–2211. IEEE (2014)
97. Grunditz, E.A., Thiringer, T., Lindström, J., Lundmark, S.T., Alatalo, M.: Thermal capability of electric vehicle PMSM with different slot areas via thermal network analysis. eTransportation **8**, 100107 (2021)
98. Chen, Q., Shao, H., Huang, J., Sun, H., Xie, J.: Analysis of temperature field and water cooling of outer rotor in-wheel motor for electric vehicle. IEEE Access **7**, 140142–140151 (2019)
99. Chen, B., Wulff, C., Etzold, K., Manns, P., Birmes, G., Andert, J., Pischinger, S.: A compre-hensive thermal model for system-level electric drivetrain simulation with respect to heat exchange between components. In: 2020 19th IEEE Intersociety Conference on Thermal and Thermomechanical Phenomena in Electronic Systems (ITherm), pp. 558–567. IEEE (2020)

Testing/Calibration/Monitoring/Diagnostics

Calibration of a Real Time Cycle SI Engine Simulation Model in the Entire Engine Operating Map

Momir Sjerić, Josip Krajnović, and Darko Kozarac

Abstract A detailed experimental research is time consuming and expensive, which increases the use of numerical simulations in all engine development stages, especially for extensive analysis and optimization of various operating parameters that affect the engine operation. In such cases the so-called cycle-simulation models, based on 1-D/0-D approach, can be very helpful by shortening the process of engine's operating map calibration. In recent years, such models are more and more intended to be real-time capable to enable their application in HiL (Hardware-in-the-Loop) simulations. Therefore, the challenge is to find the optimal balance between predictability and calculation speed of the model. This chapter presents a calibration procedure of a real-time cycle-simulation model with the aim of achieving good predictability of the model with minimal requirements for experimental investigations necessary for calibration of model parameters. The procedure includes calibration of both gas exchange and combustion model constants with the help of Proportional Integral (PI) controllers for each of the experimentally investigated operating points, and afterwards the parametrization of the selected constants based on the calibration results. The resulting parametrized model features computational speed compatible for HiL simulations, and the validation of the model showed that calculation results are in close agreement with the experimentally obtained results in the entire operating map of a naturally aspirated 1.4 L spark ignited engine.

Keywords IC engine · Calibration · Real time simulation · Spark ignited · Combustion

M. Sjerić · J. Krajnović · D. Kozarac (✉)
Faculty of Mechanical Engineering and Naval Architecture, University of Zagreb, 10002 Zagreb, Croatia
e-mail: darko.kozarac@fsb.hr

© The Author(s), under exclusive license to Springer Nature Switzerland AG 2022   293
T. Parikyan (ed.), *Advances in Engine and Powertrain Research and Technology*,
Mechanisms and Machine Science 114,
https://doi.org/10.1007/978-3-030-91869-9_12

1 Introduction

Experimental testing and calibration is an indispensable process in engine development. However, the rapid increase in computational power of modern computers enabled the use of numerical simulations in variety of applications during the entire engine development process, thus reducing the need and shortening the process of experimental investigations. Depending on the desired application, simulation models of different fidelity can be applied. Compared to the experimental results, detailed 3-D CFD models offer the most accurate results, as well as the possibility to analyze the physical phenomena that are difficult to be investigated and quantified experimentally. However, due to their complexity, they are also time consuming and therefore inconvenient for extensive analysis of various operating parameters that affect the engine operation. In such cases the so-called cycle-simulation models, based on 1-D/0-D approach, can be very helpful by offering a good compromise between computational effort and accuracy of the simulation results.

The combustion process in these models can be simulated by application of empirically based combustion models such as the Vibe function [1]. This type of combustion model is capable of real-time operation, but is not predictive since it is based on a predefined rate of heat release. Another approach is the application of predictive quasi-dimensional combustion models, such as the well-known fractal combustion model [2, 3]. Such models introduce the description of flame front geometry to the 0-D simulation environment, which can be calculated directly by the model or evaluated externally [4]. The complexity of these models can be modified in search for the optimal trade-off between calculation time and predictability. The addition of more details in such combustion models will improve model accuracy, but with the added cost of increasing the calculation time. It will also most often introduce even more calibration constants and thus the goal of such models is to ensure that for a given engine, a single set of calibration constants can be found which will result with very good accuracy of simulation results in the entire engine operating map. For example, the new quasi-dimensional combustion model based on the flame tracking approach, where the infinitely thin and smooth flame segment entrains the fresh mixture by the turbulent flame speed, was developed and validated in [5]. It was shown that the model has acceptable accuracy on the cycle-average and cycle-resolved results with a single set of calibration constants.

Reducing the complexity of quasi-dimensional models negatively impacts the predictability of the model, but offers the possibility to achieve a real-time capability with still reasonably predictive model which, if properly calibrated, can be used for extensive analysis and optimization of engine operation in a Hardware-in-the-Loop environment. Furthermore, a possibility for the use of such models in a model based engine control application is presented in [6], where a quasi-dimensional turbulent entrainment model, originally developed by Blizard and Keck [7], and later refined by Tabaczynski et al. [8] is successfully used as a basis for real-time combustion phasing prediction.

For creation of complete engine simulation models, the 0-D combustion models are most commonly coupled with 1-D flow model across the intake and exhaust pipes. The real-time capability of the entire model thus depends not only on the complexity of the models which describe the in cylinder processes, but on the gas path model as well. To reduce the computational effort of a gas path model to an appropriate level, a 0-D interpretation of the intake and exhaust system components is mandatory as well.

In this study, such a 0-D modelling approach is presented, where the combustion is modeled by a slightly simplified implementation of the Flame Tracking Model (FTM), available in AVL Cruise™ M [9], and used to investigate its potential for real-time simulation applications. Finally, the calibration method that results with a real-time capable simulation model of the complete engine system with good predictability across the entire engine operating map is defined.

2 Experimental Results

For this study, experimental results of a naturally aspirated 1.4 L spark ignited engine are used. In total, 162 experimentally tested operating points were available, but for this study, the total number of operating points that were used was reduced to 55, which still adequately represents the entire engine map, as shown in Fig. 1. During the study, 12 operating points were identified as points used for calibration of the model parameters, while the others were used in validation. Four different engine speeds were selected and for each engine speed three different loads, ranging from minimal to full load, are chosen.

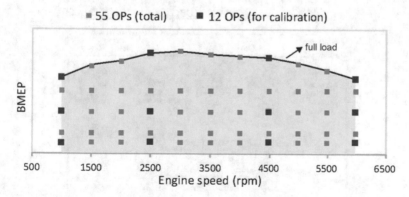

Fig. 1 Experimentally obtained engine map

3   Real Time Simulation Model

The cycle simulation model that represents a real 1.4 L spark ignited engine was made and a schematic of the model is shown in Fig. 2. Since the gas path in [9] is calculated as a zero-dimensional system, specific pipe sections, which connect the main elements of the engine, are represented by a combination of plenums and restrictions. The intake system thus consists of the air cleaner, throttle represented

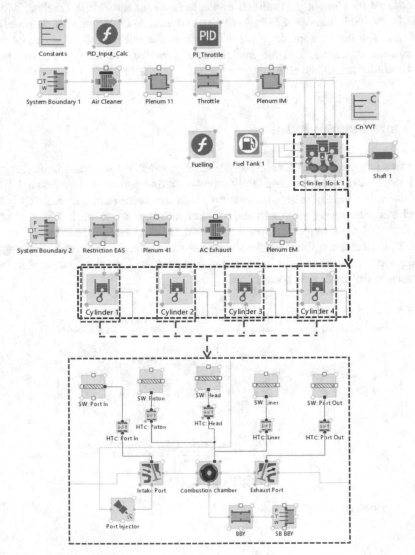

Fig. 2 Simulation model of a naturally aspirated 1.4 L spark ignited engine made in AVL Cruise™ M

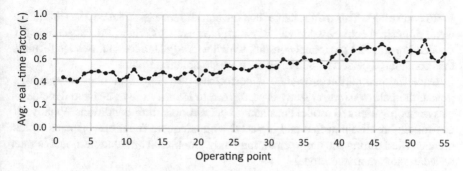

Fig. 3 Average real-time factor for each of the 55 operating points

by a restriction element and the intake manifold represented by a single plenum whose volume corresponds to that of the manifold. The exhaust system is simplified in a similar manner and consists of the exhaust manifold represented by a single plenum while all additional exhaust pipes and elements, such as muffler and resonator, are combined into a single quasi-dimensional pipe and restriction. The catalyst is represented by the air cleaner element since emission analysis is not part of this study and its purpose is only to adequately introduce the pressure drop in the system. The purpose of simplifying the intake and exhaust systems is to reduce the computational effort and thus enhance the real-time capability of the model. The cylinder block consists of four identical cylinders whose geometry and valve timings correspond to the cylinder configuration of the experimental engine.

The model also features control elements which enable the creation of specific operating conditions. In this study, the control of the intake pressure was performed with PI controller that adjusts the flow coefficient of the throttle element and thus controls the engine load. Additionally, two user defined functions are implemented, one to calculate the required inputs for the controller, and the other to control the fueling of the engine based on the cycle averaged airflow and the desired value of air excess ratio.

Figure 3 shows the averaged real-time factor of the presented simulation model during execution for all simulated operating points, and since the factor is lower than 1 in all the cases, the model is proven to be real time capable in the entire range of operating conditions. This was achieved with the global time step of 1 ms, and cylinder calculation step of 1°CA (crank angle).

4 Calibration Method and Model Parametrization

For calibration purposes additional control elements and functions are added to the model, but unlike those that control the engine load and fueling to achieve the desired operating conditions, these elements are used only for automatic calibration of model constants based on experimental results which serve as control values for PI controllers. This means that, when the convergence is achieved, the simulation result of desired control value will be exactly the same as that of the experimentally obtained result, and the model constant which was adjusted to achieve such result is thus considered perfectly calibrated for a single operating point.

Once the calibration procedure is performed on selected number of operating points, the parametrization study is performed to identify the trends and dependencies of such calibrated model constants to certain operating conditions. Finally, PI controllers are disconnected and replaced with adequate functions or maps which are identified during the parametrization study. Such final simulation model is thus called the parametrized model.

Five constants are identified as crucial for achieving a good accuracy of the simulation model and each is described in following subsections.

4.1 Calibration of Gas Exchange

As mentioned, the gas path in [9] is calculated as a zero-dimensional system, which is why gas dynamics in the intake manifold cannot be properly captured by the model. That is why the gas exchange is calibrated by adjusting the pressure offset in the intake port. This was performed by a PI controller that used the experimentally obtained result of average air mass flow as a control value. The results of calibration, as well as the subsequent parametrization are shown in Fig. 4a. It was found that the required intake pressure offset is dominantly dependent on engine speed and that, while different values were obtained for different loads no sensible trend in load-dependency was identified. That is why the parametrization was performed in such a way that for a considered engine speed an average value of 3 calculated pressure offsets was found. However, if for a single operating point a drastic deviation in

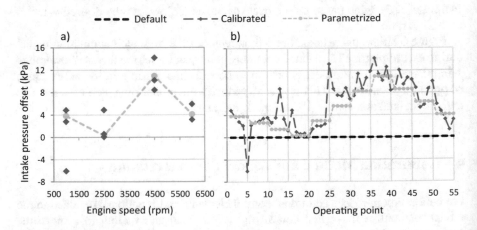

Fig. 4 Calibration and parametrization of intake pressure offset

calculated pressure offset was observed, that operating point was disregarded. As a result, a table for intake pressure offset is created and a linear interpolation is performed in-between points with defined engine speeds. Figure 4b shows the result of parametrized pressure offset for each operating point investigated in this study and, only for reference, it is shown what would be the result of calculated pressure offset if each of the 55 operating points are individually calibrated.

4.2 Calibration of Turbulence Model

In-cylinder turbulence level and turbulent eddies of different integral length scales affect the fresh mixture entrainment in spark ignited engines and the mixture formation (mixing) process in compression ignition engines. In a quasi-dimensional combustion models the burning rate mainly depends on the geometrical parameters of the combustion chamber, the in-cylinder turbulence properties and the fuel type. The mostly used zero-dimensional (0-D) turbulence models are based on the energy cascade mechanism. In this study two-equation revised K-k turbulence model was applied [10]. Two governing equations are defined, one for the kinetic energy (K) of the mean flow field and the second for the turbulent kinetic energy (k). More details about the K-k turbulence model can be found in [11].

The most credible validation of 0-D turbulence model could be made if experimental results of the in-cylinder turbulence would be available. Otherwise, the verification of the turbulence model can be performed by using three-dimensional (3-D) computational fluid dynamics (CFD) results of the cylinder domain if such simulation is made for the considered engine. If neither experimental nor 3-D CFD data of the in-cylinder flow are available, the calibration of turbulence model constants can be made by considering the turbulence velocity u' at the top dead center (TDC) which can be correlated to a mean piston speed. For the pent roof cylinder head geometry which generates tumble flow structure in the cylinder, the expected turbulence velocity at TDC is within the range of 1–2 times of mean piston speed [12]. In this study, the verification of 0-D turbulence model was performed with reference to mean piston speed and the results are plotted in Fig. 5. This step was performed manually, but simultaneously with the calibration of other model constants because the in-cylinder turbulence level has a strong impact on combustion progress.

The dotted line in Fig. 5 represents turbulence velocity that is equal to mean piston speed, while the dashed line is turbulence velocity equals 2 times the mean piston speed. These two lines represents lower and upper boundary of the expected turbulence velocity at TDC. If default values of 0-D turbulence model constants are applied, the calculated turbulence velocity u' at TDC is below the expected lower limit referenced to mean piston speed. The increase of in-cylinder turbulence level was made with the increase of integral length scale constant (from 0.06 to 0.14) and the resulted turbulence velocity u' at TDC is approx. equal to 1.3 times the mean piston speed for full load conditions on 4 considered engine speeds during calibration process (Fig. 5a). The ratio of turbulent velocity at TDC and mean piston speed for

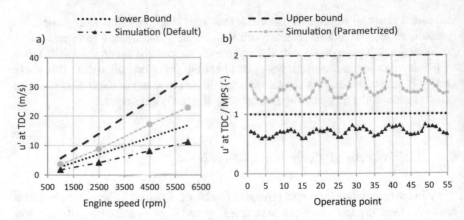

Fig. 5 Verification of 0-D turbulence model—comparison of calculated turbulence velocity u' at TDC with mean piston speed

all 55 operating points after parametrization of other model constants is shown in Fig. 5b and it can be seen that, for each operating point, the turbulence velocity u' at TDC is within the expected range.

4.3 Calibration of Ignition Delay

The combustion model used in this study originally assumes a constant ignition delay of 0.075 ms regardless of operating conditions, and this was identified from the start as something that is necessary to calibrate. The ignition delay was first evaluated from experimentally obtained pressure traces, as shown in Fig. 6. In such

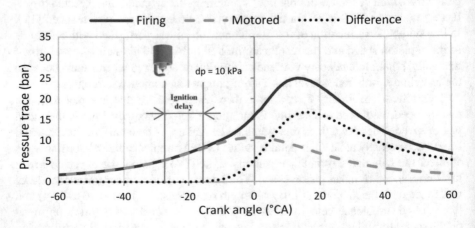

Fig. 6 Evaluation of the ignition delay from experimentally obtained pressure trace

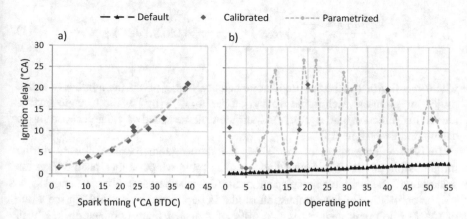

Fig. 7 Calibration and parametrization of ignition delay

an approach the criteria for definition of combustion starts must be identified. In this study, it was considered that start of combustion occurred when the increase of in-cylinder pressure, compared to a motored case, reaches 10.000 Pa.

Since a physical model for calculating the ignition delay based on in-cylinder state at the spark timing is not available in the simulation tool used in this study, the ignition delay was parametrized by a quadratic function that directly depends on spark timing. In this way, for a given engine, the model will ensure a very good accuracy of the calculated ignition delay, but the model is not predictable and will not be able to capture the influence of changes in the operating conditions without the appropriate change in spark timing adjustment.

The result of individual ignition delay calibration on 12 selected operating points is shown in Fig. 7a, while the resulting parametrized ignition delay for each of the 55 operating points is shown in Fig. 7b. It can be seen that the default ignition delay of 0.075 ms significantly differs from experiment which then has a major impact on the calibration of the combustion model.

4.4 Calibration of Combustion Model

The combustion burning rate in the applied combustion model is proportional to the entrainment of fresh mixture by the flame front that propagates across the combustion chamber. The main novelty of the FTM is that the flame surface is described by a set of discretized triangular elements whose corners shift along the predefined rays directed from the spark plug location toward the combustion chamber walls [5]. The overall mass burning rate is equal to the sum of turbulent mass burning rates over each flame front triangle:

$$\frac{dm_{BZ}}{dt} = \sum_{j=1}^{Ntria} \left[\rho_{UZ} \cdot \sum_{i=1}^{3} \left(\frac{1}{N_{j,\,act}} A_{norm,\,j} \cdot u_{t,\,i} \right) \right] \tag{1}$$

where ρ_{UZ} (kg/m^3) is the unburned zone density, $N_{j,\,act}$ $(-)$ is the number of active projection points at each flame front triangle, $A_{norm,\,j}$ (m^2) is the triangle surface area in ray direction, and $u_{t,\,i}$ (m/s) is turbulent flame speed calculated according to correlation derived in [13].

In [9] implementation of the FTM, the flame geometry is evaluated during initialization, tabulated and stored as a lookup table which is then used during the calculation. The available flame speed multiplier is added as calibration target.

The calibration of the combustion model is performed by controlling the value of CA50 (the crank angle at which 50% of the heat from combustion has been released). This was done by adjusting the FSM, while the control value was obtained from experimental pressure trace analysis. The model itself offers the possibility to parametrize the FSM with load, where the load is represented by the in-cylinder pressure at the start of combustion, as well as with engine speed. This is why it is important that the calibration of turbulence model, and even more significantly the ignition delay are calibrated prior to, or simultaneously with the FSM calibration.

The parametrization study showed that calibrated FSM is almost exclusively load dependent. Therefore, the parametrization was performed as shown in Fig. 8a. The comparison of such parametrized FSM with the values individually calibrated for all 55 operating points is shown in Fig. 8b.

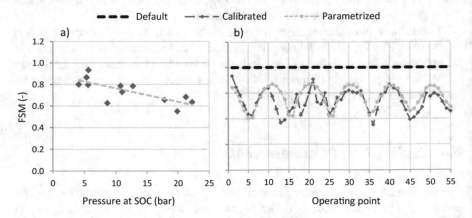

Fig. 8 Calibration and parametrization of flame speed multiplier

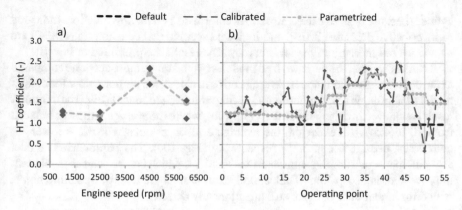

Fig. 9 Calibration and parametrization of heat transfer coefficient

4.5   Calibration of Cylinder Heat Transfer

Finally, the calibration of in-cylinder gas-to-wall heat transfer, calculated by the Woschni correlation [14], is performed. Since there were no experimentally obtained references in terms of temperatures of engine parts, the temperatures of walls in the model are assumed based on usual practice and experience. The temperatures are imposed as load dependent values from the start, and further calibration was performed by adjustment of heat transfer coefficient with indicated mean effective pressure (IMEP) as a control value. Parametrization study showed very similar trend of dependency of calibrated heat transfer coefficient on engine speed as the intake pressure offset. Similarly, significant dissipation in calculated heat transfer coefficient was observed so the parametrization was performed in the same way as for the intake pressure offset, by calculating the average value for a given engine speed while disregarding the operating points that deviate too much. It is important to emphasize that the PI controller for heat transfer calibration is activated last and the parameters of PI controller were such that it converges slowly. This enables the adjustment of air mass flow and CA50 values prior to the adjustment of the heat transfer values.

The result of calibration of heat transfer coefficient on selected 12 operating points is shown in Fig. 9a, and the resulting parametrized heat transfer coefficient for each of the 55 operating points is shown in Fig. 9b. Again, individually calibrated heat transfer coefficient for each of the 55 operating points is shown for reference.

5   Validation of Parametrized Model

The validation of the parametrized model is mainly based on the results which were used as control values during calibration. The accuracy of calculated air mass flow will directly determine if the right amount of fuel energy is introduced into the cylinder because desired excess air ratio is controlled by a fueling function. Indicated mean effective pressure (IMEP) and indicated specific fuel consumption (ISFC) are a good overall measure of model accuracy. The results are presented in Fig. 10.

It can be seen that the parametrized model is overall much more accurate compared to the default simulation model. Average error in calculated air mass flow for default model of approx. 9% was reduced to 3% for parametrized model (Fig. 10a). Similar result is achieved in calculated IMEP (Fig. 10b), where the average error was reduced from 10 to 4.5%. As expected, the average error in calculated ISFC is lower for both cases, being 8% and 3.3% respectively (Fig. 10c). The parametrized model also eliminates most of the significant inaccuracies which can be seen as significant improvement in dissipation of calculated error. A small number of operating points with error higher than 10% are still unfortunately detected.

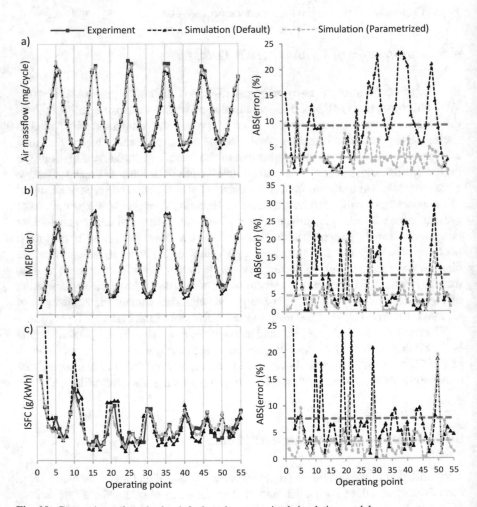

Fig. 10 Comparison of results for default and parametrized simulation models

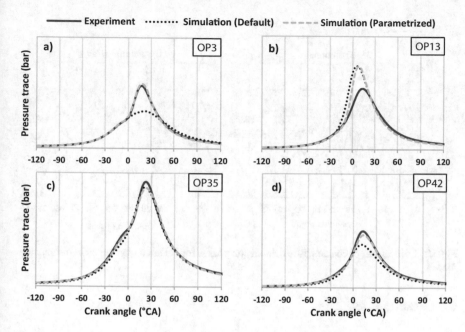

Fig. 11 Comparison of pressure traces for default and parametrized simulation models

Figure 11 shows pressure traces for a couple of investigated operating points. For OP3, which was among the 12 operating points used for calibration, it can be seen that the pressure trace of the parametrized model almost perfectly matches with the experimentally obtained one, unlike the default model which is very much different (Fig. 11a). A very good fit is also evident in Fig. 11c and d, for arbitrary selected operating points OP35 and OP42 which were not part of the calibration. OP35, being one of the full load conditions, shows that for WOT case the default model can achieve good accuracy as well. Figure 11b shows the result of OP13 as the worst case where only a small improvement was achieved compared to the default model and the pressure trace deviation is still high.

In Fig. 12, along with CA50, which was used as a control value for calibration of the FSM, CA5 and CA90 are presented as well, as a means to validate the complete combustion progress without the need to analyze pressure traces for each operating point individually. CA5 is also a good measure of the accuracy of the imposed ignition delay.

The validation of characteristic combustion phases shows significant improvement after calibration and parametrization of the model constants (Fig. 12b) in comparison to the default model (Fig. 12a). Although in small number of operating points faster combustion than the one obtained from the experiment still occurs, e.g. OP13 (Fig. 11b), over 80% of the operating points have an error in calculated CA50 lower than 2°CA, while only 4 out of 55 points are off by more than 5°CA.

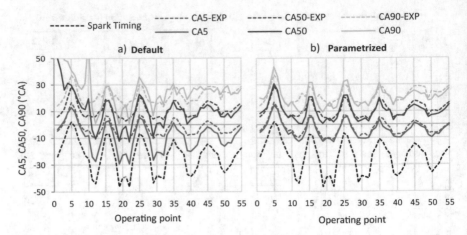

Fig. 12 Comparison of characteristic combustion phases for default and parametrized simulation models

6 Conclusions

The study presented a method for calibration of a real-time cycle-simulation model. The presented results confirm that a good predictability of the real-time capable model can be achieved and that only a small number of experimentally investigated operating points are required for the calibration of model parameters. Significant improvement is evident in combustion process prediction, which consequently reduces the average error in results by approx. 2–3 times compared to the simulation model with default parameters. The overall accuracy of the simulation results, compared to the experimental results, is above 95%.

The use of PI controllers for the automatic calibration of the model parameters significantly reduces user effort and speeds up the calibration process. Identification of model parameters and control values for calibration is a crucial part of this method. However, general guidelines presented in this chapter can be applied to different engines and are not limited to the combustion model used in this study.

References

1. Vibe, I.I.: Brennverlauf und Kreisprozess von Verbrennungsmotoren. VEB Verlag Technik Berlin, Berlin (1970)
2. Bozza, F., Gimelli, A., Merola, S., Vaglieco, B.: Validation of a Fractal Combustion Model through Flame Imaging. SAE Technical Paper 2005-01-1120 (2005)
3. Ma, F., Li, S., Zhao, J., Qi, Z., Deng, J., Naeve, N., et al.: A fractal-based quasi-dimensional combustion model for SI engines fuelled by hydrogen enriched compressed natural gas. Int. J. Hydrogen Energy **37**, 9892–9901 (2012)

4. Altın, I., Bilgin, A.: Quasi-dimensional modeling of a fast-burn combustion dual-plug spark-ignition engine with complex combustion chamber geometries. Appl. Therm. Eng. **87**, 678–687 (2015)
5. Sjerić, M., Kozarac, D., Schuemie, H., Tatschl, R.: A new quasi-dimensional flame tracking combustion model for spark ignition engines. Energy Convers. Manage. **165**, 263–275 (2018)
6. Wang, S., Prucka, R., Zhu, Q., Prucka, M., et al.: A real-time model for spark ignition engine combustion phasing prediction. SAE Int. J. Engines **9**(2), 1180–1190 (2016)
7. Blizard, N., Keck, J.: Experimental and Theoretical Investigation of Turbulent Burning Model for Internal Combustion Engines. SAE Technical Paper 740191 (1974)
8. Tabaczynski, R., Trinker, F., Shannon, B.: Further refinement and validation of a turbulent flame propagation model for spark ignition engines. Combust. Flame **39**, 111–121 (1980)
9. AVL Cruise^{TM} M.: User Manual (2019)
10. Bozza, F., Fontana, G., Galloni, E., Torella, E.: 3D-1D Analyses of the Turbulent Flow Field, Burning Speed and Knock Occurrence in a Turbocharged SI Engine. SAE Technical Paper 2007-24-0029 (2007)
11. Bozza, F., Gimelli, A.: A Comprehensive 1D Model for the Simulation of a Small-Size Two-Stroke SI Engine. SAE Technical Paper 2004-01-0999 (2004
12. Heywood, J.B.: Internal Combustion Engine Fundamentals, 1st edn. McGraw—Hill, New York (1988)
13. Frolov, S.M. et al.: Flame tracking—Particle method for 3D simulation of normal and abnormal (knocking) operation of spark-ignition automotive engines. In: International Automotive Conference JUMV 2015, Yugoslav Society of Automotive Engineers, Belgrade, Serbia (2015)
14. Woschni, G.: A universally applicable equation for the instantaneous heat transfer coefficient in the internal combustion engine. SAE Trans. **76**, 3065–3083 (1967)

Model-Based Calibration
of Transmission Test Bench Controls
for Hardware-in-the-Loop Applications

Enrico Galvagno ⓘ, **Antonio Tota** ⓘ, **Gianluca Mari,**
and Mauro Velardocchia ⓘ

Abstract Hardware-in-the-loop test rigs represent one of the most adopted experimental platforms to assess automotive transmissions dynamic performances. Even if their architectures may range among few different configurations, they share general requirements from the mechanical design to their controller implementation for a real-time deployment. This chapter is focused on a real application of a Dual Clutch Transmission Hardware-in-the-loop test rig. Two electric motors are installed to emulate the effect of the Internal Combustion Engine to the transmission input and the vehicle motion resistances to the transmission output, respectively. For a proper Hardware-in-the-loop operation, if a torque control is selected for one electric motor, the second one requires to be controlled in speed. Moreover, the controls structure cannot be usually customized since they are conventionally constrained by industrial drive limitations. This chapter includes an accurate analysis of the reciprocal influence of the two controllers since they are mechanically applied to the input and output of the same transmission. The analysis is further supported by a linear model of the transmission test rig which is able to predict the sensitivity effect of the two controller's activation on the test rig performance. An optimal tuning of the two controllers' parameters is then described to achieve the desired level of reference tracking and disturbance rejection targets.

Keywords Dynamic modelling · HiL testing · Experimental model validation · Transmission test bench · Control calibration · Frequency response functions · AC motors speed and torque control tuning · Torsional vibrations

E. Galvagno (✉) · A. Tota · G. Mari · M. Velardocchia
Politecnico Di Torino, 10129 Turin, Italy
e-mail: enrico.galvagno@polito.it

A. Tota
e-mail: antonio.tota@polito.it

G. Mari
e-mail: gianluca.mari@polito.it

M. Velardocchia
e-mail: mauro.velardocchia@polito.it

1 Introduction

Automotive transmission systems are designed to comply with strict functionality, reliability, and safety requirements to meet the desired vehicle drivability performance [1]. Usually, carmakers face the transmission development process through extensive experimental campaigns carried out to evaluate the transmission performance. A complete experimental set-up would require the connection of the transmission system to the internal combustion engine (ICE) or its direct installation on board the vehicle to reproduce with high fidelity the working operating conditions. However, this solution presents a considerable number of drawbacks in terms of costs and maintenance services that an ICE or a full vehicle system would implicate. Therefore, a more convenient solution involves the introduction of one or more electric motors to emulate the steady-state and the dynamic characteristics of the ICE as well as the load condition imposed by the vehicle motion resistances [2]. In this regard, Hardware-In-the-Loop (HiL) test benches have been very popular due to their advantages in terms of cost, flexibility, repeatability, and test automation since they adopt a model-based approach with a mix of real and emulated sensors, actuators, and vehicle subsystems to meet the cost and time constraints. HiL can reduce the testing effort up to a 90% if compared to the conventional driving test procedures, as described in [3]. Furthermore, HiL test benches have been widely used in the automotive industry to study the behavior of physical vehicle components and subsystems, e.g. the dual mass flywheel influence on transmission dynamics in [4], or to design and validate electronic control units. For example, a HiL simulation has been developed in [5, 6] for testing active brake control strategies such the Anti-lock Brake System (ABS), the Traction Control System (TCS), and the Electronic Stability Program (ESP). A similar application is also presented in [7–9], where a continuous braking pressure control strategy is implemented and validated on a HiL test rig with a conventional passenger car braking system. In [10] a HiL system is set-up for the development process of an electronic steering device to guarantee repetitive simulations and to demonstrate the efficacy with respect to on-board vehicle testing sessions. Finally, it is also convenient to integrate two or more HiL test rigs for evaluating their mutual influence, as also described by [11] where a DCT and a brake HiL systems are coupled to enhance the transmission Noise, Vibrations, Harshness (NVH) performance.

The activity described in this paper presents a real application of a DCT HiL test rig available at the Politecnico di Torino. Two electric motors are installed to emulate the effect of the ICE to the transmission input and the vehicle motion resistances to the transmission output, respectively. The control units of the electric motors adopted for HiL purposes, are usually designed with a rigid structure, typically a Proportional Integral Derivative (PID) logic, that is conventionally constrained by industrial drive limitations. The main contribution of the paper is to provide a model-based approach for analyzing and tuning the interaction between the control of the two electric motors. The torsional dynamics of the driveline installed on the test bench is simulated through a 6-degree-of-freedom (DOF) lumped parameter model, that represents an extended version of the 5-DOF model proposed in [12]. The transmission model is also enhanced by considering the closed-loop dynamics introduced by the PID algorithm for regulating the speed of one electric motor. An extensive sensitivity analysis of the model to the mechanical and control parameters is proposed to evaluate their effect on the controller reference speed tracking performance and on its disturbance rejection against the torque applied by the second electric motor. Then, the tuned speed control strategy is experimentally validated on the DCT HiL test rig.

The manuscript is organized as follows: Sect. 2 provides an overview of the hardware and software available on the DCT HiL test rig; Sect. 3 introduces the transmission model, and analyze the dynamic behavior of both open-loop and closed-loop configurations; Sect. 4 presents the sensitivity analysis of the linear model to the mechanical and control parameters, while Sect. 5 covers the experimental validation, both in frequency and time domain, of the proposed methodology; finally, conclusions are drawn in Sect. 6.

2 Transmission HiL Test Rig: Hardware (HW) and Software (SW)

The test bench here presented is meant to be used as a HiL system for automotive transmission testing. The typical loading condition associated with the usage of a mechanical transmission on a real car can be reproduced. To this aim, a simulation model running and exchanging in real time signals with sensors and actuators is implemented. Some examples of test benches sharing a similar HiL technology are reported in [13, 14]. A detailed description of the hardware and software components is given in the next two sections.

2.1 Hardware Components

Figure 1 shows a picture of the HiL transmission test bench in the Mechanical Laboratory of Politecnico di Torino. It features two electric motors (M1 and M2) and two transmissions, the first is a Dual Clutch Transmission (DCT), that is the system under investigation, and the second a Manual Transmission (MT).

As visible in Fig. 2, the connection between the two transmissions is realized through the two output shafts SA1 and SA2, i.e. the original left half shafts of the two drivelines, and a brake disk D.

More specifically, the main subsystems that are visible in Fig. 1 are commented below (starting from the left part of the picture):

(i)   M1: a 37 kW 2-pole 3-phase induction motor featuring a nominal torque of 121 Nm and a nominal speed of 2920 rpm. It is operated in the four quadrants of the torque-speed plane by an inverter with a torque overload capacity of 150%. For this application, the "torque-control with speed limitation" mode

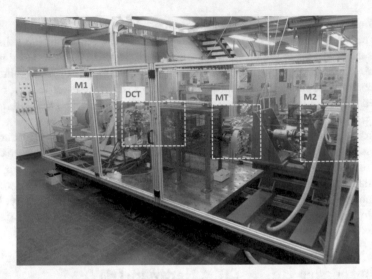

Fig. 1 Transmission test bench in the Mechanical Laboratory of Politecnico di Torino

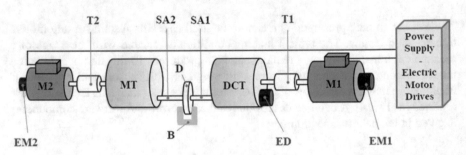

Fig. 2 Layout of the transmission test bench: electric motors (M1, M2), speed sensors (encoders EM1, EM2 and ED), torque sensors (T1, T2), disk (D), brake (B) and half shafts (SA1, SA2)

is selected for reproducing the torque delivered by the internal combustion engine of the car.

(ii) DCT: a 6-speed dry Dual Clutch Transmission (see, e.g., [15] for the kinematic and dynamic behavior of this transmission).

(iii) MT: a 6-speed Manual Transmission with one primary and two secondary shafts;

(iv) M2: a 11 kW 6-pole 3-phase induction motor with a nominal torque of 110 Nm and a nominal speed of 955 rpm. It is powered by an inverter with 150% torque overload capacity which allows the electric machine operating in four quadrants. For this application, the "encoder feedback speed-control" mode is selected and is used to simulate the vehicle load at the output shaft of the DCT, that is due to the aerodynamic and rolling resistance, the road slope, and the vehicle inertial effects.

To monitor the actual dynamic state of the transmission system, the test bench is equipped with a number of sensors that are here reported.

(i)   Two torque-meters (T1 and T2 in Fig. 2), measuring the torque provided by the two electric motors: T1 torque meter, is particularly suitable for dynamic torque measurements having a bandwidth of 3 kHz and a maximum torque of 500 Nm; T2 is a torque meter with an integrated tachometer, capable of withstanding dynamic torques of 226 Nm.

(ii) Three rotary incremental encoders: EM1 and EM2 measure the angular position and speed of the two motors while ED measures the position and speed of the DCT differential; the number of pulses per revolution of the three sensors are 1024, 4096 and 9600 respectively. It is worth noting that encoders EM1 and EM2 are the same sensors used by the electric drives for implementing the closed-loop speed control of the two motors.

(iii) Magnetic pickup sensors for monitoring the angular speed of the components inside the DCT, they are positioned: on the secondary mass of the dual mass flywheel, on the differential crown, on the two differential pinions (lower and upper secondary shaft), on the first and third gear on the lower secondary shaft and on the second and fourth gear on the upper secondary shaft.

(iv) Two thermocouples are also mounted on the DCT, one for measuring the temperature of the lubricating oil and the other one for monitoring the temperature of the hydraulic fluid of the actuation system.

(v)   A triaxial accelerometer on the gearbox housing and a microphone for further NVH analysis, e.g. gearshift noise and vibration.

The two electric motors are operated by means of electric drives, the 37 and 11 kW inverters shown in Fig. 3, which allow controlling either the actual torque or the speed in open or closed-loop, the latter through PID controllers, depending on the user's choice. Both the drives are set for operating with vector control method.

The electric cabinet of the test rig also include an Active Front End (AFE) which connects the two inverters via a direct current (DC) bus, allowing an efficient power exchange between the two motor drives. AFE may absorb or give back to the electric grid the net electric power deriving from the specific operating condition of the bench. Power regeneration is therefore enabled through this system architecture, thus

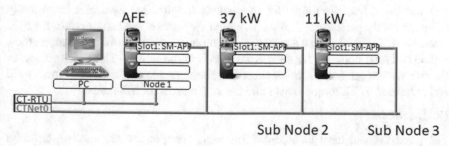

Fig. 3 Communication network between the electric drives (37 and 11 kW) and the active front end (AFE)

avoiding the dissipation, e.g. on braking resistors, of the negative electric power. For closed-loop control, the microcontrollers inside the electrical cabinet use the information from the torque transducers and the encoders mounted on the bench. The front panel of the electrical cabinet features displays for manual settings of the motor drives and many connectors for analog and digital I/O, Controller Area Network (CAN) and Ethernet communication for real-time control of the test bench and for remote monitoring and programming the inverter parameters.

It is worth underlining that a good controllability of both electric motors in terms of torque and speed reference tracking performance is mandatory for such kind of test bench. The optimization of the available parameters, in particular the PID gains of the feedback controllers, is crucial for obtaining a satisfactory trade-off between the conflicting requirements, i.e. high dynamic performance, low control effort and low noise and vibrations. The step change of some system parameters during the normal working condition of the bench, i.e. the gear ratio of the two transmissions, leads to different optimal parameter sets for different test bench configurations. This last specific characteristic requires specific analysis and considerations that are the aim of this paper.

2.2 Software Components

The HiL testing methodology requires modelling the dynamics of all the systems interacting with the transmission under test that are present on a real car, but that are not physically installed on the bench. Figure 4 highlights the main interactions between hardware and software components of the test bench. In the central part of the picture are the two models for engine and vehicle simulation, on the left the inputs signals from the Transmission Control Unit (TCU) via CAN, from encoder and torque sensors, while on the right the output signals for the TCU, the torque and the speed setpoints for the controllers of the two motors.

Suitable models for the internal combustion engine and for the vehicle longitudinal dynamics simulation must be developed and implemented in the HiL software.

Vehicle dynamics model

A 1 degree of freedom model for the vehicle longitudinal dynamics is normally adequate for the present application, in which the pure rolling condition for the tires is assumed. The model must account for the motion resistance due to aerodynamics, rolling and road slope and for the inertial effects of the vehicle. This block computes the reference speed for M2 motor according to the torque applied by the powertrain and measured by T2 torque sensor and the actual speed of M2 motor.

Engine model

The engine model must include the following functions to accomplish the tasks required by HiL transmission testing for conventional powertrains: starter simulation, engine cut-off during accelerator pedal release, redline control for maximum speed

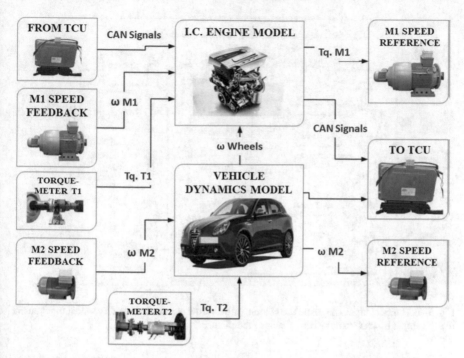

Fig. 4 HiL software scheme: on the left, the inputs from the TCU and the torque and speed sensors of the electric motors; in the middle, the main simulation blocks; on the right, the output setpoints for the electric motors and the signals for the TCU

limitation, idle speed control, torque saturation according to engine maximum and minimum torque map, combustion delay and turbo lag to account for the delay introduced by the gasoline or diesel engine torque generation system. Figure 5 shows a block diagram of the engine model, in which the said functions are reported; this model computes the torque delivered by the virtual ICE according to the actual working condition of the bench. The engine torque signal becomes a reference for the torque control of M1 motor and is sent via CAN to the corresponding drive.

Typical examples of experiments carried out on the test bench include: vehicle start-up, gear shift, engine start and stop, accelerator pedal tip-in/tip-out, run-up and coast-down, acceleration and deceleration with upshifts and downshifts. This test bench was also recently used to study the integration with other active chassis systems with the aim of enhancing the transmission NVH performance.

3 System Model and Analysis

The DCT torsional dynamics is investigated through a six-degree-of-freedom (DOF) lumped parameter model, as shown in Fig. 6.

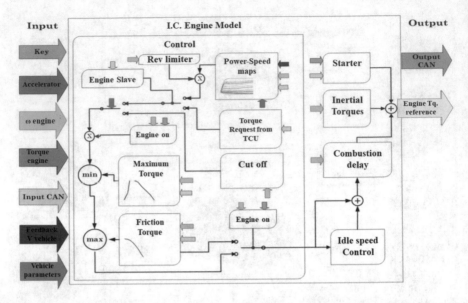

Fig. 5 ICE model block diagram. The colours of the arrow are used to identify which input enters in each single blocks from the central part of the picture

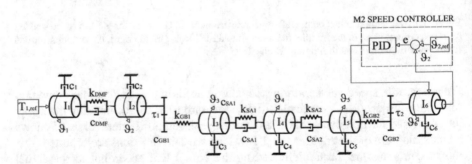

Fig. 6 Schematic of the torsional model of the transmission test bench

The test bench torsional model includes the following elements:

- The Dual Mass Flywheel (DMF) is modelled with two inertial components, I_1 and I_2, linked by a low stiffness spring k_{DMF} and a viscous damper with damping coefficient c_{DMF}. The moment of inertia I_1 includes the first DMF mass and the M1 motor inertia, meanwhile the second DMF mass and the equivalent inertia of the DCT gearbox (except its differential inertia I_3) are integrated in I_2.
- The DCT gearbox model account for the actual gear ratio τ_1 (defined as the ratio between the input and output speeds) and an equivalent torsional stiffness k_{GB1} at the transmission output shaft. The value of k_{GB1} varies with the gear ratio τ_1 due to the variation of the portion of the shaft through which the power is transmitted when the different gears are engaged.

- The MT gearbox is modelled with the actual gear ratio τ_2 and an equivalent stiffness k_{GB2} (evaluated at its output shaft) between differential inertia I_5 and inertia I_6 which includes the M2 mass moment of inertia and the equivalent inertia of the MT gearbox.
- The inertia of the brake disk I_4 connecting the DCT and MT output shafts through the two half shafts SA1 and SA2, respectively, which are modelled as a pair of spring damper elements.

The values of the said parameters used to define the transmission test bench model are reported in Table 1. While inertial and elastic parameters can be quite accurately derived from component technical drawing or data sheet, damping cannot be easily evaluated; therefore a curve fitting method based on the comparison of the simulated and experimental frequency response functions (FRFs) (see Sect. 3.3 for further information) was set up to identify the viscous damping parameters of the model.

As already mentioned in the former section, the electric motor M1 is usually set to apply a desired torque to emulate the steady-state torsional behavior of the ICE engine meanwhile the motor M2 is controlled in speed to replicate the vehicle longitudinal dynamics through the mathematical model of the motion resistance.

The system analysis is conducted in two steps that will be called.

(a) *open-loop system* if the two motors M1 and M2 apply torque to the mechanical system without using any feedback signals from the test rig.

(b) *closed-loop system* if M2 is speed controlled using feedback from EM2 sensor and M1 applies torque to the mechanical system without feedback.

3.1 Open-Loop System Equations

The equations of motion of the torsional model depicted in Fig. 6 are:

$$I_1\ddot{\vartheta}_1 + c_{DMF}(\dot{\vartheta}_1 - \dot{\vartheta}_2) + c_1\dot{\vartheta}_1 + k_{DMF}(\vartheta_1 - \vartheta_2) = T_1$$

$$I_2\ddot{\vartheta}_2 + \left(c_2 + \frac{c_{GB1}}{\tau_1^2}\right)\dot{\vartheta}_2 - c_{DMF}(\dot{\vartheta}_1 - \dot{\vartheta}_2) - \frac{k_{GB1}}{\tau_1}\left(\vartheta_3 - \frac{\vartheta_2}{\tau_1}\right) - k_{DMF}(\vartheta_1 - \vartheta_2) = 0$$

$$I_3\ddot{\vartheta}_3 + c_3\dot{\vartheta}_3 + c_{SA1}(\dot{\vartheta}_3 - \dot{\vartheta}_4) + k_{GB1}\left(\vartheta_3 - \frac{\vartheta_2}{\tau_1}\right) + k_{SA1}(\vartheta_3 - \vartheta_4) = 0$$

$$I_4\ddot{\vartheta}_4 + c_4\dot{\vartheta}_4 + c_{SA1}(\dot{\vartheta}_4 - \dot{\vartheta}_3) + c_{SA2}(\dot{\vartheta}_4 - \dot{\vartheta}_5) + k_{SA1}(\vartheta_4 - \vartheta_3) + k_{SA2}(\vartheta_4 - \vartheta_5) = 0$$

$$I_5\ddot{\vartheta}_5 + c_5\dot{\vartheta}_5 + c_{SA2}(\dot{\vartheta}_5 - \dot{\vartheta}_4) + k_{GB2}\left(\vartheta_5 - \frac{\vartheta_6}{\tau_2}\right) + k_{SA2}(\vartheta_5 - \vartheta_4) = 0$$

$$I_6\ddot{\vartheta}_6 + \left(c_6 + \frac{c_{GB2}}{\tau_2^2}\right)\dot{\vartheta}_6 - \frac{k_{GB2}}{\tau_2}\left(\vartheta_5 - \frac{\vartheta_6}{\tau_2}\right) = T_6 \quad (1)$$

where $\tau_1 = \tau_{G1}\tau_{F1}$, $\tau_2 = \tau_{G2}\tau_{F2}$ and the generalized coordinates are the angular positions of each inertial component in Fig. 6: $\boldsymbol{q} = \{\vartheta_1\vartheta_2\vartheta_3\vartheta_4\vartheta_5\vartheta_6\}^T$, and T_1 and T_6 represent the torques applied by M1 and M2, respectively.

Table 1 Parameters of the transmission test bench model

Quantity	Symbol	Value	Units
Primary DMF inertia	I_1	0.195	kgm^2
Secondary DMF inertia	I_2	0.147	kgm^2
DCT inertia (at output shaft)	I_3	0.123	kgm^2
Brake disk inertia	I_4	0.18	kgm^2
MT differential inertia	I_5	0.172	kgm^2
MT and M2 inertia	I_6	0.069	kgm^2
DCT gear ratio	τ_{G1}	[4.15, 2.27, 1.43, 0.98, 0.76, 0.62, −4]	−
MT gear ratio	τ_{G2}	[4.17, 2.35, 1.46, 0.95, 0.69, 0.55, −4.08]	−
DCT final drive ratio	τ_{F1}	4.118	−
MT final drive ratio	τ_{F2}	4.222	−
Fly-wheel stiffness	k_{DMF}	458.4	Nm/rad
DCT stiffness	k_{GB1}	[2.73, 7.84, 0.96, 3.78, 0.33, 1.82] × 10^5	Nm/rad
MT stiffness	k_{GB2}	1.34 × 10^5	Nm/rad
First driveshaft stiffness	k_{SA1}	0.12 × 10^5	Nm/rad
Second driveshaft stiffness	k_{SA2}	0.15 × 10^5	Nm/rad
DOF damping	$[c_1, c_2, \dots, c_6]$	[0.2, 2, 0.02, 0.34, 4, 0]	Nms/rad
DMF Damping	c_{DMF}	7.5	Nms/rad
DCT damping	c_{GB1}	0	Nms/rad
MT damping	c_{GB2}	4	Nms/rad
First driveshaft damping	c_{SA1}	0.1	Nms/rad
Second driveshaft damping	c_{SA2}	0.1	Nms/rad

Equation (1) can be expressed with the following matrix formulation:

$$M\ddot{q} + C\dot{q} + Kq = \underbrace{\begin{bmatrix} 1 & 0 \\ 0 & 0 \\ 0 & 0 \\ 0 & 0 \\ 0 & 0 \\ 0 & 1 \end{bmatrix}}_{E} \begin{bmatrix} T_1 \\ T_6 \end{bmatrix} \tag{2}$$

where the mass matrix M, the damping matrix C and the stiffness matrix K are defined as:

$$M = diag(I_1, I_2, I_3, I_4, I_5, I_6)$$

$$C = \begin{bmatrix} c_{DMF} + c_1 & -c_{DMF} & 0 & 0 & 0 & 0 \\ -c_{DMF} & c_{DMF} + c_2 + \frac{c_{GB1}}{\tau_1^2} & 0 & 0 & 0 & 0 \\ 0 & 0 & c_3 + c_{SA1} & -c_{SA1} & 0 & 0 \\ 0 & 0 & -c_{SA1} & c_{SA1} + c_4 + c_{SA2} & -c_{SA2} & 0 \\ 0 & 0 & 0 & -c_{SA2} & c_5 + c_{SA2} & 0 \\ 0 & 0 & 0 & 0 & 0 & c_6 + \frac{c_{GB2}}{\tau_2^2} \end{bmatrix}$$

$$K = \begin{bmatrix} k_{DMF} & -k_{DMF} & 0 & 0 & 0 & 0 \\ -k_{DMF} & k_{DMF} + \frac{k_{GB1}}{\tau_1^2} & -\frac{k_{GB1}}{\tau_1} & 0 & 0 & 0 \\ 0 & -\frac{k_{GB1}}{\tau_1} & k_{GB1} + k_{SA1} & -k_{SA1} & 0 & 0 \\ 0 & 0 & -k_{SA1} & k_{SA1} + k_{SA2} & -k_{SA2} & 0 \\ 0 & 0 & 0 & -k_{SA2} & k_{SA2} + k_{GB2} & -\frac{k_{GB2}}{\tau_2} \\ 0 & 0 & 0 & 0 & -\frac{k_{GB2}}{\tau_2} & \frac{k_{GB2}}{\tau_2^2} \end{bmatrix} \tag{3}$$

The system is then expressed with the state-space representation:

$$\dot{x} = \underbrace{\begin{bmatrix} 0_{6x6} & I_{6x6} \\ -M^{-1}K & -M^{-1}C \end{bmatrix}}_{A_{OL}} x + \underbrace{\begin{bmatrix} 0_{6x2} \\ -M^{-1}E \end{bmatrix}}_{B_{OL}} u \tag{4}$$

where $x = \{ q \ \dot{q} \}^T$ is the state vector and $u = \{ T_1 \ T_6 \}^T$ is the input vector.

Modal Analysis

By solving the eigenvalue problem associated with the system in Eq. (4), the damped natural frequency $\omega_{n,r}$, the damping ratio ζ_r, the amplitude and phase of the complex eigenvector ψ_r associated to the r_{th} mode are computed and shown in Figs. 7 and 8 for different values of the MT gears g_2. The eigenvectors are normalized so that

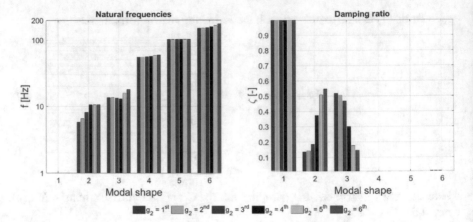

Fig. 7 Natural frequencies and damping ratio of the open-loop system with different MT gears and DCT gear $g_1 = 5th$

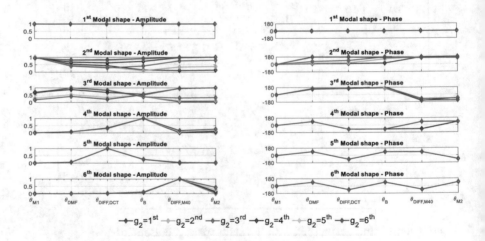

Fig. 8 Modal shape amplitude and phase of the open-loop system with different MT gears and DCT gear $g_1 = 5th$

the modulus of the maximum element is unitary and, due to the presence of the two gear ratios τ_1 and τ_2, they are reduced to the shaft of the electric motor M1. As an example, the r_{th} eigenvector $\boldsymbol{\psi}_{r,M1}$ is scaled as follows:

$$\boldsymbol{\psi}_{r,M1} = \left\{ \psi_1 \ \psi_2 \ \psi_3 \tau_1 \ \psi_4 \tau_1 \ \psi_5 \tau_1 \ \psi_6 \frac{\tau_1}{\tau_2} \right\}_r^T \tag{5}$$

The first mode represents a rigid body mode while the second mode represents the first real torsional mode of the system. The effect of the MT gear ratio mainly influences the second and third modes whereas it does not show a similar influence for higher vibration modes. The last three modes show a very similar characteristic: a single inertia vibrates with respect to the remaining part of the driveline, which stays practically stationary.

A similar sensitivity analysis is conducted for the DCT gear g_1, thus obtaining the results shown in Figs. 9 and 10.

Even in this case, the DCT gear only influences the second and the third modal shapes while higher frequency modes are not affected. Furthermore, the DCT architecture with different torque paths through the transmission, depending on the gear engaged, leads to a non-monotonous trend of the natural frequency with the gear

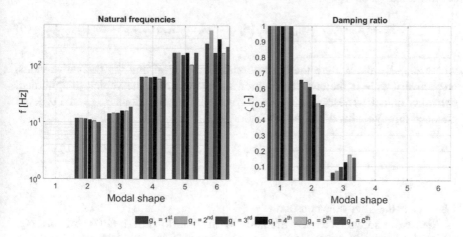

Fig. 9 Natural frequencies and damping ratio of the open-loop system with different DCT gears and MT gear $g_1 = 5th$

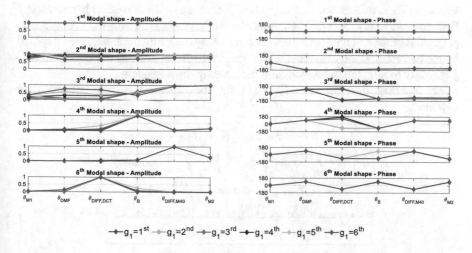

Fig. 10 Modal shape amplitude and phase of the open-loop system with different DCT gears and MT gear $g_1 = 5th$

ratio. Indeed, the even and odd gears of DCT are installed on a different input shafts if compared to the MT architecture where each gear is sequentially mounted on the unique input shaft.

3.2 Closed-Loop State-Space Model

The closed-loop system is obtained by adding the following M2 speed control logic, as shown in Fig. 6:

$$T_6 = K_p(\dot{\vartheta}_{ref} - \dot{\vartheta}_6) + K_i(\vartheta_{ref} - \vartheta_6) + K_d(\ddot{\vartheta}_{ref,F} - \ddot{\vartheta}_{6,F}) \tag{6}$$

where K_p, K_i and K_d are the proportional, the integral and the derivative gains, respectively. ϑ_{ref} is the angular position calculated from the vehicle longitudinal dynamics model, as described in Sect. 2.2. Angular accelerations $\ddot{\vartheta}_{ref,F}$ and $\ddot{\vartheta}_{6,F}$ are obtained through a band-limited derivation of $\dot{\vartheta}_{ref}$ and $\dot{\vartheta}_6$, respectively:

$$\begin{cases} t_F \dddot{\vartheta}_{ref,F} + \ddot{\vartheta}_{ref,F} = \ddot{\vartheta}_{ref} \\ t_F \dddot{\vartheta}_{6,F} + \ddot{\vartheta}_{6,F} = \ddot{\vartheta}_6 \end{cases} \tag{7}$$

where t_F is the filter time constant.

Alternative solutions exist in literature for the definition of the PID structure, e.g. the proportional and derivative terms can be placed in the feedback signal rather in the feedback error [16]. Indeed, this alternative structure provides a good disturbance rejection and removes zeros from the closed-loop transfer function thus reducing the overshoot for reference tracking response. To obtain a similar result with the standard PID structure of Eq. (6), a reference prefilter is often used. However, the PID structure selected for the presented activity is constrained by the hardware limitations imposed by the two electric motors drives with a standard control logic as in Eq. (6) with the possibility to tune the three gains and the filtering time constant.

By inserting Eqs. (6) and (7) in (1), the following 8 dynamic equations are obtained:

$$I_1\ddot{\vartheta}_1 + c_{DMF}(\dot{\vartheta}_1 - \dot{\vartheta}_2) + c_1\dot{\vartheta}_1 + k_{DMF}(\vartheta_1 - \vartheta_2) = T_1$$

$$I_2\ddot{\vartheta}_2 + \left(c_2 + \frac{c_{GB1}}{\tau_1^2}\right)\dot{\vartheta}_2 - c_{DMF}(\dot{\vartheta}_1 - \dot{\vartheta}_2) - \frac{k_{GB1}}{\tau_1}\left(\vartheta_3 - \frac{\vartheta_2}{\tau_1}\right) - k_{DMF}(\vartheta_1 - \vartheta_2) = 0$$

$$I_3\ddot{\vartheta}_3 + c_3\dot{\vartheta}_3 + c_{SA1}(\dot{\vartheta}_3 - \dot{\vartheta}_4) + k_{GB1}\left(\vartheta_3 - \frac{\vartheta_2}{\tau_1}\right) + k_{SA1}(\vartheta_3 - \vartheta_4) = 0$$

$$I_4\ddot{\vartheta}_4 + c_4\dot{\vartheta}_4 + c_{SA1}(\dot{\vartheta}_4 - \dot{\vartheta}_3) + c_{SA2}(\dot{\vartheta}_4 - \dot{\vartheta}_5) + k_{SA1}(\vartheta_4 - \vartheta_3) + k_{SA2}(\vartheta_4 - \vartheta_5) = 0$$

$$I_5\ddot{\vartheta}_5 + c_5\dot{\vartheta}_5 + c_{SA2}(\dot{\vartheta}_5 - \dot{\vartheta}_4) + k_{GB2}\left(\vartheta_5 - \frac{\vartheta_6}{\tau_2}\right) + k_{SA2}(\vartheta_5 - \vartheta_4) = 0$$

$$I_6\ddot{\vartheta}_6 + \left(c_6 + \frac{c_{GB2}}{\tau_2^2} + K_p\right)\dot{\vartheta}_6 - \frac{k_{GB2}}{\tau_2}\vartheta_5 + \left(\frac{k_{GB2}}{\tau_2^2} + K_i\right)\vartheta_6 + K_d\ddot{\vartheta}_{6,F} - K_d\ddot{\vartheta}_{ref,F} = K_p\dot{\vartheta}_{ref} + K_i\vartheta_{ref}$$

$$I_6{}^t{}_F\ddot{\vartheta}_{6,F} + \left(c_6 + \frac{c_{GB2}}{\tau_2^2} + K_p\right)\dot{\vartheta}_6 - \frac{k_{GB2}}{\tau_2}\vartheta_5 + \left(\frac{k_{GB2}}{\tau_2^2} + K_i\right)\vartheta_6 + (I_6 + K_d)\ddot{\vartheta}_{6,F} - K_d\ddot{\vartheta}_{ref,F} = K_p\dot{\vartheta}_{ref} + K_i\vartheta_{ref}$$

$$t_F\ddot{\vartheta}_{ref,F} + \ddot{\vartheta}_{ref,F} = \ddot{\vartheta}_{ref} \tag{8}$$

The introduction of the two additional differential equations in Eq. (7) increases the state vector dimension of the closed-loop system by including the two additional states $\ddot{\vartheta}_{6,F}$ and $\ddot{\vartheta}_{ref,F}$. The tuning of proportional, integral and derivative gains modifies the equivalent damping, stiffness and inertial characteristics of the 6-DOF model. The state-space representation of the closed-loop system is then given by:

$$\dot{x}_a = A_{CL}x_a + B_{CL}u_a \tag{9}$$

where $x_a = \left\{x^T \; \ddot{\vartheta}_{6,F} \; \ddot{\vartheta}_{ref,F}\right\}^T$ is the augmented state vector and $u_a = \left\{T_1 \; \vartheta_{ref} \; \dot{\vartheta}_{ref} \; \ddot{\vartheta}_{ref}\right\}^T$ is the augmented input vector. Matrices A_{CL} and B_{CL} of the closed-loop system are defined in the Appendix.

Modal Analysis

The complex modal analysis for the closed-loop system in Eq. (9) was carried out and the results for MT gear influence are shown in Figs. 11 and 12 while the DCT gear effect is reported in Figs. 13 and 14. It is important to remark that all simulation results shown in the rest of this subsection are obtained by setting the PID gains to their optimal value reported in Table 3; the tuning procedure that led to those controller parameters will be described in Sect. 4.2.

By comparing the closed-loop results in Fig. 12 with the open-loop analysis in Fig. 8, the presence of the speed controller on the motor M2 introduces an additional vibration mode which is represented by the closed-loop second modal shape

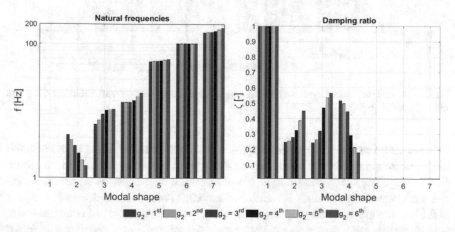

Fig. 11 Natural frequencies and damping ratio of the closed-loop system with different MT gears and DCT gear $g_1 = 5th$

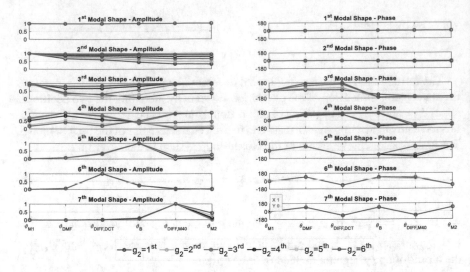

Fig. 12 Modal shape amplitude and phase of the closed-loop system with different MT gears and DCT gear $g_1 = 5th$

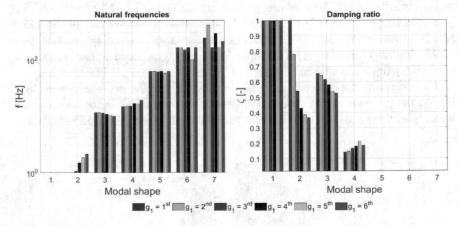

Fig. 13 Natural frequencies and damping ratio of the closed-loop system with different DCT gears and MT gear $g_2 = 5th$

at lower frequencies. The closed-loop modal shapes from the third to the seventh modes corresponds to the open-loop modal shapes from the second to the sixth, respectively. The speed control logic has also a small impact on the natural frequencies while the damping factor of the closed-loop third mode is increased with respect to the correspondent open-loop second mode. It is interesting to note that the speed controller does not influence the modal shape, natural frequency, and damping factor of the last three modes.

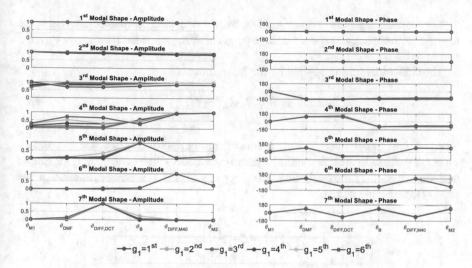

Fig. 14 Modal shape amplitude and phase of the closed-loop system with different DCT gears and MT gear $g_2 = 5th$

The effect of the DCT gear g_1 on the open-loop and the closed-loop modal analysis comparison is analyzed in Figs. 13 and 14.

The DCT gear ratio has a lower influence on the modal shape of the additional mode (the second) introduced by the closed-loop system. Even if the closed-loop is still beneficial in increasing the damping factor of the third and fourth modes, respect to the equivalent second and third open-loop modes, the variation of the DCT gear ratio is not as effective as observed for the variation of the MT gear.

3.3 Model Experimental Validation

A torsional vibration test is executed on the DCT test rig to obtain the experimental data and validate the linear model described and analyzed in the previous sections. During the test, one electric motor (e.g. M2) is controlled to apply a sinusoidal torque with a constant amplitude and a continuously variable frequency, while the second motor (M1) is controlled to keep a constant speed. Before the torsional vibration test is started, a preliminary phase is required to bring the transmission test rig into a steady-state condition, identified by constant torque and speed, where the behavior of the system can be linearized through Eq. (1). The preliminary phase is fundamental to keep approximately constant the system parameters, e.g. torsional stiffnesses, and to avoid any inversion of torque sign thus preventing nonlinear phenomena such as the impact between rotating components due to backlash.

Time histories of the torques applied by the two electric motors are measured together with the angular speeds in three points of the transmission line ($\dot{\theta}_1 = \dot{\theta}_{M1}$, $\dot{\theta}_3$, and $\dot{\theta}_6 = \dot{\theta}_{M2}$).

The sinusoidal torque applied by the M2 motor is:

$$T_6 = \overline{T} + T_0 \sin(2\pi f(t)t) \tag{10}$$

where $\overline{T}$ is a mean value of torque constantly applied during the test to create a torsional preload in the transmission system (to avoid nonlinearities), T_0 is the torque amplitude and $f(t)$ the excitation frequency calculated as a power function of the time t (logarithmic chirp):

$$f(t) = f_0 \left(\frac{f_1}{f_0} \right)^{\frac{t}{t_1}} \tag{11}$$

where f_0 is the initial frequency and f_1 is the final frequency at time t_1.

Experimental measurements are then processed resulting in the Frequency–Response Functions (FRFs) between the input torque T_6 applied by M2 and the three rotational speeds. Under the assumption that the system is linear (at least in the neighborhood of the equilibrium point around which the system vibrates) and characterized by time-invariant parameters, the estimation of a FRF from experimental data can be performed by calculating the Power Spectral Density (PSD) and the Cross-Power Spectral Density (CSD) through the Welch's method [17].

Figures 15 and 16 show the estimated FRFs from the measurements together with their coherence functions to evaluate the frequency range for which the estimation algorithm is considered reliable.

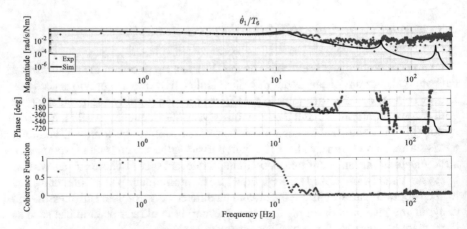

Fig. 15 FRF $\dot{\theta}_1/T_6$ resulting from the experimental measurements and calculated using the open-loop linear model with $g_1 = 3rd$ and $g_2 = 4th$

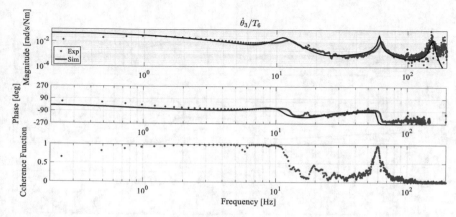

Fig. 16 FRF $\dot{\theta}_3/T_6$ estimated from the experimental measurements and calculated from the open-loop linear model with $g_1 = 3rd$ and $g_2 = 4th$

Since the excitation for the dynamic system is generated by the electric motor M2, the amplitude of the $\dot{\theta}_3/T_6$ and $\dot{\theta}_6/T_6$ responses are high enough to be measured accurately by the encoder up to more than 150 Hz. On the other hand, the lower amplitude of oscillation for the other M1 motor limits the range of reliability of $\dot{\theta}_1/T_6$ to 20 Hz.

The results concerning the first FRF ($\dot{\theta}_1/T_6$) are affected by a relevant phase difference in the high frequency range, and this is mainly due to very low signal to noise ratio in the experimental measures, as also confirmed by the coherence function that drops to zero very quickly after 20 Hz.

On the other hand, the experimental FRF of $\dot{\theta}_3/T_6$ is very well estimated from the encoder measurements in the whole frequency range, thanks to the high peak amplitudes and the high encoder resolution. Furthermore, the linear model can capture both the experimental magnitude and phase up to more than 200 Hz.

Finally, the FRF of $\dot{\theta}_6/T_6$ is shown in Fig. 17, where the accuracy of the experimental estimation decreases significantly immediately after the third peak in magnitude. However, the FRF magnitude and phase is also well captured by the linear model within the validation range of the experimental estimation.

4　Sensitivity Analysis of the Linear Model

The experimentally validated DCT torsional model represents an important tool for predicting and calibrating the effect of the two motor controllers on the test rig dynamic behavior. The present section aims at analyzing the DCT model sensitivity to the parameters that are subject to calibration, e.g. the speed control logic gains, or to optimization for accomplishing the desired operative condition, e.g. the DCT or MT gear ratio.

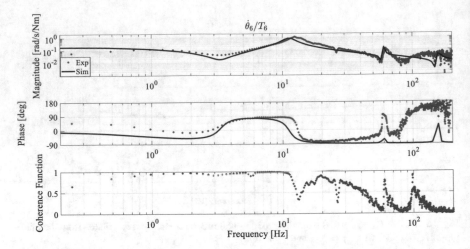

Fig. 17 FRF $\dot{\theta}_6/T_6$ estimated from the experimental measurements and calculated from the open-loop linear model with $g_1 = 3rd$ and $g_2 = 4th$

The sensitivity analysis is carried out in the frequency domain by considering the frequency response functions between the closed-loop system input (M1 motor torque T_1 and reference speed $\dot{\theta}_{ref}$) and the desired output quantities selected for evaluating the DCT test rig performance. One of the main tasks for the considered HiL test rig is to regulate the brake disk speed $\dot{\theta}_4$ to track the reference angular speed calculated from the vehicle model. However, the speed control logic in Eq. (6) is applied to the M2 motor speed $\dot{\theta}_6$. Based on these considerations, four FRFs are evaluated from the closed-loop system described by the state space in Eq. (9): two Reference Speed Tracking (RST) FRFs for evaluating the controller tracking performance and two Disturbance Rejection (DR) FRFs for analyzing the M2 controller sensitivity to the application of the M1 motor torque.

The two RST FRFs are obtained by considering a harmonic excitation for the reference speed $\dot{\theta}_{ref}$ with amplitude $\dot{\theta}_{ref,0}$ and frequency Ω by fixing the M1 torque $T_1 = 0$:

$$\boldsymbol{\alpha_{RST}}(\Omega) = \left[G_{RST}^4(\Omega) \; G_{RST}^6(\Omega) \right]^T = \boldsymbol{C}(j\Omega\boldsymbol{I} - \boldsymbol{A_{CL}})^{-1}\boldsymbol{B_{CL}}\left[0 \; 1/(j\Omega) \; 1 \; j\Omega \right]^T \tag{12}$$

where $\boldsymbol{C} = \left[0\,0\,0\,0\,0\,0\,0\,0\,0\,1\,0\,1\,0\,0 \right]^T$ is the output matrix; $G_{RST}^4 = \tau_2\dot{\theta}_{4,0}/\dot{\theta}_{ref,0}$ and $G_{RST}^6 = \dot{\theta}_{6,0}/\dot{\theta}_{ref,0}$ are the RST FRFs from the reference speed to the disk brake $\dot{\theta}_4$ and to the M2 motor speed $\dot{\theta}_6$, respectively.

The desired shape for both the RST FRFs is $G_{RST}^4(\Omega) = G_{RST}^6(\Omega) = 1$ for $\Omega < \Omega_B$ where Ω_B represents the closed-loop bandwidth frequency above which it is required to drop for reducing the sensitivity to high frequency noises.

The two DR FRFs are computed by considering a harmonic excitation for the M1 torque T_1 with amplitude $T_{1,0}$ and frequency Ω by fixing the reference speed $\theta_{ref} = \dot{\theta}_{ref} = \ddot{\theta}_{ref} = 0$:

$$\boldsymbol{\alpha}_{\mathbf{DR}}(\Omega) = \begin{bmatrix} G^4_{DR}(\Omega) & G^6_{DR}(\Omega) \end{bmatrix}^T = \boldsymbol{C}(j\Omega\boldsymbol{I} - \boldsymbol{A}_{CL})^{-1}\boldsymbol{B}_{CL}\begin{bmatrix} 1 & 0 & 0 & 0 \end{bmatrix}^T \quad (13)$$

where $G^4_{DR} = \dot{\theta}_{4,0}/T_{1,0}$ and $G^6_{DR} = \dot{\theta}_{6,0}/T_{1,0}$ are the DR FRFs from the M1 torque T_1 to the disk brake $\dot{\theta}_4$ and to the M2 motor speed $\dot{\theta}_6$, respectively.

The desired shape for both the DR FRFs is $G^4_{DR}(\Omega) = G^6_{DR}(\Omega) = 0$ for the whole frequency range.

4.1 Sensitivity to Gear Ratios

The first analysis concerns the model sensitivity to the variation of the DCT gear ratio, which is a parameter imposed by the TCU having a strong influence on the test bench dynamics and the motor controller performance. The PID gains of the speed controller are set to the nominal values reported in Table 3.

Figure 18 shows the trend of the RST and DR FRFs when the DCT gear ratio is changed, and the MT gear ratio is kept constant to $g_2 = 5th$.

Both RST FRFs show a clear resonance peak at the first natural frequency that increases with the DCT gear ratio, as already shown in Fig. 13. The amplitude of the first resonance peak increases with the DCT gear ratio thus reflecting the

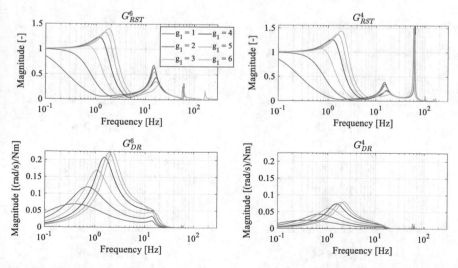

Fig. 18 DR and RST FRFs magnitude for the closed-loop system with different DCT gears and MT gear $g_2 = 5th$

damping ratio characteristics obtained from the modal analysis in Fig. 13. A second resonance peak is also visible for the G^6_{RST} FRF whose frequency increases with the DCT gear but its amplitude decreases for higher gears. A similar shift of the resonance frequencies is obtained for the G^4_{RST} FRF but the amplitude of the second and the third peaks are more pronounced than their correspondents in G^6_{RST}. This result agrees with the modal shape of the fifth mode (~60 Hz) in Fig. 14 where the brake disk inertia vibrates with respect to the remaining part of the driveline, which stays practically stationary.

The effect of the motor M1 torque is analyzed through the two DR FRFs plotted in Fig. 18. The influence of the M1 torque on $\dot{\theta}_6$ and $\dot{\theta}_4$ dynamics is more pronounced for G^6_{DR} than G^4_{DR} and amplifies the first peak for higher DCT gears.

The main conclusion for the sensitivity analysis on the DCT gear variation is that the dynamic behavior of the M2 speed controller is improved in terms of disturbance rejection from M1 when the first gear is engaged in the DCT. However, the reference tracking attitude for $g_1 = 1$ does not represent the best selection since the closed-loop bandwidth is drastically reduced if compared to higher gears RST FRFs.

The second sensitivity analysis involves the variation of the MT gear ratio, considered as a tunable parameter for extending the test rig operative speed and torque range based on the maneuver selection. The effect of τ_2 on the RST FRFs is shown in Fig. 19 by engaging the fifth DCT gear.

The effect of MT gear on the first peak of both G^6_{RST} and G^4_{RST} is opposite if compared to the DCT sensitivity analysis in Fig. 18: the resonance frequency and the peak amplitude decreases with MT gear. Moreover, the first and second peaks tend to get closer for lower gears. The sixth gear would represent a good choice in terms of reference tracking performance due to the lower amplitude of the first peak, even if

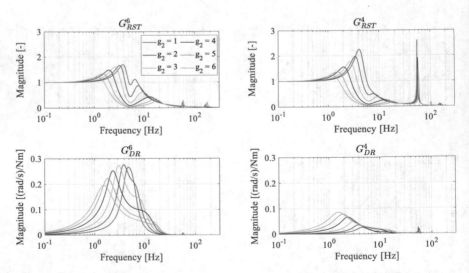

Fig. 19 RST and DR FRFs magnitude for the closed-loop system with different MT gears and DCT gear $g_1 = 5th$

this advantage is obtained with a lower closed-loop bandwidth, if compared to other gears. Finally, the DR FRFs modification to the MT gear ratio variation is reported in Fig. 19. The influence of M1 torque on the M2 speed controller dynamics is still more predominant on G_{DR}^6 than G_{DR}^4. The selection of higher gears reduces the first peak amplitude for G_{DR}^6 while a lower gear is beneficial for attenuating G_{DR}^4.

The sensitivity analysis on the MT gear ratio variation led to the conclusion that the sixth gear represents an optimal solution to always guarantee the desired reference tracking performance and external disturbance rejection. However, the actual usage of the HiL test bench requires to run the desired maneuver considering also the constraints imposed by the power and torque limitations of the two electric motors, that may require a different gear selection from the optimal solution. Therefore, the performance of the test bench must be carefully verified with all the possible combinations of gear ratios.

4.2 Sensitivity to PID Gains

Differently from the mechanical parameters such as the DCT and MT gear ratios, the PID gains as well as the filter time constant t_F, can be tuned and eventually adapted to the test rig operative conditions with more flexibility.

The effect of the speed proportional gain K_p on the RST FRFs is evaluated in Fig. 20 where the fifth gear is set for both DCT and MT.

The proportional speed gain cannot shift none of the resonance frequencies referring to both G_{RST}^6 and G_{RST}^4 but it can modify their amplitude values. The reason

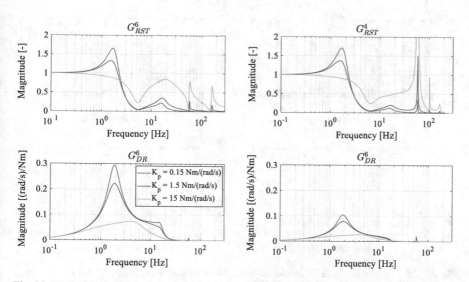

Fig. 20 RST and DR FRFs magnitude for the closed-loop system with different proportional gains K_p, by setting $K_i = 75$ Nm/rad, $K_d = 0.75$ Nm/(rad/s^2), $t_F = 100$ s, $g_1 = 5th$ and $g_2 = 5th$

behind that is well explained by Eq. (8) where K_p modifies the multiplicative coefficient of $\dot{\theta}_6$. An increment of K_p would improve the reference tracking performance by reducing the first peak amplitude and by extending the closed-loop bandwidth. However, the benefits achieved with a high K_p gain on the G^6_{RST} FRF produces a negative effect on the G^4_{RST} FRF for the high frequency range. This aspect aims at remarking that even a fine tuning of the M2 speed controller may provoke excessive oscillations in other driveline parts and then transmitted to the test rig supports in terms of vibrations or noises perceived by the user. A similar sensitivity analysis for the disturbance rejection FRFs is shown in Fig. 20 where an increment of K_p is positive for both G^6_{DR} and G^4_{DR} in terms of resonance peak attenuation.

The effect of the integral gain K_i on the RST and DR FRFs is then evaluated in Fig. 21.

Differently from the proportional gain, the integral contribution can shift the resonance frequency of the first peak since it modifies the stiffness contribution in Eq. (8) for the last degree of freedom (motor M2). The main benefit of increasing the integral gain is the extension of the closed-loop bandwidth but at the cost of a more pronounced peaks amplitude in the whole frequency range. An increment of K_i also produces a shift of the DR FRFs first resonance peak towards higher frequencies and it shrunk the frequency band around the peak, as visible for both G^6_{DR} and G^4_{DR}. Another benefit from the increment of the integral contribution is the attenuation of the G^6_{DR} peak amplitude meanwhile G^4_{DR} amplitude is not influenced by the variation of this parameter.

Furthermore, the effect of the derivative PID gain on both RST and DR FRFs is shown in Fig. 22.

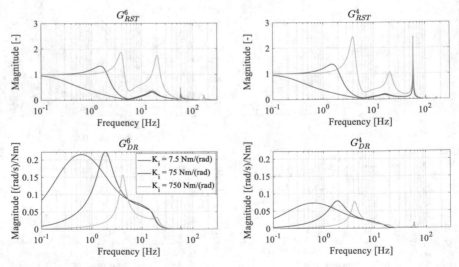

Fig. 21 RST FRFs magnitude for the closed-loop system with different integral gains K_i, by setting $K_p = 1.5$ Nm/(rad/s), $K_d = 0.75$ Nm/(rad/s²), $t_F = 100$ s, $g_1 = 5th$ and $g_2 = 5th$

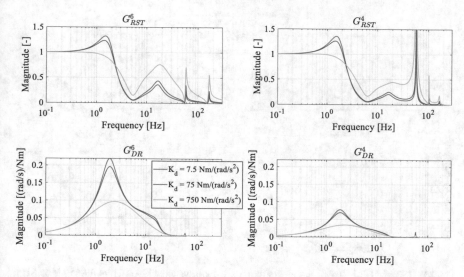

Fig. 22 RST and DR FRFs magnitude for the closed-loop system with different derivative gains K_d, by setting $K_p = 1.5$ Nm/(rad/s), $K_i = 75$ Nm/rad, $t_F = 100$ s, $g_1 = 5th$ and $g_2 = 5th$

The derivative gain can modify the peaks amplitude, but it cannot shift any resonance frequency. The positive benefit of a high derivative gain is the reduction of the first peak magnitude for both G^6_{RST} and G^4_{RST} but at the cost of an increment in the peak amplitudes for higher frequencies, especially for G^4_{RST}. A similar conclusion is observed for the DR FRFs, where an increment of K_d is beneficial for the whole frequency range in terms of peak attenuation but at the cost of a larger frequency band pass.

Finally, the impact of the filter time constant t_F on both RST and DR FRFs is illustrated in Fig. 23. It can be seen that, the effect of the filter time constant on system dynamics in the low-mid frequency range is non monotonic: for low values of t_F, an additional damping to the first resonant mode is obtained; on the contrary, for high values, a further increase reduces the damping and increases the frequency of the main peaks of all the considered FRFs. On the other hand, at high frequencies, the effect of the time constant increment is always to attenuate the response of both RST. This last effect is particularly beneficial in this application, where high frequency disturbances are generated by the staircase encoder speed signals that requires to filter out this high frequency content to avoid NVH issues during the test.

The results obtained through the sensitivity analysis of the M2 speed controller parameters, represent a valid tool for calibrating the PID gains thus achieving a closed-loop FRF response which can be shaped according to the desired requirements reported in Table 2. The desired shape for both G^6_{RST} and G^4_{RST} FRFs magnitudes should be constrained as close as possible to 1 to satisfy the reference tracking performance. However, this can be achieved only within the frequency bandwidth range, that is the frequency range where the magnitude of a closed-loop FRF is greater than -3 dB. The sensitivity analysis shows that the most effective parameter

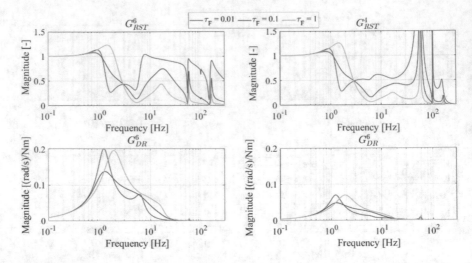

Fig. 23 RST and DR FRFs magnitude for the closed-loop system with different filter time constant t_F, by setting $K_p = 1.5$ Nm/(rad/s), $K_i = 75$ Nm/rad, $K_d = 0.75$ Nm/(rad/s^2), $g_1 = 5th$ and $g_2 = 5th$

Table 2 Desired FRF requirements for PID calibration

Property	Requirements
Peak amplitude of G^6_{RST}	≤ 1.3
Peak amplitude of G^4_{RST}	≤ 1.5
Bandwidth of G^6_{RST} (at half-power point)	≥ 2 Hz
Bandwidth of G^6_{DR} (at half-power points)	≤ 4 Hz
Peak amplitude of G^6_{DR}	≤ 0.25(rad/s)/Nm

to extend the frequency bandwidth is the integral contribution K_i. An excessive selection of the integral gain produces an increment of all resonance peak amplitudes, thus reducing the reference tracking performance. This negative consequence can be in part compensated by increasing the proportional and the derivative gains to smooth the peaks, at least in the lower frequency range. The amplitude of the G^4_{RST} peaks also require a constraint to avoid an undesired vibration level in any other point of the driveline. Indeed, since the closed-loop system is feedbacked on the M2 motor speed, the magnitude constraint for G^4_{RST} peaks is set higher than G^6_{RST} FRF.

To keep the desired reference tracking performance achieved for RST FRSs, the influence of the M1 motor torque should be also mitigated by shaping G^6_{DR} and G^4_{DR}. The sensitivity analysis proved that G^6_{DR} represents the more critical FRF in terms of resonance peaks amplitude and on the peak frequency band which can be shrunk with an increment of the integral gain and a reduction of proportional and derivatives gains.

Table 3 Final values of PID gains	Gain	Value
	Proportional gain K_p	1.5 Nm/(rad/s)
	Integral gain K_i	75 Nm/(rad)
	Derivative gain K_d	0.75 Nm/(rad/s^2)
	Derivative time constant t_F	100 s

To achieve the required dynamic performance of the closed-loop system reported in Table 3, a brute force method is applied to search for a suitable set of PID gains among a predefined range of values. The lower and the upper limits of each PID parameter range are defined based on the sensitivity analysis carried out in the present section, thus constraining the brute force algorithm to test only the set of gains that provoke a perceptible variation of the RST and DR frequency responses. The final gains computed by the brute force algorithm are then reported in Table 3:

The final set of PID parameters clearly shows that the solution computed by the brute force algorithm provides a low derivative gain with a high filtering time constant, whose combined effect produces a negligible intervention of the derivative term. Indeed, Fig. 22 clearly demonstrates that only a derivative gain of two orders of magnitude greater than the nominal value would modify the frequency response of the closed-loop system. The calibrated speed controller is then verified through the time domain response of the closed-loop system, by providing a delayed step input between $\dot{\theta}_{ref}$ and T_1. Time histories results of $\dot{\theta}_6$, $\dot{\theta}_4$ and T_6 are shown in Fig. 24.

The speed controller on M2 provides satisfactory response in terms of reference speed tracking performance with an overshoot of 5.6%, a rise time of 0.2 s and a settling time of 0.4 s. The step torque on T_1, applied 2 s later after the imposition of

Fig. 24 Time response of $\dot{\theta}_6$ and $\dot{\theta}_4$ to a delayed step input between $\dot{\theta}_{ref}$ and T_1 with $g_1 = 5th$ and $g_2 = 5th$

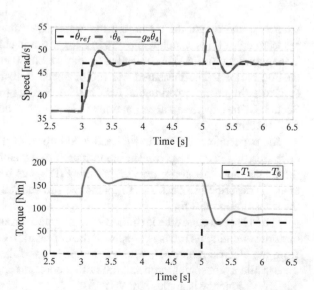

the step on the reference speed, provides useful information regarding the capability of the controller to react to external disturbance. Even if the M1 torque T_1 produces a 16% overshoot on $\dot{\theta}_6$ and $\dot{\theta}_4$, the closed-loop system can reject the torque disturbance after 1 s.

5 Controller Experimental Validation

The PID gains selected through the sensitivity analysis described in the previous section, are then implemented to the transmission test rig for validating the controller performance. The controller performance is validated both in frequency and time domain to prove its efficiency in terms reference tracking accuracy and disturbance rejection to external torques. To guarantee that the transmission works as much as possible in a linear operating range, before the execution of any experiments, the test rig is brought into a steady-state condition identified by a constant angular speed, imposed by the speed controller of M2, and by a transmission torsional preload through the application of a constant torque from the motor M1. This procedure is necessary to avoid the inversion of the torque sign which is responsible for extremely non-linear gear teeth impact due to the torsional backlashes between transmission rotating components [11, 18].

During the first test, the M2 electric motor is controlled in speed through the PID logic in Eq. (6) and parametrized according to Table 3. A sinusoidal speed with a constant amplitude and a continuously variable frequency is generated for $\dot{\theta}_{6,ref}$:

$$\dot{\theta}_{ref} = \dot{\theta}_m + \dot{\theta}_a \sin(2\pi f(t)t) \tag{14}$$

where $\dot{\theta}_m$ and $\dot{\theta}_a$ are the mean and the amplitude values of the sinusoidal signal, respectively. $f(t)$ is calculated to obtain a logarithmic chirp as described in Eq. (11). The second motor (M1) is controlled to apply a constant torque T_1 to guarantee the desired preload. The measured M2 motor speed is than elaborated, together with $\dot{\theta}_{ref}$, to estimate the experimental RST FRF, according to the algorithm described in Sect. 3.3. The experimental estimation of G_{RST}^6 is then compared to the linear model RST FRF in Fig. 25.

The experimental G_{RST}^6 is well estimated up to 40 Hz, above which the coherence function drops to zero. The good match between the experimental FRF with the linear model response proves that the PID gains tuned through the sensitivity analysis described in Sect. 4, can guarantee the desired RST performance also for the transmission test rig.

The M2 speed controller is furtherly investigated through a second experiment that aims at validating its robustness against the application of a dynamic external torque from the electric motor M1. The M2 motor is controlled to keep a constant speed meanwhile a sinusoidal torque signal with a constant amplitude and a continuously variable frequency is applied by M1:

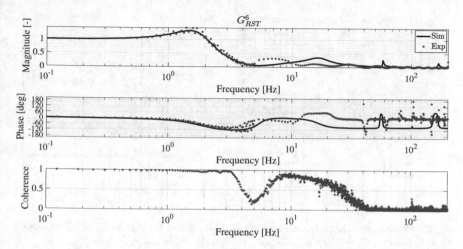

Fig. 25 Experimental estimation and model evaluation of G^6_{RST} by fixing $K_p = 1.5$ Nm/(rad/s), $K_i = 75$ Nm/rad, $K_d = 0.75$ Nm/(rad/s²), $t_F = 100$ s $g_1 = 5th$ and $g_2 = 5th$

$$T_1 = T_{1,m} + T_{1,a} \sin(2\pi f(t)t) \qquad (15)$$

where $T_{1,m}$ and $T_{1,a}$ are the mean and the amplitude values of the sinusoidal signal, respectively. $f(t)$ is calculated to obtain a logarithmic chirp as described in Eq. (11). The M1 torque is measured, as well as the M2 speed $\dot{\theta}_6$, to elaborate the estimation of G^6_{DR} which are compared against the linear model DR FRF in Fig. 26.

Within the frequency validation range of the experimental G^6_{DR} (up to 40 Hz), the M2 speed controller shows disturbance rejection properties even better than the

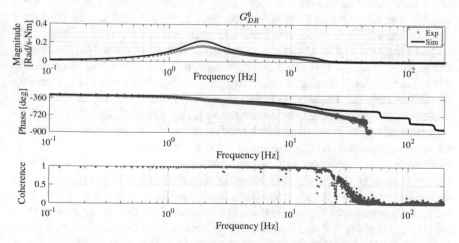

Fig. 26 Experimental estimation and model evaluation of G^6_{DR} by fixing $K_p = 1.5$ Nm/(rad/s), $K_i = 75$ Nm/rad, $K_d = 0.75$ Nm/(rad/s²), $t_F = 100$ s $g_1 = 5th$ and $g_2 = 5th$

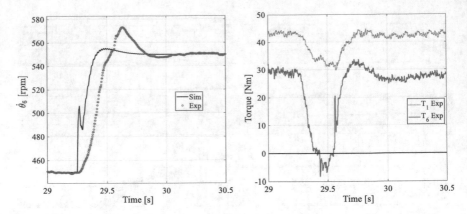

Fig. 27 Experimental and analytical time response of $\dot{\theta}_6$, T_1 and T_6 by fixing $K_p = 7.5\,\text{Nm/(rad/s)}$, $K_i = 75\,\text{Nm/rad}$, $K_d = 0.75\,\text{Nm/(rad/s}^2)$, $t_F = 100\,\text{s}$, $g_1 = 5th$ and $g_2 = 5th$

desired requirements imposed in Table 2 as proved by the lower peak amplitude around 2 Hz.

Finally, the speed controller is verified in the time domain by imposing a step signal to the reference signal $\dot{\theta}_{ref}$. The experimental time histories of the M2 motor speed are compared against the linear transmission model step response in Fig. 27.

The experimental M2 motor speed is characterized by a quick response during the initial phase of the step application, mainly due to the presence of the second resonance peak at ~ 20 Hz visible in Fig. 25, followed by a slower dynamics dictated by the first resonance peak at ~ 2 Hz. The mismatch observed between the experimental and simulation results in Fig. 27 is larger than what expected from the frequency-domain comparison. This discrepancy is mainly due to the intervention of the nonlinear transmission behavior during the test execution. The main nonlinearities in an automotive transmission system are backlashes and angular clearances between the rotating components (meshing gears and synchronizers). Indeed, the second subplot of Fig. 27 shows a typical nonlinear trend of the torque entering the MT gearbox. A torque sign inversion characterized by a dead band is clearly visible; it is related to a loss of contact between internal rotating components of MT gearbox due to load reversal. The loss of contact after zero crossing is followed by torsional impacts which generate peaks in the instantaneous torque trend (see also [11, 19] for further details). The effect of backlash nonlinearities is to make the MT working in an operating condition far from the linear hypothesis assumed for system modelling. Nevertheless, the performance of the controlled system in time domain is still satisfactory even if the system dynamic behavior is quite different from the one used for controller parameter tuning. On the other hand, the sine-sweep test used for elaborating the experimental FRFs in Fig. 26, is executed with a constantly applied mean torque $T_{1,m}$ (see Eq. 10) that guarantees a minimum level of torsional preload that avoid any backlashes recovery. A similar conclusion is also valid for the sine-sweep results of Fig. 25, where the combination of a low sine amplitude $\dot{\theta}_a$ with a constant preload T_1 constrains the transmission test rig to work in a linear operating condition thus justifying a better match with the simulated FRF.

6 Conclusions

This paper provided a model-based tuning procedure for controlling the electric motors of a transmission HiL test rig with the aim of achieving the desired closed-loop requirements with the following main conclusions.

- HiL transmission test rigs require an accurate calibration of the torque and speed feedback controllers of the actuators, e.g. electric motors, used to emulate the external load applied to the system under investigation, e.g. a DCT.
- The dynamic behavior of the driveline mounted on the test bench is well described by the proposed linear 6-DOF transmission model, for both open-loop and closed-loop configurations, as long as the transmission test rig operates in linear conditions, e.g. avoiding torque sign inversion.
- The step change of the gear ratio of the transmission under test during normal operation of the bench has a relevant effect on system torsional dynamics. Potential future developments could involve an adaptation of the control parameters to achieve satisfactory performance in all the operating conditions. However, most of industrial drives typically do not allow a gain scheduling of the motor controllers during the test execution.
- A novel methodology for analyzing and tuning the reference speed tracking and disturbance rejection performance of the motor controllers, both in frequency and time domains has been explained. The sensitivity analysis helps to understand the effect of each mechanical and control parameter on the system dynamics thus guiding the calibration process.
- The application of this method to the DCT HiL test rig installed at Politecnico di Torino provided a good tradeoff between the conflicting requirements (reference speed tracking and disturbance rejection). The torsional vibration analysis of the system revealed the presence of an underdamped mode that could be excited by high frequencies noises related to speed or torque feedback signals. This requires a speed controller design that must include the analysis of the internal dynamics of the whole transmission and driveline to avoid NVH issues during normal operation of the bench.

Appendix—Closed-Loop State Space Matrices

$$A_{CL} = \begin{bmatrix} A_{11} & A_{12} & A_{13} \\ A_{21} & A_{22} & A_{23} \end{bmatrix}$$

where:

$$A_{11} = O_{6X6}; \; A_{12} = I_{6X6}; \; A_{13} = O_{6X2}$$

$$A_{21} = \begin{bmatrix} -\dfrac{k_{DMF}}{I_1} & \dfrac{k_{DMF}}{I_1} & 0 & 0 & 0 & 0 \\[2mm] \dfrac{k_{DMF}}{I_2} & -\dfrac{\left(k_{DMF}+\frac{k_{GB1}}{\tau_1^2}\right)}{I_2} & \dfrac{k_{GB1}}{\tau_1 I_2} & 0 & 0 & 0 \\[2mm] 0 & \dfrac{k_{GB1}}{\tau_1 I_3} & -\dfrac{(k_{GB1}+k_{SA1})}{I_3} & \dfrac{k_{SA1}}{I_3} & 0 & 0 \\[2mm] 0 & 0 & \dfrac{k_{SA1}}{I_4} & -\dfrac{(k_{SA1}+k_{SA2})}{I_4} & \dfrac{k_{SA2}}{I_4} & 0 \\[2mm] 0 & 0 & 0 & \dfrac{k_{SA2}}{I_5} & -\dfrac{(k_{SA2}+k_{GB2})}{I_5} & \dfrac{k_{GB2}}{\tau_2 I_5} \\[2mm] 0 & 0 & 0 & 0 & \dfrac{k_{GB2}}{\tau_2 I_6} & -\dfrac{\left(\frac{k_{GB2}}{\tau_2^2}+K_i\right)}{I_6} \\[2mm] 0 & 0 & 0 & 0 & \dfrac{k_{GB2}}{\tau_F \tau_2 I_6} & -\dfrac{\left(\frac{k_{GB2}}{\tau_2^2}+K_i\right)}{\tau_F I_6} \\[2mm] 0 & 0 & 0 & 0 & 0 & 0 \end{bmatrix}$$

$$A_{22} = \begin{bmatrix} -\dfrac{(c_{DMF}+c_1)}{I_1} & \dfrac{c_{DMF}}{I_1} & 0 & 0 & 0 & 0 \\[2mm] \dfrac{c_{DMF}}{I_2} & -\dfrac{\left(c_{DMF}+c_2+\frac{c_{GB1}}{\tau_1^2}\right)}{I_2} & 0 & 0 & 0 & 0 \\[2mm] 0 & 0 & -\dfrac{(c_3+c_{SA1})}{I_3} & \dfrac{c_{SA1}}{I_3} & 0 & 0 \\[2mm] 0 & 0 & \dfrac{c_{SA1}}{I_4} & -\dfrac{(c_{SA1}+c_4+c_{SA2})}{I_4} & \dfrac{c_{SA2}}{I_4} & 0 \\[2mm] 0 & 0 & 0 & \dfrac{c_{SA2}}{I_5} & -\dfrac{(c_5+c_{SA2})}{I_5} & 0 \\[2mm] 0 & 0 & 0 & 0 & 0 & -\dfrac{\left(c_6+\frac{c_{GB2}}{\tau_2^2}+K_p\right)}{I_6} \\[2mm] 0 & 0 & 0 & 0 & 0 & -\dfrac{\left(c_6+\frac{c_{GB2}}{\tau_2^2}+K_p\right)}{\tau_F I_6} \\[2mm] 0 & 0 & 0 & 0 & 0 & 0 \end{bmatrix}$$

$$A_{23} = \begin{bmatrix} O_{5X2} \\[2mm] -\dfrac{K_d}{I_6} & \dfrac{K_d}{I_6} \\[2mm] -\dfrac{(I_6+K_d)}{\tau_F I_6} & \dfrac{K_d}{\tau_F I_6} \\[2mm] 0 & -\dfrac{1}{\tau_F} \end{bmatrix}$$

$$B_{CL} = \begin{bmatrix} O_{6X4} \\[2mm] \dfrac{1}{I_1} & 0 & 0 & 0 \\[2mm] O_{4X4} \\[2mm] 0 & \dfrac{K_i}{I_6} & \dfrac{K_p}{I_6} & 0 \\[2mm] 0 & \dfrac{K_i}{\tau_F I_6} & \dfrac{K_p}{\tau_F I_6} & 0 \\[2mm] 0 & 0 & 0 & \dfrac{1}{\tau_F} \end{bmatrix}$$

References

1. Schnabler, M., Stifter, C.: Model-Based Design Methods for the Development of Transmission

Control Systems. In: Conference Paper, SAE International, 0148-7191 (2014)
2. Castiglione, M., Stecklein, G., Senseney, R., Stark, D.: Development of Transmission Hardware-in-the-Loop Test System. SAE Technical Paper, 2003-01-1027 (2003)
3. Bagalini, E.,Violante, M.: Development of an automated test system for ECU software validation: An industrial experience. In: 15th Biennial Baltic Electronics Conference (BEC), pp. 103–106, Tallinn, Estonia (2016)
4. Galvagno, E., Velardocchia, M., Vigliani, A., Tota, A.: Experimental Analysis and Model Validation of a Dual Mass Flywheel for Passenger Cars. SAE Technical Paper, 2015-01-1121 (2015)
5. Hwang, T., Roh, J., Park, K., Hwang, J., Lee, K.H., Lee, K., Lee, S.-J., Kim, Y.-J.: Development of HIL systems for active brake control systems. In: SICE-ICASE International Joint Conference (2006)
6. Lee, S.J., Kim, Y.J.: Development of hardware-in-the-loop simulation system for testing multiple ABS and TCS modules. Int. J. Veh. Des. 36(1), 13–23 (2004)
7. Tota, A., Galvagno, E., Velardocchia, M., Vigliani, A.: Passenger car active braking system: Model and experimental validation (Part I). Proc. Inst. Mech. Eng. C J. Mech. Eng. Sci. 232(4), 585–594 (2018)
8. Tota, A., Galvagno, E., Velardocchia, M., Vigliani, A.: Passenger car active braking system: Pressure control design and experimental results (part II). Proc. Inst. Mech. Eng. C J. Mech. Eng. Sci. 232(5), 786–798 (2018)
9. Galvagno, E., Tota, A., Vigliani, A., Velardocchia, M.: Pressure following strategy for conventional braking control applied to a HIL test bench. SAE Int. J. Passenger Cars-Mech. Syst. 10(2017-01-2496), 721–727 (2017)
10. Lee, M.H., et al.: Development of a hardware in the loop simulation system for electric power steering in vehicles. Int. J. Automot. Technol. V12, 733–744 (2011)
11. Galvagno, E., Tota, A., Velardocchia, M., Vigliani, A.: Enhancing Transmission NVH Performance through Powertrain Control Integration with Active Braking System. SAE Technical Paper 2017-01-1778 (2017)
12. Galvagno, E., Velardocchia, M., Vigliani, A.: Torsional oscillations in automotive transmission: Experimental analysis and modelling. Shock Vibration 2016 (2016)
13. Mendes, A., Meirelles, P.: Application of the hardware in-the-loop technique to an elastomeric torsional vibration damper. SAE Int. J. Engines 6(4), 2004–2014 (2013)
14. Bracco, G., Giorcelli, E., Mattiazzo, G., Orlando, V., Raffero, M.: Hardware-in-the-loop test rig for the ISWEC wave energy system. Mechatronics 25, 11–17 (2015)
15. Galvagno, E., Velardocchia, M., Vigliani, A.: Dynamic and kinematic model of a dual clutch transmission. Mech. Mach. Theory 46(6), 794–805 (2011)
16. Leonhard, W.: Control of Electrical Drives. Springer Science & Business Media (2001)
17. Welch, P.D.: The use of fast Fourier transform for the estimation of power spectra: a method based on time averaging over short, modified periodograms. IEEE Trans. Audio Electroacoust. 15(2), 70–73 (1967)
18. Guercioni, G.R., Galvagno, E., Tota, A., Vigliani, A., Zhao, T.: Driveline Backlash and Half-Shaft Torque Estimation for Electric Powertrains Control. SAE Technical Paper 2018-01-1345 (2018)
19. Osella, G., Cimmino, F., Galvagno, E., Vafidis, C., Velardocchia, M., Vigliani, A., Antonio, T.O.T.A.: U.S. Patent No. 10946855. U.S. Patent and Trademark Office, Washington, DC (2021)

Turbomachinery Monitoring and Diagnostics

Aly El-Shafei (ORCID)

Abstract Turbomachinery condition monitoring and diagnostics has come a long way in the last few decades. The main tool for monitoring and diagnosis of turbomachinery is the use of vibration measurement and analysis. Along with machine operating parameters, machine vibration analysis provides a complete picture of turbomachinery condition and can be used to perform a complete analysis and a diagnostic result can be subsequently obtained. Usually diagnostics is performed after an anomaly has been detected through monitoring. This is the approach taken by ISO 13373 series of diagnostic standards. This chapter discusses methods of vibration measurement for machinery monitoring and diagnostics. The sensors used, their operating principles and installation methods are described, as well as the required instrumentation and measurement set-up. The main tool for vibration diagnosis is the spectrum, however, for a complete and accurate diagnosis other vibration tools may have to be considered, including the overall amplitude, time waveform, the phase, the orbit and in some cases an operating deflection shape analysis (ODS) may need to be performed. Finally recent trends shall be briefly discussed, including the use of wireless sensors, cloud monitoring, expert diagnosis, the use of artificial intelligence in diagnosis, as well as the very recent use of vibration video motion magnification technology.

Keywords Turbomachinery · Condition monitoring · Machinery diagnostics · Operating parameters · Vibration measurement · Signal processing · Diagnostic process · Remote monitoring and diagnosis

A. El-Shafei (✉)
Cairo University, Giza 12316, Egypt
e-mail: elshafei@ritec-eg.com

RITEC, Cairo 11431, Maadi, Egypt

© The Author(s), under exclusive license to Springer Nature Switzerland AG 2022
T. Parikyan (ed.), *Advances in Engine and Powertrain Research and Technology*,
Mechanisms and Machine Science 114,
https://doi.org/10.1007/978-3-030-91869-9_14

1 Introduction

This chapter summarizes the state-of-the-art in turbomachinery condition monitoring and diagnostics. The condition monitoring market has expanded tremendously in the last few years [1], and many technologies have been promoted for machinery condition monitoring including infra-red thermography, oil analysis, motor current analysis and ultrasound. However, vibration condition monitoring is by far the leader in this market, with more than 65% market share [1]. In fact, this author has spent more than 30 years in this field and has monitored (along with his teams) tens of thousands of machines and it is the opinion of this author that it is sufficient to monitor vibration and operating parameters to fully diagnose any rotating machine and many reciprocating machines. Thus, this chapter is dedicated almost exclusively to the vibration condition monitoring and diagnosis of machines. The other condition monitoring technologies are useful and have specific usage, for example infra-red thermography is very useful for monitoring of electric panels or furnaces, ultrasound is extremely useful for diagnosis of electric discharge in transformers, while oil analysis is generally useful in confirming machine condition and the state of the lubricant. Motor current signature analysis is specific for motors, although motor diagnostics can be done by vibration analysis through experienced users.

Over the years the vibration measurement and analysis tools have evolved. In 1993 [2], the author summarized the then state-of-the-art of vibration measurement for machinery monitoring and diagnosis. The basic machinery vibration analysis tools are described in detail by Eshleman [3] and Peters and Eshleman [4]. A coherent methodology for machinery vibration monitoring is described by El-Shafei and Rieger [5], while a whole series for vibration machinery diagnostic standards, ISO 13373, were developed by experts for the International Organization for Standardization (ISO) [6–12]. The ISO 13373 series is based on the author's work [13, 14] in developing a flow-chart approach for machinery diagnostics. An explanation of the use of ISO 13373 standards is given in [15, 16]. Examples of Diagnosis of specific machines is given in [17, 18]. It should be noted that the diagnosis of gearboxes is actually quite elaborate and may require more advanced analysis tools [18].

This chapter starts by reviewing the vibration measurement tools and the instrumentation used in machinery monitoring and diagnostics, followed by a review of the vibration analysis tools used for machinery diagnostics. This is followed by the description of the principles of machinery condition monitoring. Next, the process for machinery diagnostics is described with some examples. Finally, the chapter ends with recent trends for machinery vibration monitoring and diagnostics.

2 Vibration Measurement and Instrumentation

In order to use vibration as a monitoring and diagnosing tool, vibration needs to be measured successfully. Vibration needs to be sensed (i.e. measured and transformed into an electric signal) then processed and converted to a digital signal for analysis. All of these steps are important and should be done correctly. If any of these steps fail, then the whole measurement process is jeopardized. It is extremely important to measure correctly. Guidance on the vibration measurement process for condition monitoring and machinery diagnosis is given in [2–4, 7].

2.1 Vibration Sensors

There are many types of vibration sensors. The most common sensor for vibration analysis is an accelerometer, which, as its name implies, measures absolute (seismic) acceleration, most commonly through a piezoelectric crystal that converts acceleration into electric charge [2].

Accelerometer mounting is very important. There are many ways for mounting an accelerometer (depending on the application), but the most common mounting methods are by a stud, mounting pad or magnet [3]. The mounting method affects heavily the useful frequency range of an accelerometer, which can be reduced from 6 kHz for stud-mounting to 2 kHz for magnet mounting, for a typical accelerometer [3]. This is due to the need to avoid the mounting resonance and to use the accelerometer in the region of constant sensitivity [2]. However, if the accelerometer is used for envelope analysis [4], then usually the acceleration data are processed at the mounted resonance to use the resonance magnification of the measured signal for further processing [4].

General machines are usually monitored in units of velocity. In this case, the acceleration signal sensed by the accelerometer is integrated to velocity (usually electronically). However, a high-pass filter at a very low corner frequency (around 0.5 Hz) is used before integration to avoid the inevitable magnification near zero frequency. This is the most common method for measuring vibration velocity. Laser vibration velocity sensors are used in the market, but are expensive, thus are of limited use in monitoring. However, they are used in particular diagnosis measurements, especially for torsional vibration.

As a general practice, acceleration is used to monitor machines in the frequency range above 1 kHz, and velocity is used in the range from 20 Hz up to 1 kHz, while vibration displacement is used below 20 Hz [3].

To measure displacement, an accelerometer signal can be double integrated, with the same signal processing described above. This is for absolute displacement. For relative displacement a Linear Variable Differential Transformer (LVDT) [2] can be used to measure vibration (example on turbine casing expansion). However, the LVDT is a contact measurement and is a relative measurement. The LVDT consists of two energized coils acting as a transformer. When the intermediate rod is centered, the differential voltage from the transformer coils is zero. Upon rod displacement, the differential voltage measured is proportional to the rod displacement. The LVDT has a maximum frequency of 400 Hz [2].

For a contactless relative measurement, a proximity probe is used. This is most commonly used on rotors mounted on fluid film bearings. The proximity probe is essentially a coil of wire inserted in a polymer probe and installed in proximity of the rotating shaft. The coil in the probe is energized by an oscillator supplied with a carrier frequency in the radio frequency range, and when the rotating shaft (made of magnetic conducive material) approaches the proximity probe, the resulting magnetic field produces eddy currents on the surface of the rotating shaft, which represent losses in the magnetic field. As the shaft moves closer to the probe, the eddy current losses increase. Therefore, the proximity probe actually measures the gap between the probe and the shaft [2]. The measured signal is then processed through a demodulator that removes the radio frequency carrier signal, leaving the measured vibration displacement.

Other applications may require the use of strain gages for measurement of vibration displacement. This is particularly important in the diagnosis of machines, as the measurement is done at the source (not transmitted through machine casing). Strain gage measurement has the advantage of being able to measure both lateral and torsional vibration. For measurements on a rotating shaft, the measured signal is transmitted through telemetry or slip rings [2].

2.2 Instrumentation

Instrumentation is used to process the signal sensed by the vibration transducer. The most important aspect in the measurement loop is to consider impedance matching of all instruments in the loop [2], from the sensor to the measurement and display instrument.

Many intermediate instruments are used: amplifiers are used to improve the signal to noise ratio, integrators are used to transform the signal from one domain to the other, while filters are used to select certain frequency ranges. As discussed, a high pass filter is used before integration, while a low pass filter with a high corner frequency is used before sampling as an anti-aliasing filter. Band pass and notch filters are used in particular applications. Tracking filters use band pass filters to measure transient response of rotating machinery during start-up and shut-down [2].

The final instrument in a measurement loop is essentially a voltmeter capable of displaying the signal in the time domain (oscilloscope) or frequency domain (spectrum analyzer) [3].

Virtual instruments and virtual instrument loops can be created using specialized software. However, when virtual instruments are used, care should be taken in the analog signal processing before sampling to ensure that problems such as aliasing are avoided [4].

2.3 Digital Data Processing

In our digital world, there are many aspects of digital signal processing that has to be considered for vibration condition monitoring and diagnostics. The most important issue is that Shannon's sampling theorem has to be respected. This theorem states that the sampling rate has to be twice as fast as the highest frequency of the signal. This is to avoid frequency folding [3]. Since we do not know the vibration signal beforehand (that is why we are measuring it), but we know the frequency range that we are interested in, which is F_{max} (from the machine knowledge) [3], we can have a low pass anti-aliasing filter applied to the vibration signal at this maximum frequency to eliminate any unwanted noise above F_{max}. Actually, because of the characteristics of the anti-aliasing filter, the sampling rate is 2.56 F_{max}. This translates in the number of samples actually is also 2.56 times the number of lines in a spectrum [3].

Two very important equations, which are quite useful in setting-up a vibration condition monitoring and diagnosis system are [3]:

$$T_s = \# \text{ of Lines}/F_{max} \tag{1}$$

and

$$Res = 2 \, WF \, F_{max}/\# \text{ of Lines} \tag{2}$$

where F_{max} is the frequency range, # of Lines is the number of lines in the spectrum, T_s is the measurement time, Res is the resolution, and WF is the window factor. The resolution is the ability to distinguish between two frequencies in a spectrum [3] and is very important in the diagnosis of electric motors [4, 7, 12]. The window factor is used when a window is applied to the vibration measurement. The purpose of the window is to force the beginning and end of the measurement to zero, thus ensuring periodicity and reducing leakage in a spectrum [2, 3]. Leakage is a spectral phenomenon where spectral energy leaks from one spectral line to neighboring lines because of lack of periodicity in the vibration signal [2, 3]. The most common windows are uniform (with a WF of 1.0), Hanning (with a WF of 1.5) and Flat Top (with a WF of 3.8). In almost all vibration condition monitoring and diagnosis applications, a Hanning window should be used. However, for certain applications, other windows may be used. For example, a uniform window is used for bump tests, and flat-top window is used for accurate and specific amplitude measurement [4].

Due to noise, a single measured spectrum is usually not indicative of the actual machine vibration. Generally multiple spectra are measured at the same point and are averaged out to reduce noise. 8 averages are recommended, but this may significantly increase the measurement time. The minimum acceptable number of averages is 4 averages. For low F_{max} or for a high # of lines, the measurement time according to Eq. (1) would be high. In this case, overlap averaging can be used [4], where part of the previous data block is used in the subsequent data block.

Specialized averaging procedures are sometimes used. Peak-hold averaging is sometimes used in lieu of a tracking filter to track the transient behavior of a machine. Synchronous time averaging is used to average the time waveform, triggered by a certain event, to accentuate this particular event [4].

In summary, to set-up a vibration condition monitoring and measurement system, it is important to do the following:

- The sensor frequency range is suitable for the application. The sensor sensitivity should be considered in set-up.
- The F_{max} should be selected based on the expected frequencies generated by the machine [3].
- The # of lines should be selected to ensure adequate resolution and suitable measurement time.
- Hanning window should almost always be selected.
- Averaging should be selected to reduce the noise effect. Overlap processing may be used to reduce measurement time.

Finally, if a compromise set-up cannot be achieved, then two measurements should be considered at the same point with two different measurement set-ups to ensure accurate, useful and meaningful measurement.

3 Tools for Vibration Monitoring and Diagnosis

There are many tools for vibration monitoring and diagnosis. The most important of which are the time waveform and the spectrum.

3.1 Vibration Time Waveform

The vibration time waveform is a plot of vibration data versus time. This is the most basic vibration data. The peak overall value is obtained from the time waveform. It is actually a very useful tool for visualizing the machine behavior. Directional forces, truncation and pulses are clearly seen in the time waveform [4]. Some effects such as beating and modulation need to be distinguished in a spectrum. Figure 1 shows the time waveform showing directional forces due to misalignment [18].

3.2 Vibration Spectrum

The vibration spectrum is the FFT of the vibration data shown in the time waveform. Even though it contains the same data, yet the spectrum display is a very powerful tool for vibration diagnosis, as it illustrates the frequency content of the signal. The

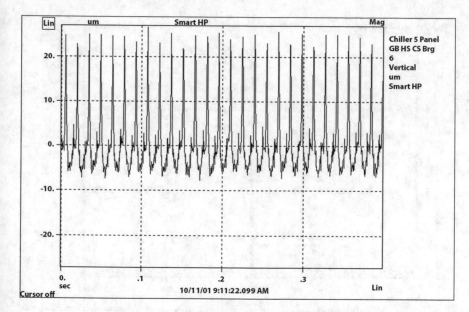

Fig. 1 Time waveform showing a directional force due to misalignment [18]

role of diagnosis is to relate each significant frequency line in the spectrum to a particular machine fault [4]. Figure 2 shows a spectrum on a cement mill showing beating between two frequencies [18].

3.3 Vibration Waterfall

Vibration waterfall plot is simply a three dimensional plot of spectra as a certain operating parameter is changed, usually speed. Figure 3 shows a waterfall plot showing progressive looseness with time [4]. For rotating machinery, the waterfall plots may be in the form of a full spectrum, where positive and negative frequencies are plotted, indicating forward and backward whirls.

3.4 Envelope Spectrum

The envelope spectrum is a special measurement to detect impacts in a signal by accentuating the signal envelope [4]. Usually envelope measurements are used to detect early rolling element bearing faults. Figure 4 shows an envelope spectrum for a highly loaded gearbox [18].

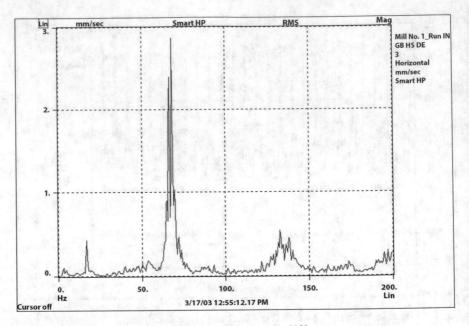

Fig. 2 Spectrum showing beating between two frequencies [18]

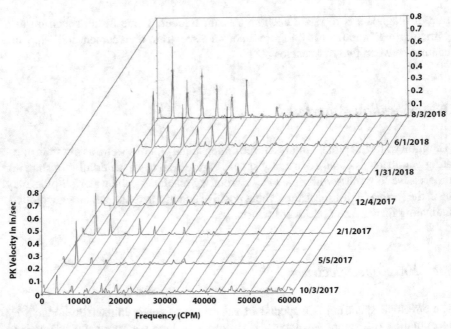

Fig. 3 Waterfall plot showing progressive looseness with time [4]

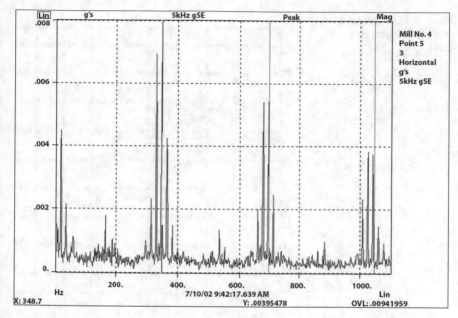

Fig. 4 Envelope spectrum of a highly loaded gearbox [18]

3.5 Phase

Phase is defined as the time difference between two signals, at the same frequency, proportioned to their period (which is a complete cycle of 360°). Absolute phase of a signal is defined relative to a reference point. For rotating machinery, the reference point is usually a key way on the shaft, or a reflective tape installed on the shaft, sometimes termed keyphasor. To measure absolute phase a proximity probe is used to monitor the keyway, or a laser probe is used to monitor the reflective tape. A pulse is generated by the proximity probe whenever the keyway passes in front of it, and similarly a pulse is generated when the reflective tape passes in front of the laser probe. This pulse is then the reference signal for phase at the running speed. Phase is then defined by the angle between the generated pulse and the first peak in the measured vibration signal [4].

Phase is very important in machinery diagnostics. Many faults show vibration at the synchronous frequency, i.e. a frequency equal to the running speed, usually termed 1x frequency. This renders the spectrum, which is the main diagnostic tool, useless. In order to distinguish between these faults, phase is then used [13]. For example, to differentiate between unbalance and misalignment, phase is measured across the coupling. For rigid rotors, if the vibration signals across the coupling are in-phase, then the problem is unbalance, if they are out-of-phase, then the problem is misalignment. Many more examples are given in the ISO 13373 series [6–12].

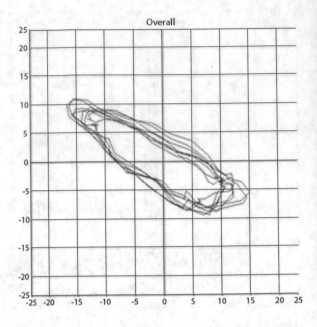

Fig. 5 Typical unfiltered orbit for a rotating machine (μm)

3.6 Orbit

The orbit is a plot of two vibration signals from two proximity probes 90° apart monitoring a shaft (usually mounted on fluid film bearings). Time is eliminated, and the two signals are plotted against each other, thus providing the actual trajectory of the rotor in the bearing. The orbit is a very useful tool for visualizing the shaft vibration. A typical unfiltered orbit for a rotor in a fluid film bearing is shown below in Fig. 5. A filtered orbit at 1x would show as an ellipse.

3.7 Shaft-Centerline Plot

The Shaft Centerline Plot (SCP) is also obtained from two proximity probes 90° apart monitoring a shaft on fluid film bearings. However, for an SCP plot, the gap voltage (i.e. the static component) is plotted as the machine speed is changed. This is in contrast to the orbit plot where the vibration signal (i.e. the dynamic component) is plotted. It may be considered that the SCP is the locus of the orbit center under speed changes.

The SCP is a very important tool in the evaluation of fluid film bearing condition. Usually the shaft is sitting at the bottom of the bearing at start-up. As the speed increases, the shaft centerline lifts and follows a certain trajectory. For heavily loaded bearings the shaft centerline remains near the bottom of the bearing, which can lead to excessive wear or rubbing under certain conditions. If however the bearing is

Fig. 6 Shaft centerline plot for a rotating machine on fluid film bearings

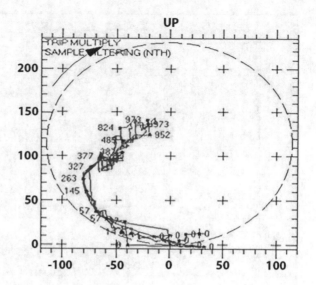

lightly loaded, and the shaft centerline approaches the center of the bearing, then this machine would be prone to instability. Figure 6 shows a typical SCP for a rotating machine during start-up.

3.8 Frequency Response Function

The Frequency Response Function (FRF) is a system characteristic function that represents the system behavior in the frequency domain. It is the Fourier Transform of the Impulse Response Function (IRF), and is closely related to the Transfer Function obtained by the Laplace Transform of the IRF, which is used frequently in control system analysis.

All the tools used in this section from 3.1 to 3.7 are signal analysis tools, meaning that a vibration signal is measured and is analyzed. However, the FRF is a system analysis tool. It does not describe a vibration signal, but it actually represents the system behavior, in the frequency domain, to particular excitation, often in a controlled test. It is possible from a measured FRF to identify the natural frequencies and their associated damping ratios, and even the mode shapes (through further analysis) in a certain measured frequency range. Figure 7 shows the magnitude plot of a typical FRF. Peaks in the magnitude FRF represent natural frequencies.

This might be straightforward for structures, however for rotating machines with speed dependent characteristics, this requires special attention and deeper understanding. Consider a rotating machine with speed dependent characteristics, as shown in Fig. 8 [19]. The natural frequencies change with speed. Consider that the machine is operating at a certain speed. If the machine is excited (for example by a magnetic bearing), then the measured FRF would show two resonant peaks at the two natural

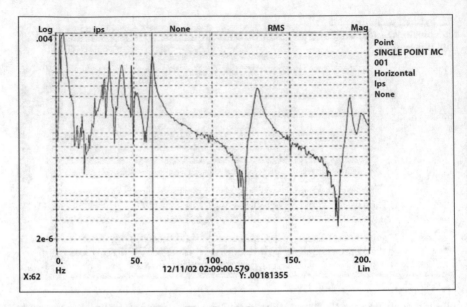

Fig. 7 Typical magnitude FRF for a Girth Gear [18]

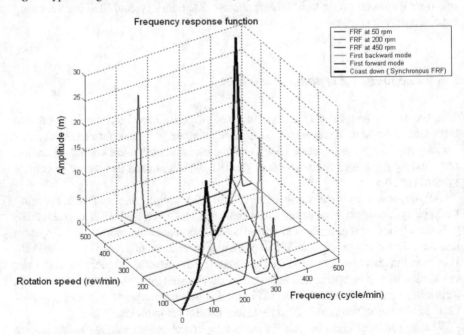

Fig. 8 FRFs for a rotating machine with speed dependent characteristics [19]

frequencies shown on the blue FRF. If however, the machine is operating at a different speed, and the same test is repeated, then we would get the green FRF shown in Fig. 8. Notice that the peaks of the green FRF (and thus the natural frequencies) are different from those of the blue FRF. Similarly the red FRF shown in Fig. 8, at a higher speed, shows different natural frequencies, clearly illustrating the speed dependent characteristics of rotating machinery. Notice that all of these are natural frequencies, but excited at different speeds.

If however, the machine is just excited by the unbalance, and the measured signal is filtered at the running speed, where the measured vibration is synchronous with the running speed (i.e. 1x), then we get a synchronous FRF, which is the black FRF in Fig. 8. Note that the synchronous FRF is just the well-known Bode plot used frequently for transient analysis of rotating machinery. Notice that the peaks of the synchronous FRF (black FRF in Fig. 8) consists of one peak from the green FRF and one peak from the red FRF, illustrating the change in characteristics and natural frequencies with speed. The peaks in a synchronous FRF are called critical speeds, and are very important for the analysis and diagnosis of rotating machinery.

Bode Plot. The FRF is a complex number, which can be presented as magnitude and phase or real and imaginary. For rotating machinery, the Bode plot is a plot of the magnitude of the synchronous FRF versus speed and phase of the synchronous FRF also versus speed. Figure 9 shows the magnitude and phase plots of a Bode plot at the horizontal and vertical directions of the drive end bearing of a rotor. The upper Fig. 9a and b represent the magnitude Bode plot, while the Fig. 9c and d represent the phase Bode plot. Notice that balancing does not affect the phase Bode plot.

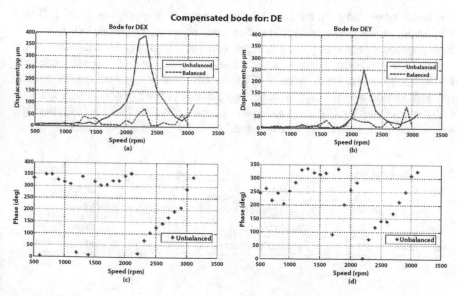

Fig. 9 Bode plot of a rotor before balancing and after balancing

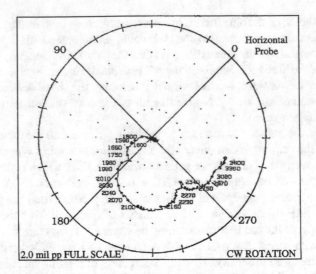

Fig. 10 Polar plot of a synchronous FRF during a turbine shutdown

Polar Plot. The Polar plot is another representation of an FRF, where the imaginary component of an FRF is plotted versus the real component, and speed (or frequency depending on the FRF type) is implicit. Figure 10 shows a Polar plot of a synchronous FRF during a turbine shutdown. Each circle represents a resonant condition. The critical speeds can be identified by a 90° phase shift in each circle.

Real and Imaginary Plots. The real and imaginary components of an FRF can be plotted versus frequency (or speed in case of synchronous FRFs). It may be advantageous to use this particular representation for proper identification of the system characteristics. Table 1 shows the advantageous use of real and imaginary FRFs. For example, for a mobility FRF, the real plot of the FRF can be used to obtain the mode shape, while the imaginary plot can be used to obtain more accurate estimates of damping. Figure 11 shows an example of the real FRF plot for a mobility measurement on a pump rotor, which was later used to confirm the mode shape.

Table 1 Real and imaginary FRFs

FRF	Measures	Real	Imaginary
Dynamic Compliance	Displacement/Force	Damping	Mode
Mobility	Velocity/Force	Mode	Damping
Accelerance	Acceleration/Force	Damping	Mode

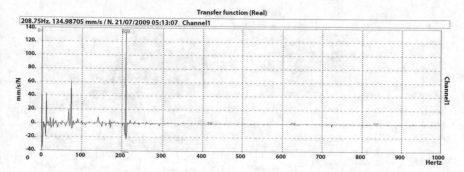

Fig. 11 Real plot of a mobility FRF [17]

3.9 Operating Deflection Shape (ODS)

The Operating Deflection Shape (ODS) analysis is a powerful tool for visualization of machinery and structural vibration. Often used with animation, ODS can be a powerful tool for visualizing machine vibration. To measure ODS, the machine or structure is divided into grid points and measurement of amplitude and phase of vibration at a particular frequency is then conducted and plotted, which can be a time consuming process. The result is usually an impressive figure clearly identifying the root cause of the problem. Figure 12 shows the 6x ODS of a cooling fan base with excessive horizontal deflection at the blade passing frequency causing shaft failures [14], while Fig. 13 shows 3D 1x ODS plots of steam turbine bearings, where the high pressure rotor was experiencing excessive directional steam induced vibration. The axes represent the actual physical dimensions of the structure. In both figures, the deformed structure is compared to the undeformed skeleton to clearly visualize the deformation at that particular frequency. ODS is increasingly being replaced by vibration video motion magnification. However, ODS still has the advantage of being a 3D measurement, while vibration video motion magnification is not.

Fig. 12 ODS of cooling fan base [14]

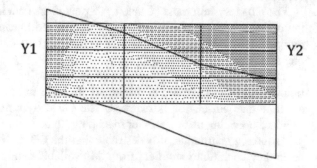

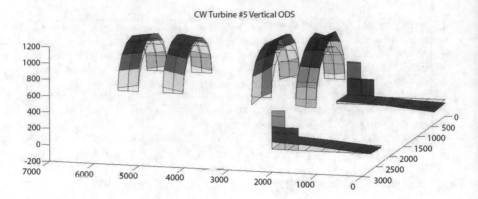

Fig. 13 3D ODS of steam turbine bearings

4 Vibration Condition Monitoring

The purpose of vibration condition monitoring is to be able to determine the condition of a machine and to be able to use this information to make a decision on the functionality of the machine and to plan for any required maintenance of the machine. This is the basic premise of Condition-Based Maintenance (CBM).

4.1 Condition Based Maintenance System

The general objective of a Condition-Based Maintenance System can be summarized as the ability to view multiple datasets associated with a specific system or component set, in an environment that facilitates better operation and maintenance decisions [5].

This means that the CBM System must have the ability to access all data sets in a timely manner, and the system must be designed for scalability, from relatively small data sets to substantial databases.

The CBM System must be able to display these data in a useful manner, so that the operator can determine the state of monitored machinery relative to time and process demand level. It is important to represent attributes such as condition and rate of change.

The goal of the CBM System is to aid in forecasting or predicting possible conditions for any selected piece of machinery. In order to successfully achieve this goal, accurate data correlation is required. The CBM System should allow baseline or standard operating parameters for machinery to be input from an external data analysis source. In particular, it is desirable that the CBM System can be configured to select key condition identifiers from all available condition identifiers for display and/or correlation. Data sets should be possible to be correlated against each other in a meaningful manner. For example, pressure versus temperature, and specific vibration frequency parameters versus load etc. are examples of simple correlated sets. Any correlation analysis should take into account the asynchronous nature of the data to be correlated, and provide algorithms that allow accurate cross correlation.

It is possible with such a CBM System to receive meaningful data that can be incorporated by plant operators to provide accurate condition assessment of the machine, as well as possible fault diagnosis. It is therefore important to differentiate between the requirements of condition monitoring and those of fault diagnosis. A plant operator should be able to utilize the condition data received from the CBM System to accurately assess the condition of the machine and provide a diagnosis for the possible machine fault.

4.2   Condition Monitoring Versus Machinery Diagnostics

At this point, it is important to differentiate between the requirements of condition monitoring versus those of machinery diagnostics. Consider Fig. 14, which is the famous bathtub curve for probability of failure of machines with time [5]. In the run-in period, the probability of failure is high, until the machine goes into normal operation, which (if the maintenance work is done correctly) should be a long period of time. Upon the inception of a fault, point A on the bathtub curve, the probability of failure starts to rise again.

We will not concern ourselves with the run-in period, since usually careful human attention is given to new or repaired equipment until they pass the run-in period. However, it is important to clearly define the requirements for the jobs of monitoring and diagnosis. In normal operation, the purpose of monitoring is to detect any change in machine condition that would lead to a machine fault (point A on Fig. 14), while the purpose of the diagnosis is to find the cause of the fault and time to failure (beyond point A on Fig. 14).

For plants with large numbers of machines, the task for monitoring machines can be quite burdensome, particularly that vibration data (the main tool for CBM Systems) are dynamic in nature, and usually require measurements on a large number of points on the machine, in different domains, and with different spectral ranges

Fig. 14 Bathtub curve

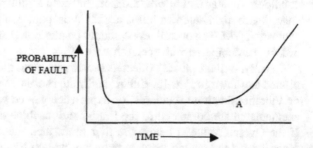

and resolutions [2]. Moreover, most of the time, condition monitoring data are not providing new information on the condition of the machine, but actually the condition data should be telling the user most of the time that the machine is in good condition (otherwise that plant has a major problem). Thus, it makes a lot of sense to reduce the amount of data collected on a machine during monitoring to simple numbers that can clearly indicate the level of vibration on the machine.

On the other hand, for diagnostics, the CBM system already has determined that the machine is faulty, and it is required to diagnose the cause of the fault. This usually requires the full analysis of the dynamic data measured on the machine. This means that all the data on all points on the machine, on all directions, and with different formats, spectral ranges and resolutions, are to be used in the diagnosis [2]. Moreover, special testing, such as Operating Deflection Shape (ODS) analysis, may be required to accurately diagnose the cause of the fault, as well as the time to failure.

4.3   Condition Evaluation Based on Statistical Data

Statistical analysis is usually used as a condition evaluation tool rather than as a diagnosis tool. This is because statistical analysis can help reduce the condition evaluation burden, and accurately determine the condition of the machine from a large amount of data.

Vibration data usually come in different formats, and contain a wide set of information. The vibration data can be presented in spectrum, time waveform, orbit, amplitude, phase, and many other data formats. In order to be able to judge a machine condition, the vibration data are usually condensed to a single number. This single number is termed the Overall Value, and can be quantified as peak value or root mean square (rms) value, and either as displacement, velocity or acceleration data [20]. This Overall Value is then trended and compared to limits.

Using vibration measurement to evaluate machine condition, computerized maintenance systems can evaluate machine condition based on overall levels of measures. Overall levels of a measure are typically judged in terms of limits. Two or three alarm values are typically used in the trending process. An alert limit may result in initiating collection of a spectrum or a time waveform, or changing the frequency of periodic measurement to be taken, and usually signifies a deterioration in machine condition. An alarm limit would signify the need to shut down the machine. Normal, surveillance and shutdown limits are the most commonly used limits for condition evaluation [20]. The overall levels should be monitored and recorded for a period, in order to be compared with each other.

Usually, a trend plot is collected in velocity data, and the monthly data can be plotted as shown in Fig. 15. Either the limits chosen can be selected based on ISO or Vibration Institute standards, or experience gained with the machine. The data accumulated are compared to the limits, and machine condition can be evaluated. If the Overall Value is below the first limit, then the machine is normal; on the other hand if the overall value is between the two limits, then the machine should

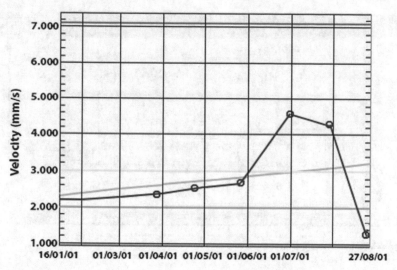

Fig. 15 Trend plot

be under close surveillance, and maintenance action should be planned. However, if the vibration level is above the second limit then the machine should be shut down and maintenance action performed immediately.

In the trend plot, some statistical analysis, such as a regression analysis or correlation analysis may be useful. This will aid in understanding and evaluating the trend of all points on a machine, as well as forecast the future behavior based on its history. Figure 15 shows actual trend data (Navy lines) for a point on a machine, with three limits shown (green line for Alert limit, yellow line for Alarm limit and red line for Danger limit), and a regression line (light blue line). Note that the machine condition was normal, but a fault developed some time during June 2001. Maintenance action during August 2001 reduced the Overall value within normal condition.

Basing maintenance action on the Overall Value can be useful, but is incomplete. The reason is that the Overall Value may not always be sensitive to the change in machine condition. This is particularly true for rolling element bearings [21]. The change in bearing condition and initiation of bearing fault rarely affect the wide band measurement embedded in the Overall Value. Rather small changes in the amplitude of particular bearing frequencies indicate deterioration in bearing condition. Usually different techniques such as ultrasonic detection or spectral analysis of vibration data would be required. However, spectral analysis is an involved process. There are two simple techniques that utilize spectral data to trend the vibration, that can be used to determine the condition of machines, and are much more sensitive (than overall values) to the change of individual defects in a machine.

These two techniques for trending of the vibration data are band alarms and spectrum enveloping. In the band alarm technique, limits are defined for preset frequency bands not for the entire frequency span. This technique is powerful in monitoring bearings; since the vibration levels of bearing fault frequencies are always lower than

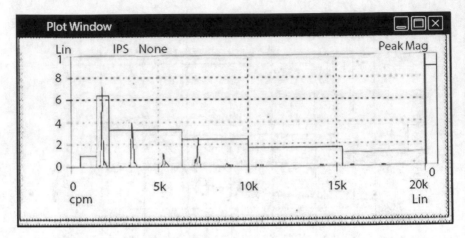

Fig. 16 Band alarm using 6 bands

other frequencies appearing in the spectrum. Figure 16 represents a sample of a band alarm plot.

With the band alarm technique, a single value is measured for each band of interest and these band amplitudes are trended and subjected to limits, same as with the overall value. However, with the band amplitudes, these measures of the machine condition are each sensitive to a particular fault in the machine. Thus it is possible to detect early faults with band amplitudes, before they affect the overall value of the measurement. Thus, the condition of the machine, and some basic diagnostics, can be evaluated through the trending of band amplitudes, and compared to band limits.

The second technique, the spectrum enveloping technique, uses a spectrum mask for each peak in the spectrum, such that each frequency component has its own alarm and shutdown levels. Using this technique, it is possible to monitor the changes that occur to each vibration component and to determine whether this component exceeded the previously set shutdown level, or not. Figure 17 illustrates a sample for the spectrum enveloping. Basically, the spectrum enveloping technique does not provide any particular amplitude, but rather it provides a spectrum mask. This spectrum envelope is usually based on the current normal spectrum, and both an alert and a danger envelope are initiated, such that each narrow band in the spectrum has an alert limit of twice the normal value. If in any subsequent measurement a single line reaches one of these envelope limits at any frequency, the user can immediately be alerted of the deterioration of the condition of the machine.

5   Vibration Diagnosis

It has been debated for a long time whether the diagnosis of rotating machinery could be standardized. In fact the debate always centered on whether the diagnosis process

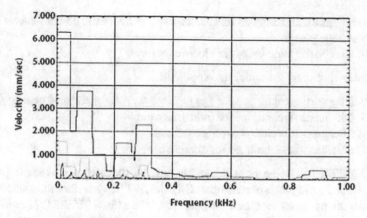

Fig. 17 Spectrum enveloping

is an art, where the knowledge of a human being in diagnosing machine faults is a personal process that relies on the knowledge and experience of an analyst, or whether it is a science that can be explained and documented.

However, the International Organization for Standardization (ISO), through the ISO 13373 series, has adopted an approach that relies on Fault Tables to document available knowledge on machine faults, and the use of step-by-step flow charts or process tables to document the structured procedure that is used by many analysts in reaching an appropriate diagnosis of a machine. This process, particularly the use of flow charts for diagnosis, was pioneered by the author in his papers on the diagnosis of installation faults [13] and the diagnosis of fans [14].

The author of this chapter has participated in developing the ISO 13373 series as the Convenor of ISO TC108/SC2/WG10 and as the project leader for many of its standards. This section summarizes the author's opinion and many of the approaches that are accepted now as international standards.

5.1 Standards Overview

The ISO 13373 series consists of the following parts, under the general title *Condition monitoring and diagnosis of machines—Vibration condition monitoring*:

- Part 1: General procedures
- Part 2: Processing, analysis and presentation of vibration data
- Part 3: Guidelines for vibration diagnosis
- Part 4: Diagnostic techniques for gas turbines and steam turbines on fluid film bearings
- Part 5: Diagnostic techniques for fans and blowers

- Part 7: Diagnostic techniques for machine sets in hydraulic power generating and pump-storage plants
- Part 9: Diagnostic techniques for electric motors.

The following parts are under development:

- Part 6: Diagnostic techniques for gas turbines on rolling element bearings
- Part 8: Diagnostic techniques for industrial pumps
- Part 10: Diagnostic techniques for generators
- Part 11: Diagnostic techniques for gearboxes.

ISO 13373-3 [8] is the base document for the diagnosis process. It provides guidelines for the diagnosis of machines. ISO 13373-1 [6] describes the measurement procedures on machines for diagnosis purposes, while ISO 13373-2 [7] describes the processing and analysis of data for diagnosis purposes. The other parts [9–12] are machine specific standards that describe the unique diagnostic procedures for each machine type.

ISO 13373-3 [8] is described in the standard as: "has been developed as a set of guidelines for the general procedures to be considered when carrying out vibration diagnostics of machines. It is intended to be used by vibration practitioners, engineers and technicians and it provides them with useful diagnostic tools. These tools include diagnostic flowcharts, process tables and fault tables. The material contained herein presents a structured approach of the most basic, logical and intelligent steps to diagnose vibration problems associated with machines. However, this does not preclude the use of other diagnostic techniques." [8]

The diagnosis process is described in ISO 13373-3 [8] as consisting of flowcharts, process tables and fault tables. The flowcharts and the process tables are essentially a step-by-step question and answer procedure that guides the user in the diagnosis process. The flowcharts are used for an overview of the vibration events and characterize the features, while the process tables are used for more in-depth analysis. The fault tables are used to illustrate common machinery events and how they manifest themselves [8].

5.2 Initial Analysis

The diagnostic process starts by asking some basic questions, both on the safety of the humans and machines and to ascertain the history and current status of the machine. This is included in the normative Annex A in 13373-3 [8], which specifies the systematic approach to the vibration analysis of machines:

(a) Annex A.1 is used to gather background information regarding the machine, nature and severity of the vibration.
(b) Annex A.2 is used to answer a set of questions aimed at arriving at a probable diagnosis of such common faults as unbalance, misalignment and rubs.

(c)   Annex A.3 is used to set out certain considerations when recommending actions following a probable diagnosis.

5.3   Common Analysis

The diagnosis of faults common to a wide range of machines are shown in other annexes in 13373-3 [8]:

- Installation faults and examples are described in Annex B.
- Radial hydrodynamic fluid-film bearing faults and examples are described in Annex C.
- Rolling element bearing faults and examples are described in Annex D.

The use of ISO 13373-3 [8] is necessary with all machine specific diagnostics standards [9–12], and should be used in conjunction with these machine specific standards, as it describes the initial analysis that must be applied to the diagnosis of all machines [8] as well as common faults to all machines including installation faults, fluid film bearing faults and rolling element bearing faults.

5.4   Fault Table

A sample fault table is given below for the diagnosis of fans and blowers [10]. The systematic approach to vibration analysis of fans and blowers is given by the fault table in Table 2. The fault table includes mainly installation faults. For faults regarding fan or blower bearings, see ISO 13373-3 [8], Annexes C and D. Several faults can give similar indications and further investigation would be necessary to distinguish between them.

The fault table (Table 2) above includes the most common faults in fans and blowers including the descriptors of their vibration characteristics, as well as other fault descriptors. To complement the fault table, ISO 13373-5 [10] for fan diagnostics introduces the concept of a symptom table (Table 3 below). The symptom table includes the observable symptoms of typical faults, in the form of elevated vibration signals (subharmonic, 1x, 2x, ..., etc.) as well as vibration phase and other diagnostic discriminating factors. The symptom table distinguishes between symptoms almost certain to be seen if fault occurs and symptoms that may or may not be seen.

5.5   Symptom Table

An example of symptom table used by ISO 13373-4 [9] for steam and gas turbine diagnostics, is given below in Table 3.

Table 2 Fault table for fans and blowers

Fault	Vibration characteristics	Other descriptors	Comment
Shaft misalignment/concentricity errors	1x, or 1x and 2x, sometimes 1x and 2x and 3x	Directional force 180° phase shift across coupling. Offset misalignment tends to produce phase shift across the coupling in the radial direction, while angular misalignment tends to produce the phase shift in the axial direction	There are two types of misalignments: parallel and angular, and in most cases there would be a combination of the two
Looseness	Usually a series of peaks at rotational speed and integer harmonics of rotational speed, generally the amplitude of these peaks decreasing with higher harmonic numbers	Looseness can be at bearings or skid, or anchor bolts. Check for difference in amplitude and/or phase at the interface to discern position of looseness	Looseness can be at the bearing housing (sometimes due to the bearing installation), and/or at the pedestal or the skid
Excessive bearing clearance	1x. With low amplitude harmonics	Directional	Can be due to wear, in both fluid film and rolling element bearings
Piping strain	1x	Directional, wave clipping in time waveform	Piping flanges should match without jacking
Soft foot	1x, plus 2x line frequency in the electric motor	Soft foot test	Soft foot is the condition that exists when all feet are not correctly supporting the machine. See also ISO 13373-9
Shaft rubbing	Clipping in time waveform, with 1x and multiple harmonics in spectrum. Light rubbing can cause rotating vectors (spiral vibration)		Not commonly observed on fans

(continued)

Table 2 (continued)

Fault	Vibration characteristics	Other descriptors	Comment
Unbalance	1x	Phase shift across coupling depends on the mode. Cylindrical modes tend to have 0° phase shift across the coupling, while conical modes tend to have 180° phase shift Usually, 90° phase shift between the horizontal and vertical measurements at the same bearing location	Unbalance is often due to erosion, or deposits on blades. Overhung fans may require a couple balance, while centre-hung fans can generally be balanced in a single-plane
Bent shaft	1x similar to unbalance, manifests itself at slow roll speed	Can cancel with unbalance at particular rotational speeds	Rarely seen on fans
Casing distortion	1x, sometimes 2x	180° phase shift from end to end	Only important where bearings are integral with the casings
Resonance	High vibration at a particular frequency	Resonance testing indicates natural frequency	Avoid operating close to a resonant frequency e.g. by changing speed, or by changing resonant frequency, e.g. by stiffening machine or adding mass. Sometimes damping may be needed
Tilting foundation	High 1x vibration levels that cannot be explained by unbalance, misalignment, bent shaft or eccentricity	Rocking motion in 1x ODS	ODS study to analyse problem in more depth
Aerodynamic forces	Blade passing frequency	Can have high noise	Usually caused when fan is operating off best efficiency point

(continued)

Table 2 (continued)

Fault	Vibration characteristics	Other descriptors	Comment
Belt faults	Belt Passing frequency	Less than 1x	Typically due to belt wear, misalignment and/or incorrect tension
Belt resonance	Belt resonance frequency	Usually less than 1x	Usually due to lack of belt tension
Excessive belt tension	1x	Directional	Similar symptoms to misalignment
Belt pulley eccentricity	Usually directional 1x, sometimes 1x and 2x	Sometimes visually observed as wobbly motion	

Note *ODS* operational deflection shape

5.6   Flow Chart

The use of flow charts in machine diagnosis is quite useful, and presents a step-by-step approach to machine diagnosis that can be applied in the field. The diagnosis of Installation Faults in ISO 13373-3 [8] is actually based on reference [13]. The standard methodology illustrated in Fig. 18 starts by visual inspection but considers spectral analysis as the main component of the testing of the installed machine (M/C). In addition, resonance testing, time waveform analysis, phase analysis and operational deflection shape (ODS) analysis are used if and when judged necessary.

Basically, a visual inspection of the machine and the site should be completed before any testing of installed machinery be performed. In many cases, the presence of skid looseness and/or piping strain would be evident to the naked eye. Actually, it is suggested that all skid and anchor bolts be tightened before testing an installed machine. Also, all piping connections should be checked before testing. All flanged connections should be checked to make sure that connecting bolts pass through the flanges without any restriction, thus causing no piping strain.

Spectral analysis is the core of the diagnosis of rotating machinery. These spectral data should be measured on all bearings on the driver and driven machine, in all three directions, horizontal, vertical and axial. Complete knowledge of the machine should be available to identify characteristic frequencies. The purpose of the spectral analysis is to identify the frequencies causing the machine to vibrate. If all vibration amplitude levels are within acceptable limits, then the machine would be accepted as normal. However, if any of the spectral components has high amplitude, then spectral analysis is used to correlate the frequency of the high amplitude vibration to a machine frequency.

The result of the spectral analysis of the high amplitude vibration is one of three cases: (a) at a known frequency, (b) at an unknown frequency, or (c) at the running speed. By a known (unknown) frequency it is meant that the reason of the presence

Table 3 Observable symptoms of typical faults

Fault type	Elevated vibration signals				Time				Critical speed changed	Barring			Varies with load	Repeatable	Comments
	Sub 1x	1x	2x	> 2x	Transient	Sudden appearance	Gradual increase	Steady state		Audible rubbing	Increased slow roll	Barring not possible			
Shaft unbalance (generic)		●				●	●	●						●	Immediately evident
Shaft unbalance (loss of material)		●			○	●	○	●	●	○					Most effect at bearings of affected rotor
Bearing Elevation change	○	●			○	○	●	●	○					○	Occurs following transient
Permanent Shaft Bend		●				○	●	●	●		●	○		●	
Transient shaft bend—no rubbing		●			●	○			○		○			○	E.g. During temperature changes
Transient shaft bend—hard rubbing		●	○	○	●				○		○				During speed or load change
Oil whirl (or whip)	●				○	○								○	Whip locks rotor to 1st critical speed

(continued)

Table 3 (continued)

Fault type	Elevated vibration signals				Time				Critical speed changed	Barring			Varies with load	Repeatable	Comments
	Sub 1x	1x	2x	> 2x	Transient	Sudden appearance	Gradual increase	Steady state		Audible rubbing	Increased slow roll	Barring not possible			
Steam induced vibration		●			○	○								●	Load dependent. Modify admission sequence; repair diaphragms; install nozzle blocks properly
Steam whirl / steam induced vibration	●				○	○								●	Load dependent. Reducing steam flow rapidly removes problem

(continued)

Table 3 (continued)

Fault type	Elevated vibration signals				Time				Critical speed changed	Barring			Varies with load	Repeatable	Comments
	Sub 1x	1x	2x	> 2x	Transient	Sudden appearance	Gradual increase	Steady state		Audible rubbing	Increased slow roll	Barring not possible			
Rotating stall in GT	●				●	●								●	Return to correct flow immediately removes problem
Differential creep on rotors		●					●	●			●				
Rotor Crack		●	○			●	●	●	○						Critical speed may reduce and show two peaks
Looseness in bearing or pedestal	●	●	●	○			●	●	○						

(continued)

Table 3 (continued)

Fault type	Elevated vibration signals				Time				Critical speed changed	Barring			Varies with load	Repeatable	Comments
	Sub 1x	1x	2x	> 2x	Transient	Sudden appearance	Gradual increase	Steady state		Audible rubbing	Increased slow roll	Barring not possible			
Morton Effect/Newkirk Effect (light rubbing)	●	●					●							○	Periodic variation in amplitude or spiralling, approx. 30 – 120 mins per cycle

This table is not exhaustive but contains the most prevalent faults associated with steam and gas turbines with fluid film bearings

● Indicates symptom almost certain to be seen if fault occurs

○ Indicates symptom may or may not be seen

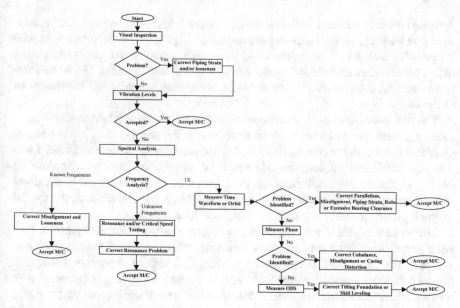

Fig. 18 Flow chart for installation faults [13]

of this frequency in the spectrum is known (unknown). The first case is the easiest to analyze. If the high amplitude vibration is at a known frequency, then the problem is correlated to that known frequency. For example, if 2x vibration is identified then this is usually correlated to misalignment. If decreasing harmonics of the running speed are present in the spectrum, then this spectrum shape is usually correlated with looseness in the bearings or the skid. If however, there were unknown frequencies in the spectrum, then additional testing would be required to determine the source of the unknown frequencies. Amongst the additional testing that may be required are: resonance testing (including impact (bump) test and critical speed test), modal testing, and flow characteristics testing. The purpose of the resonance testing is to correlate the unknown frequency to natural frequencies (stationary components) or critical speeds (rotating components) of the machine. Modal testing is a more advanced form of resonance testing, where all the modal characteristics of the machine are determined, including natural frequencies, damping ratios, and mode shapes. Modal testing is rarely used in the field, as it is an elaborate testing method, and is usually time consuming and costly. However, when justified, it can be a very powerful tool to obtain the machine characteristics and both identify clearly the unknown frequency in the spectrum, and suggest a solution to the problem. As for the flow characteristics testing it is always a good idea to make sure that the rotating machine is operating at or near the best efficiency point, otherwise high amplitude vibration is to be expected. This is the case for recirculation and cavitation in pumps, and stall in compressors.

The most difficult case occurs when the spectral analysis reveals high 1x vibration. There are many faults, related to installation problems, that lead to high 1x vibration.

Amongst these faults are unbalance, misalignment, casing distortion, tilted foundation, skid leveling, piping strain and excessive bearing clearance. In this case, special
vibration measurements have to be conducted on the machine to describe the nature
of this 1x vibration, and to distinguish between the different 1x faults. These measurements include: time waveform measurement, phase measurement, and measurement
of the ODS.

The time-waveform measurement can be used to distinguish between misalignment, piping strain and excessive bearing clearance. For piping strain, it will be quite
clear that the forcing on the machine is directional, usually in the horizontal direction, and this directional force would be acting on the whole machine. Inappropriate
bearing clearance also results in directional forces, however this would be localized
at the bearing with the inappropriate clearance. This is particularly true for special
geometry bearings, such as lemon bore or multi-lobe bearings.

The phase analysis is quite important to diagnose unbalance, misalignment, bent
shaft and casing distortion. In many cases, misalignment (the main installation
anomaly) manifests itself as vibration at 1x only. One of the best ways to distinguish
between 1x vibration due to unbalance and 1x vibration due to misalignment is to
measure phase across the coupling. If there is a 180° phase shift across the coupling,
then the problem is misalignment. If no phase shift occurs across the coupling, then
the problem is unbalance. Casing distortion can be easily identified by 180° phase
shift across the machine (side-to-side or end-to-end) in the horizontal, vertical and/or
axial directions. A cocked bearing can be identified by measuring phase around the
bearing housing and noticing the phase shift due to the wobbly action of the cocked
bearing. In many cases, a coupled time-waveform-phase analysis is quite useful in
visualizing the vibration pattern and identifying the problem.

If the 1x vibration problem is still not solved after the time waveform and phase
analysis, then an Operational Deflection Shape (ODS) should be measured. The ODS
is useful in identifying problems of tilted foundation, skid leveling, skid looseness,
and shaft parallelism. In the ODS measurement, phase-referenced 1x vibration is
measured at grid points on the machine structure or skid. This reveals the actual
deflection shape of the machine under the operating load, and at the operating speed.
Note that the ODS is not a mode shape of the machine or structure, unless the machine
is in resonance, but it can be considered as a summation of the contribution of all of
the modes of vibration. ODS analysis can be quite useful in identifying installation
problems, as it provides a visualization of the actual vibration pattern of the machine
and/or skid. In particular, if a machine skid exhibits a node in its ODS, then this is
a clear indication of a tilted foundation or a leveling problem in the skid. Accurate
measurement of skid and/or foundation levels would then be required to confirm the
results of the ODS analysis.

Both reference [13] and ISO 13373-3 [8] provide many examples of the application
of this methodology in diagnosing installation faults.

6 Recent Trends in Vibration Monitoring and Diagnosis

The vibration monitoring and diagnosis technologies are at an inflection point. The advent of low cost Micro-Electro-Mechanical-Systems (MEMS) accelerometers, with wireless sensors (thus eliminating costly cabling) provides for readily available data at reasonable cost. This means that eventually handheld data collectors shall be replaced by a wide range of low cost wireless accelerometers that provide vibration data if and when needed. These data can then be transferred to a cloud repository for further analysis. Condition monitoring tools can be used as described in Sect. 4 for condition evaluation. Moreover, historian software are being developed to store and analyze Big Data. This can be further used by Artificial Intelligence (AI) techniques such as deep learning to provide useful condition evaluation and recommended actions.

Automatic diagnosis [5] is still further away. Some rule-based expert systems are readily available in the market, and vary in their effectiveness. AI based diagnosis is still at its infancy but will probably make its way to the market. However, a very important tool for diagnosis emerged in the last few years which is vibration video magnification. This technology is starting to replace ODS as a visualization tool, and fits well with the current trends as it provides Big Data, that can plausibly be integrated in future automated diagnostic software.

6.1 MEMS Accelerometers

MEMS accelerometers are revolutionizing vibration measurement. Their low-cost will eventually allow many more machines to be permanently monitored online.

The principle of operation of MEMS accelerometers is simple [22]. They are typically based on capacitors with a structure that uses two capacitors formed by a moveable plate held between two fixed plates. Under zero net force the two capacitors are equal, but a change in force will cause the moveable plate to shift closer to one of the fixed plates, increasing its capacitance, and further away from the other fixed plate, reducing that capacitance. This difference in capacitance is detected and amplified to produce a voltage proportional to the acceleration. The dimensions of the structure are of the order of microns [22].

The author has used MEMS accelerometers. The user needs to be aware of their characteristics. An unannounced 4 kHz sensor resonance had to be mitigated by signal processing. The technology is still under development, but as it matures it will certainly replace piezoelectric accelerometers in many applications due to their low cost.

6.2   Wireless Sensors

Accelerometers with wireless connections send their acceleration signals through Wi-Fi or Bluetooth technologies to a gateway that is connected through Ethernet with the Internet. The wireless network should be designed such that it does not overload the gateway(s).

Wireless sensors tend to be battery powered, some with impressive battery lives of 5 years. However, battery life depends on the use. Frequently using the sensor drains the battery power. A common set-up is to have the sensor collect data for 10 s every 8 h. This tends to be a suitable setup to preserve battery life, but has to be evaluated for providing suitable monitoring data as described in Sect. 2 of this chapter. Finally, whatever strategy is used for data collection, care needs to be exercised to wake the sensors with a certain time lag in order not to clog the gateways.

6.3   Cloud Monitoring

Cloud vibration monitoring has been available for quite some time now. Not only online systems send their data to the cloud, but also handheld data collector systems, through their software, can upload their routinely collected data to the cloud. The cloud monitoring software has evolved and now can provide different levels of data access and display, allowing the technicians to monitor the technical data, plant managers to monitor plant equipment status and program management, and executives to do asset management.

Cloud Monitoring allows enterprises an opportunity to better manage their plants. Opportunities of centralizing the analysis or remote monitoring, or even outsourcing the condition monitoring services, provide unique strategies that can benefit the whole organization.

6.4   The Use of Artificial Intelligence in Monitoring and Diagnosis

The presence of Cloud Big Data on condition monitoring and process parameters is a ripe market for the application of AI technology for condition monitoring. In fact several Asset Performance Management (APM) software are now available in the market, using AI, particularly deep learning. These AI-based software require plant data to be assembled in historian software to be easily accessible by the condition monitoring software. This seems to be a successful application of AI in condition monitoring and it appears to be gaining momentum in the market.

However, using AI in vibration diagnostics is still at its infancy. The author ventured in using neural networks some time ago [23, 24] but developing a complete automated vibration diagnosis is still not available. Major efforts are currently being exerted in this direction [25], and we shall probably see the success of these technologies in AI-based vibration diagnostics in the future.

6.5 Vibration Video Motion Magnification

A recent technology has evolved that is a video-processing method that uses every pixel in the camera and detects vibration motion. The process decodes light to pull out information that is indicative of motion. The process involves measuring a video of the machine or structure, identifying points of high vibration, analyzing the spectrum of points of high vibration, identifying frequencies of interest, then amplifying and filtering at the selected frequencies. The result is video of motion at the particular frequency selected, showing the machine or structure behavior at that particular frequency. The result is very similar to ODS, but much more rapid in measurement and analysis. The author has used it many times in analyzing the vibration of machine casings and supporting skids, and the analysis of bridges and condition monitoring of bridges, updating finite element models based on measured bridge natural frequencies and mode shapes. Note that vibration video motion magnification (VVMM) is a 2D measurement, and it may be necessary to take multiple videos in multiple directions to get a true 3D representation, and it needs to be emphasized that appropriate lighting may be needed for in-door measurements.

An example is shown below for VVMM. A high speed test rig for testing turbocharger bearings was experiencing high unexplained vibration at particular frequencies [26]. Measurement of VVMM videos of the test rig identified problems at 100 and 180 Hz, while the test rig was operating at 42,000 rpm. VVMM was used to filter at 100 Hz frequency, showing that the shaft is exhibiting an axial motion as shown in Figs. 19 and 20, which is obvious in the coupling contraction and expansion, while there is an independent 180 Hz frequency of motor bracket rocking motion as depicted in both Figs. 21 and 22. Of course, these figures are more impressive in video format.

Fig. 19 Test rig shot at coupling contraction (100 Hz)

Fig. 20 Test rig shot at coupling expansion (100 Hz)

Fig. 21 Test rig shot at motor bracket original state (180 Hz)

Fig. 22 Test rig shot showing motor bracket is tilted towards the shaft (180 Hz)

The use of VVMM technology was quite useful in troubleshooting this test rig [26] which allowed for further testing of turbocharger bearings.

It is conceivable that VVMM technology shall be used with Big Data and AI to provide future automatic diagnostic capabilities.

7   Conclusion

This chapter reviewed the vibration measurement tools and the instrumentation used in machinery monitoring and diagnostics. Also, a review of the vibration analysis tools used for machinery diagnostics was presented. The description of the principles of machinery condition monitoring were elaborated, and the process for machinery diagnostics was described with some examples. Finally, the chapter discussed the recent trends for machinery vibration monitoring and diagnostics.

References

1. Global Condition Monitoring Services Market, Forecast to 2025, Frost & Sullivan, July (2018)
2. El-Shafei, A.: Measuring vibration for machinery monitoring and diagnostics. Shock Vibr. Digest **25**(1), 3–14 (Jan, 1993)
3. Eshleman, R.L.: Basic Machinery Vibration. The Vibration Institute, USA (2002)
4. Peters, J., Eshleman, R.L.: Machinery Vibration Analysis. The Vibration Institute, USA (2020)
5. El-Shafei, A., Rieger, N.: Automated diagnostics of rotating machinery, in Proceedings of ASME Turbo Expo, Atlanta, Georgia, USA, ASME paper GT-2003-38453 (June, 2003)
6. ISO 13373-1 Condition Monitoring and Diagnosis of Machines—Vibration Condition Monitoring: Part 1: General Procedures (2002)
7. ISO 13373-2 Condition Monitoring and Diagnosis of Machines—Vibration Condition Monitoring: Part 2: Processing, Analysis and Presentation of Vibration Data (2005)
8. ISO 13373-3 Condition Monitoring and Diagnosis of Machines—Vibration Condition Monitoring: Part 3: Guidelines for Vibration Diagnosis (2015)
9. ISO 13373-4 Condition Monitoring and Diagnosis of Machines—Vibration Condition Monitoring: Part 4: Diagnostic Techniques for Gas and Steam Turbines with Fluid-Film Bearings (2021)
10. ISO 13373-5 Condition Monitoring and Diagnosis of Machines—Vibration Condition Monitoring: Part 5: Diagnostic Techniques for Fans and Blowers (2020)
11. ISO 13373-7 Condition Monitoring and Diagnosis of Machines—Vibration Condition Monitoring: Part 7: Diagnostic Techniques for Machine Sets in Hydraulic Power Generating and Pump-Storage Plants (2017)
12. ISO 13373-9 Condition Monitoring and Diagnosis of Machines—Vibration Condition Monitoring: Part 9: Diagnostic Techniques for Electric Motors (2017)
13. El-Shafei, A.: Diagnosis of installation problems of turbomachinery, in Proceedings of ASME Turbo Expo, Amsterdam, The Netherlands, ASME paper GT-2002-30284 (June, 2002)
14. El-Shafei, A.: Fan Diagnosis in the Field, in 9th International Conference on Vibrations in Rotating Machinery, Institution of Mechanical Engineers, Exeter, UK, pp. 719–730 (2008)
15. El-Shafei, A.: Machine Diagnosis as a Standardized Approach, In: Cavalca K., Weber H. (eds) Proceedings of the 10th International Conference on Rotor Dynamics – IFToMM. IFToMM 2018. Mechanisms and Machine Science, vol 61. pp. 207–221, Springer, Cham. https://doi.org/10.1007/978-3-319-99268-6_15 (2018)
16. El-Shafei, A.: Standardized fan diagnostics according To ISO 13373-5, in Proceedings of the Second Vibration Institute Middle East Conference, RITEC, October (2020)
17. El-Shafei, A.: Diagnosis of a recurring pump shaft failure, in 37th Annual Meeting of the Vibration Institute, San Antonio, Texas, pp. 269–281, June (2014)
18. El-Shafei, A., Diagnosis of gearbox faults in the field, in 8th International Conference on Vibrations in Rotating Machinery, Institution of Mechanical Engineers, Swansea, UK, pp. 479–490 September (2004)
19. Younan, A.A., El-Shafei, A.: Model calibration of anisotropic rotordynamic systems with speed dependent parameters. ASME J. Eng. Gas Turbine Power **130**(July), 042502-1–042502-10 (2008)
20. Eshleman, R.L.: Machinery condition analysis. Vibrations **4**(2), 3–11 (1988)
21. Berggren, J.C.: Diagnosing faults in rolling element bearings—Part III electronic data collector applications. Vibrations **5**(2), 8–19 (1989)
22. http://wikid.io.tudelft.nl/WikID/index.php/MEMS-based_accelerometer. Accessed on 10 Apr 2021

23. Hassan, T., El-Shafei, A., Zeyada, Y., Rieger, N.: Comparison of neural network architectures for machinery fault diagnosis, in: ASME Turbo Expo 2003, collocated with the 2003 International Joint Power Generation Conference, American Society of Mechanical Engineers Digital Collection, pp. 415–424 (2003)
24. El-Shafei, A., Hassan, T.A., Soliman, A.K., Zeyada, Y., Rieger, N.: Neural network and fuzzy logic diagnostics of 1x faults in rotating machinery. ASME J. Eng. Gas Turbines Power **129**(3), 703–710 (2007)
25. Nath, A.G., Udmale, S.S., Singh, S.K.: Role of artificial intelligence in rotor fault diagnosis: A comprehensive review. Artif. Intell. Rev. **54**, 2609–2668 (2021)
26. Ibrahim, M.S., Dimitri, A.S., El-Shafei, A.: Troubleshooting of a high-speed turbocharger test rig, in Proceedings of the Second Vibration Institute Middle East Conference, RITEC, October (2020)

Manufacturing

Manufacturing of Structural Components for Internal Combustion Engine, Electric Motor and Battery Using Casting and 3D Printing

Ilias Papadimitriou⬤ and Michael Just⬤

Abstract The manufacturing of powertrain components for light vehicles became more challenging during the last years, due to the powertrain diversity caused by the powertrain's electrification. Based on this fact, the conventional process should be adapted to the use of different types of components and low production volumes while keeping the production costs on acceptable level. One of the extensively used manufacturing processes is High-Pressure Die Casting (HPDC) for aluminum and magnesium alloys, characterized by high complexity and high investments cost. Besides the casting processes, the additive manufacturing techniques introduced in the last years, offer new possibilities mainly due to the wide range of alloys and the possibility to shape up complex geometries. This chapter gives a comprehensive overview on the manufacturing methods used in the production of structural parts of powertrain and treats some advanced subjects of manufacturing issues and future trends.

Keywords Manufacturing · High-pressure die cast (HPDC) · Additive manufacturing · Simulation · 3D printing

1 Introduction

The trends around future manufacturing, concentrate mainly on digitalization and virtual process optimization.

With the digitalization and Industry 4.0 becoming more present, information technology is used to connect the different process steps and enable the communication between them. At the same time, it offers the possibility to store high volumes of acquired data in data centers or clouds. The connection of the process steps that include the network of machines offers the possibility to optimize the manufacturing cycle and reduce radically the manufacturing costs. The stored data can improve

I. Papadimitriou (✉) · M. Just
GF Casting Solutions AG, Amsler-Laffon-Strasse 9, 8201 Schaffhausen, Switzerland
e-mail: ilias.Papadimitriou@georgfischer.com

© The Author(s), under exclusive license to Springer Nature Switzerland AG 2022
T. Parikyan (ed.), *Advances in Engine and Powertrain Research and Technology*,
Mechanisms and Machine Science 114,
https://doi.org/10.1007/978-3-030-91869-9_15

future decisions using machine learning techniques and optimize the machine park health monitoring.

Virtual process optimization, based on simulation of the manufacturing process combined with the product development requirements, is one of the promised features of the manufacturing optimization. Compared to other simulation techniques like multibody dynamics, computer fluid dynamics or finite elements, the simulation of the manufacturing processes has not yet reached similar maturity. One of the most complex simulation tasks is the simulation of high-pressure die casting (HPDC) called filling solidification simulation (FSS). Although the simulation itself is based on the well-known Navier–Stokes equation, it can be challenging to get a good correlation with measurements manly because of statistical distributed process parameters and the alloy characterization in fluid, semisolid and solid condition under temperature up to 600 °C for aluminum. Despite the significant achievements during the last years, the correlation for casting alloys in the quasi-homogeneous state remains a challenge.

2 Overview of Power Unit Casting Components

The casting process is one of the main manufacturing processes of power unit components. Some typical components are described below.

Cylinder block. Cylinder blocks are produced mainly in aluminum or cast iron. In the last years, the production of cast iron decreased drastically compared to aluminum. Cast iron remains the primary material for track engines produced in sand casting. Aluminum cylinder blocks are produced mainly in high-pressure die-cast and permanent mold casting.

Cylinder head. Cylinder heads are produced almost exclusively from aluminum in permanent mold casting or sand casting. This is because the cooling circuit cannot be produced in high-pressure die-cast technology. The introduction of 3D-printed cores during the last years allows the production of complex and thin cores giving the possibility to realize optimized geometry for the cooling in sand casting processes.

Crankshaft. The crankshaft is one of the moving parts in the engine that is used to be produced in cast iron for years in the past. Today, cast iron is used for some industrial or track engines. The majority of the crankshafts are produced out of forged steel. Reasons for this are the lower material properties of cast iron.

Exhaust manifold and turbine housing. Engine manifolds and turbine housings of internal combustion engines are produced in sand casting using cast iron or casting steel. Exhaust manifolds of diesel engines are mainly produced in cast iron. For the gasoline engines with higher gas temperatures it is necessary to use casting steel—being more expensive—with alloy costs comparable to Nickel. The disadvantage of using cast iron is the high weight of the component, which is positioned in most cases on the top of the engine. For this reason, there are some other solutions to reduce the weight, like hydroforming of sheet metal in particular for sports cars or luxury segment cars. Another solution introduced within the last ten years is the integration of the exhaust manifold in the cylinder head using surface treatment in the gas outlet.

Oil pan/covers. The oil pan and engine covers are produced mainly in casting or in sheet metal. The use of high-pressure die-cast is very popular as it allows the realization of very thin walls and integration of functions like the oil filter possible. For some engines, covers are produced in magnesium in the high-pressure die-cast process for lightweight reasons.

Gearbox housing. Gearbox housings are typical cast parts. Today, they are mainly made of aluminum in high-pressure die cast or permanent mold casting. The production in high-pressure die-cast allows the realization of thin wall thickness that is very important in terms of lightweight, particularly for large gearbox housings.

Electric engine housing. The electric engine housing for vehicle applications is a typical casting component and produced mainly in high-pressure die-cast or permanent mold casting. The most of the high power electric engines are water-cooled and the cooling cavity is realized using sand (or salt) cores in permanent molding or as two assembled parts in high-pressure die-cast.

Battery housing. Battery housings are produced mainly in high-pressure die-cast or sheet meatal. The use of high-pressure die-cast is very popular with some limitations in terms of component dimensions. One main challenge is the cooling circuit that in some cases should be integrated into the housing's component design.

Turbine blades. Turbine blades are often made from a titanium-based alloy. These are used in both power plant turbines and aircraft turbines. One of the most effective methods of increasing the power density and thermal efficiency of a gas turbine is to raise the turbine inlet temperature. Thus, the necessary cavity design of the blade can be produced exclusively used investment casting or additive technology.

3 Casting Process

3.1 Use of Casting Process

The main reasons for the use of casting processes are:

Realization of complex geometry. A typical example is the water jacket of a cylinder head, which is not possible to produce using other process. The cylinder head is usually produced in mold or sand casting using low pressure/gravity casting.

Close to final contour. One of the characteristics of casting processes is the possibility to produce parts close to final geometry, which is released as negative geometry in cores or molds. This is very important regarding the machining effort that follows the casting process. The surface of the final part may consist of machined and

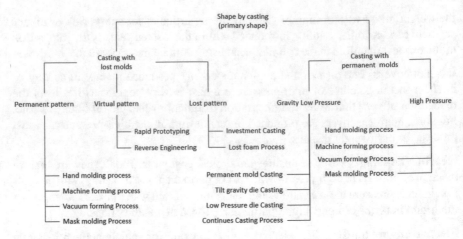

Fig. 1 Overview of casting methods [1]

non-machined areas. The roughness of non-machined areas depends on the casting process with **high roughness** in sand casting, **middle roughness** in mold casting and **low roughness** in high-pressure die-cast.

High reproduction quality. The casting process itself is a complex process with many parameters that need to be adjusted at the beginning but is stable and can be automatized for high-volume production.

Some other important characteristics of the casting process are that it can be used for almost all available technically important materials, the high level of economic efficiency and the usage of recycled material. Figure 1 shows an overview and a classification of typical casting methods.

3.2 Investment Casting

The investment casting process is the most complex and thus most labor-intensive casting process, as automated processes are rarely realized here, depending on the quantities involved. The first step is the production of a positive wax or thermoplastic preform, which is often created in a negative profile aluminum tool. In the next step, the wax preforms are joined together forming a cluster. Subsequently, this cluster is wrapped several times using a slurry to coat with sand and then dried. This process might have a duration of more of a week, depending on the granularity of the sand and the specifications of the product. The scope is to build up a ceramic layer that can later withstand the casting pressure. Depending on the ceramic type chosen, aluminum, iron or titanium fusion can be applied. Subsequently, the wax or plastic is melted out of this cluster, the mold is calcined and filled with the previously liquefied metal. Removal takes place by destroying the mold with possible subsequent mechanical

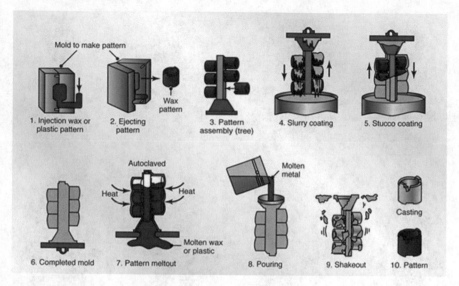

Fig. 2 Process steps of investment casting [2]

processing and/or finishing. Due to the high surface qualities and the representation of finest details, the process has gained a firm market position. Typical products are turbine blades for aviation or for the power energy sector. Figure 2 shows a typical process steps of investment casting. Figure 3 shows an example of turbine blades produced in investment casting process.

3.3 Sand Casting

In the sand casting process, the liquid metal is poured into a sand mold by gravity and solidified under normal air pressure. This is a slowly solidifying casting process. The most commonly used process is the *green sand process*. Nowadays, it is increasingly being replaced by the additive mold manufacturing process (3D-Printed sand core). This provides a design flexibility and therefore almost any complex geometry can be produced. Depending on the geometry of the product, a core can be used, which is firmly stored in the sand mold and is therefore one of the few processes that can form hollow structures. The sand mixture usually consists of quartz sand, a binder (natural or chemical) and water. A distinction is made between molds with a pattern (*green sand process*) and molds without a pattern, which are produced generatively using a 3D printer.

These are a one use molds that are reprocessed afterwards in a closed-loop process. This process is assigned to the group of lost molds. Typical components are engine cylinder heads, electric engine housings, hollow cast components, heat exchangers and many more. Figure 4 shows the process of a gravity sand casting.

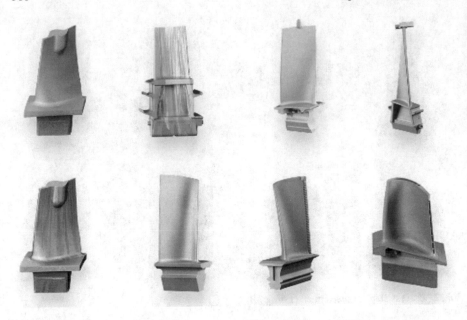

Fig. 3 Blades, example of IGT airfoils equiaxed, DS and single crystal [3]

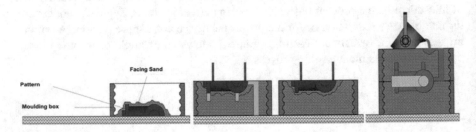

Fig. 4 Sequence gravity sand casting [4]

3.4 Permanent Mold Casting/Tilting Mold/Low-Pressure Die Casting

In the permanent mold casting process, liquid metal is poured into a reusable permanent mold usually made of metal. Due to the higher thermal conductivity of the metallic mold, more heat can be dissipated in a shorter time compared to sand casting. The faster the cooling behavior, the finer the microstructure and the better the strength properties of the casting. Permanent mold casting can be carried out by hand for small quantities and automatically for larger quantities. This means that the individual steps such as core insertion, casting, cooling and opening of the mold as well as ejection can be carried out continuously. The type of filling low pressure or atmospheric pressure or the type of filling of the permanent mold itself such as horizontal or

Fig. 5 Cross section of a
low pressure casting
machine [4]

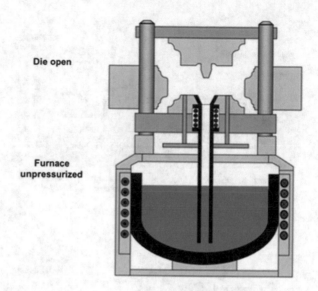

vertical has an influence on the quality and cycle time. Especially in low-pressure casting can be achieved high densities and thus excellent properties by the calmed mold filling. Cycle times are generally between 3 and 7 min. Typical components are safety components such as chassis components, wheels, swing arms for motorcycles, cylinder head, and cylinder block. Figure 5 shows section of a typical low pressure casting equipment.

3.5   High-Pressure Die Casting (Cold Chamber/Hot Chamber Process)

In this casting process, the molten metal is introduced into a metal mold under high pressure, cooled and then removed from the mold by means of ejectors. The usual cycle time is between 50 and 120 s. The advantages of this casting process are that large quantities can be produced in a short time and that the cast parts can be produced close to final contour, i.e. with tight tolerances. This leads to an optimum economic efficiency. This is only one of the main reasons why the HPDC process has become so popular in recent years. The area of application is often in the powertrain sector (engine blocks, converter housings, electric motor housings, etc.) as well as in all areas of the Body in White "BiW" (strut domes, side members, doors, instrument panel, etc.). The structural parts are characterized by their light weight, thin walls and good mechanical properties. The maximum casting pressure applied in the die casting process depends on the geometry and size of the component. This pressure is typical machine indication. Figure 6 show the layout of die typical casting machine.

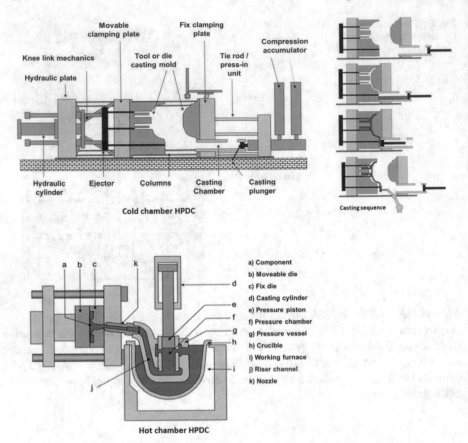

Fig. 6 Principle hot chamber/cold chamber die casting, difference is the melt flow [4]

3.6 Trend of the Modern Times: Giga Casting

Since the patent publication of US Company Tesla two years ago regarding the construction of a big die-cast machine to cast the complete Body in White, the 'American dream' arrived at the rest of the foundry world. According to the idea "bigger is better" there have been intensive activities among the suppliers of die-cast machines, the peripheral equipment manufacturers and the competitors of Tesla. Although by the patent release, there were many critical reviews regarding the feasibility of such big die-cast machines, in the meantime, both OEMS's and suppliers take into consideration concepts that follow a similar idea.

This direction has unexpectedly accelerated the innovated way of thinking of casting suppliers. Since then, they have been working diligently on checking the cost-effectiveness of such a large die-cast component, adapting/redeveloping casting alloys, scaling the changed properties to the entire vehicle, joining processes, tolerance linkages in the vehicle and the logistics that such components entail. Such large die-casting machines also pose special challenges for the fundament of a foundry, for the die-casting tools and the subsequent steps such as machining and coating systems. For example, a shop portal crane must double or triple its lifting capacity in order to be able to lift future die casting tools at all. In addition, the service life of these tools, as they wear similarly to the smaller tools, which means that these quantities of steel have to be made available and of course recycled. And this leads to more intense sustainability considerations. Also the quality and the cost of a Giga casting tool seems to be an important determining factor to evaluate.

Conventional processing machines have to be redesigned as well to be able to take e.g. 1800 × 1200 × 600 mm parts and to be able to process them with the same accuracy. Metal Alloying will be change from high performance precipitation hardening to lower performance natural hardening alloy systems to reduce the effect of components distortion on components due to heat treatment. However, no alloy system on the market has only advantages. It is therefore necessary to consider the secondary-primary alloy ratio in the case of natural hardening alloy systems and to adapt the joining technology accordingly. Figure 7 shows an example of modular construction of battery housing.

The vehicle concepts themselves are also being given new perspectives, because cast alloys behave differently in the field of crash than sheet steel or wrought alloy solutions does. One focus here, for example, is on repair approaches in the event of damage to the rear end structure. Figure 8 shows the extract of Tesla's Patent of giant machine.

Fig. 7 Multi scalable and functional battery trace [5, 6]

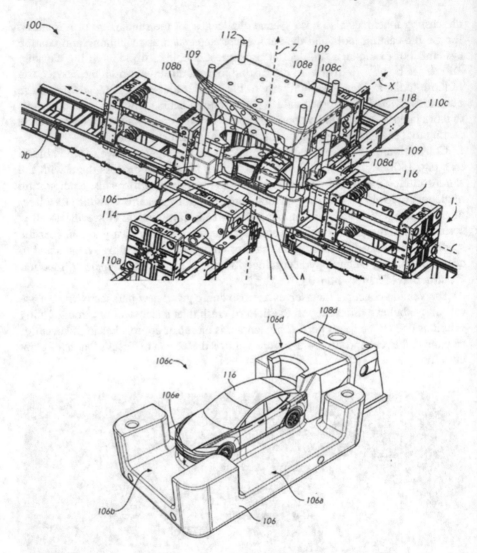

Fig. 8 Tesla's patent giant machine to produce the frame in almost one piece [7]

3.7 Quality and Tolerances of Casting Components

The quality and tolerances of casting components are classified according to standards and norms which are widely recognized in the industry. They are defined to facilitate cooperation, especially in the technical field. These used as guideline to specify in a comprehensible way what can be expected of a product or service, as well as the rules and conditions to be observed in production or services.

Regarding materials made of aluminum and its alloys, similar to all other alloying systems, can be distinguished basically in two major groups: wrought aluminum alloys and cast aluminum alloys. Both have a defined designation system according to European standards.

European standardization in the field of foundry technology is being steered by the Technical Committee CEN/TC 190 Foundry Technology [8]. Specific material topics are dealing by the CEN/TC 132 Aluminum and Aluminum Alloys [8] Committee.

European standards must be adopted by the CEN member states and transformed to national standards.

In the European norm EN 1706:2010 chemical composition and mechanical properties of aluminum casting alloys are specified.

Some specific topics are described in other standards, such as:

- DIN 1680 general tolerances for castings GTA/GTB,
- DIN 50148 tensile specimens for non-ferrous metals,
- DIN 50049 or EN 10204 Factory certificate and Certificate of Acceptance,
- VDG P202 Volume deficits of castings,

or others partly agreed on separately between institutions, customers and suppliers.

In addition, these standards describe and regulate the determined material composition, the specimen dimensions, the yield strength, tensile strength and elongation. Some of these standards apply across the board (sand casting, permanent mold casting and high-pressure die casting) to all casting processes and some are adapted to the processes. However, there are certain differences, e.g. in the required dimensional tolerances, but on the other hand similar requirements apply to casting defects.

4 Materials

4.1 Alloys

Sustainability is now one of the defining issues in the world of materials. The focus is not only on the CO_2 footprint in the production of the materials, but increasingly also on the logistics routes and the actual processing into the final component.

Other influencing factors are

- Secondary-primary content of the alloys
- Recycling of the recycled material
- Minimization of heat treatment for the alloys (changeover from artificial aging to cold-hardening system "naturally hardened").

Material selection for the automotive sector is influenced by various aspects such as cost, weight, required structural stiffness, crash relevance and, as mentioned above, the CO_2 footprint. Figure 9 shows a typical classification of aluminum alloys.

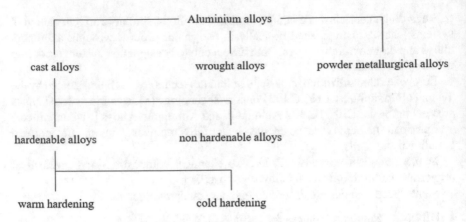

Fig. 9 Typical classification of aluminum alloys [1]

Aluminum and magnesium have excellent light weighting potential, which is particularly important considering the lifetime of a product or vehicle. However, these materials have lower mechanical properties compared to steel alloys, which can be compensated for by appropriate component design, especially if they are produced using the casting process. Compared to steel, the materials aluminum and magnesium have a higher CO_2 footprint due to the manufacturing process but it can be compensated due to the reduced weight calculated in-vehicle lifetime.

One of the advantages of both materials is their good recyclability. The aim with any material is to cover the material requirement exclusively with recycled material, if possible. For many years, the efforts of automotive manufacturers have been directed toward improving existing aluminum alloys or developing new alloy variants that meet the complex requirements of automotive construction and contain a high proportion of recycled material. In addition, many OEMs are often not open to consider the alloy developments available on the market, as on the one hand they want to source the alloy worldwide and on the other hand they do not want to place the sourcing with one supplier. This attitude of course has an impact on innovation in aluminum and magnesium alloy development (This is particularly evident when they regularly inquire what is new on the market!). The low initial strength of aluminum is increased by various alloying elements as additives. The main alloying elements are silicon, magnesium, copper, manganese and zinc (Fig. 10). Silicon (Si) is one of the most important alloying elements. It mainly improves the casting properties of the alloy. Silicon in combination with magnesium causes good hardenability. Figure 10 shows the element effects on hardening of aluminum alloys.

Magnesium (Mg) plays the role of a strengthening element. Magnesium in combination with silicon forms the Mg2Si phase, which ensures strengthening during heat treatment. Magnesium improves corrosion properties, but increases the tendency to oxidation and hydrogen absorption. Copper (Cu) influences the strength and hardness of an aluminum alloy. On the other hand, copper worsens the corrosion resistance.

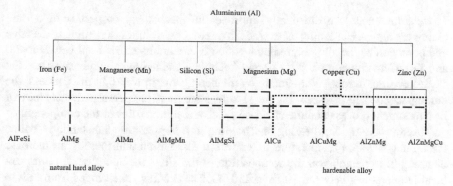

Fig. 10 Effect of the elements on the hardening of aluminum alloys [1]

Copper-based alloy has lower followability and hot cracking resistance compared to other alloys.

Iron (Fe) exerts a negative influence on the toughness of the alloy at contents above approx. 0.2%. It is present as a very brittle AlFe(Si) compound in the form of plates.

The iron-manganese ratio in particular is important when permanent steel molds are used, since it influences the tendency to stick in the casting mold. Too much iron, as already mentioned, makes the material more brittle, too little iron increases the tendency to stick and the melt sucks it out of the steel mold, which can lead to excessive aging of the casting tools.

4.2 Aluminum Alloys

The term ductility refers to the property of formability. It is particularly important in case of aluminum alloys for further processing (failure behavior in the case of a crash or for joining processes such as riveting). In the case of certain aluminum alloys, subsequent hardening is now necessary in order to achieve corresponding ductile properties. This process is called precipitation hardening. Examples of aluminum alloys that can be precipitation hardened are:

- Aluminum–silicon-magnesium alloys
- Aluminum-magnesium-silicon alloys
- Aluminum-copper alloys
- Aluminum-zinc alloys.

Today, the hardening of aluminum alloys is carried out according to two principles. One is cold aging and the other is warm aging. In the case of cold aging, the aluminum alloy is allowed to "rest" at room temperature for 5–8 days before further processing until it reaches its final strength.

Heat treatment can drastically improve the mechanical properties of certain aluminum alloys. Hardening takes place in three steps: Solution annealing to dissolve and homogenize the alloy ingredients which is essentially carried out between 400 and 530 °C and for approx. 30–180 min (technically/economically relevant), and the step of cooling to room temperature by quenching e.g. in water. In this process, the homogenized state is frozen. Figure 11 shows the time–temperature dependency of heat treatment process and the effect on mechanical properties of the component.

The Step Aging, i.e. heating and holding at a temperature substantially below the homogenization temperature to achieve a dissolution and thereby an increase in strength. Depending on the composition of the alloy, the aging temperature can be at temperatures between 100 and 250 °C. The holding time ranges from ½ h to one week, depending on the alloy composition. Table 1 shows the heat treatment

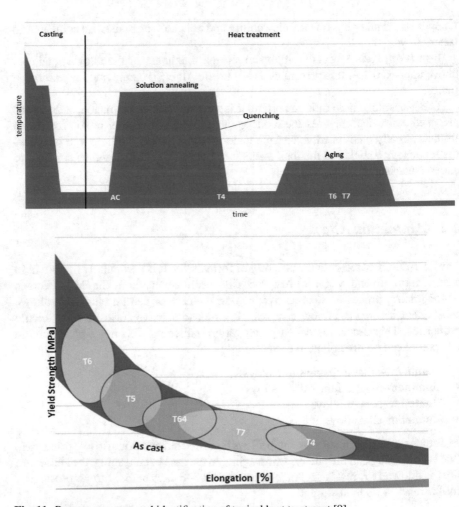

Fig. 11 Process sequence and identification of typical heat treatment [9]

Table 1 Codes and description of the typical heat treatments [1]

Sort code			Heat treatment
EN	ASTM	Int	
F	F	AC	Manufacturing condition, as-cast
	W		Solution annealed (unstable condition)
O			Soft annealed
T1	T2		Controlled cooling after casting, cold aged
	T3		Quenched to high temperature forming process, cold formed and cold aged in stable condition
T4	T4		Solution annealed and cold aged
T5	T5		Only aged (without solution annealing)
T6	T6		Solution annealed, quenched and aged
T64	T64		Solution annealed, quenched and not completely aged
T7	T7		Solution annealed, quenched and overaged

designation and description according to EN, ASTM and International Standards.

Table 2 shows the mechanical properties of aluminum and Magnesium alloys that can be reached in High Pressure Die Cast Process.

Similar tables are commercially available on the market for the various processes. In general, the manufactures or the companies processing the various metals have an in-depth understanding of the material they use and are therefore often better sources than the minimum requirements described in the DIN.

4.3   Magnesium Alloys

More than 90% of the magnesium alloys used today are processed in die casting. There are good and also cost-effective alloys available here for use of the parts at room temperature, especially AZ91 and AM50 or AM60. The former has slightly better strength, the AM alloys more ductility.

All die-casting alloys contain >4% aluminum, which improves castability. Other die-casting alloys are available for use at elevated temperatures, such as AS21/AS41, AE42/AE44 or AJ62. Gearbox housings or engine blocks are manufactured in large quantities from these alloys. Corrosion resistance is increased by adding alloying elements that form passive layers (e.g. Al, Be, Si, Ca, Y, P, etc.). Limiting the amounts of Fe, Ni, Co and Cu added reduces the formation of cathodic precipitates. In recent years, the addition of Ca, among other elements, has been increasingly investigated in order to develop so-called fire-retardant alloys, but as always, it can be observed that the elements have a favorable effect on one property and a negative effect on others (too much Ca reduces the flowability).

Table 2 Extract of mechanical properties of Al and Mg [5]

Properties		Unit	Most common aluminium high pressure die casting alloys						
			AlMg5Si2Mn	AlSi10Mg(Fe)	AlSi10MnMg	AluSiDur© +GF+ patented	AlSi9Cu3(Fe)	AlSi12Cu1(Fe)	AlSi12(Fe)
DIN designation			EN AC-51500	EN AC-43400	EN AC-43500	Non standardized	EN AC-46000	EN AC-47100	EN AC-44300
Alternative				239D		AlSi10MnMg+	226D	231D	230D
Melting point	T_m	°C	580–618	550–600	550–590		490–600	530–580	570–580
Ultimate tensile strength	R_m	N/mm²	290–370 (F)	260–300 (F)	250–295 (T6)	240–315 (T6)	≥240 (F)	≥240 (F)	≥240 (F)
0.2% yield strength	$R_{p0.2}$	N/mm²	175–215 (F)	150–170 (F)	130–180 (T6)	135–235 (T6)	≥140 (F)	≥140 (F)	≥130 (F)
Fracture elongation	A	%	5–20 (F)	2.5–6 (F)	4–9 (T6)	6–13 (T6)	≥1.0 (F)	≥1.0 (F)	≥1.0 (F)
Compression yield strength	$\sigma_{d0.2}$	N/mm²	177	169	174				
Young's modulus	E	kN/mm²	72	74	76		75	76	75
Shear modulus	G	kN/mm²	27	29	28				
Poisson's ratio	ν	–	0.333	0.327	0.332				

(continued)

Table 2 (continued)

Properties		Unit	Most common aluminium high pressure die casting alloys						
			AlMg5Si2Mn	AlSi10Mg(Fe)	AlSi10MnMg	AluSiDur© +GF+ patented	AlSi9Cu3(Fe)	AlSi12Cu1(Fe)	AlSi12(Fe)
Impact bending toughness	@20 °C	J	85.6 (20 °C)	11.7 (20 °C)	17.3 (20 °C)				
Brinell hardness		HB	88–90 (F)	87–90 (F)	85–91 (T6)		min. 80 (F)	min. 70 (F)	min. 60 (F)
Density	ρ	g/cm³	2.6	2.65	2.7		2.75	2.65	2.65
Thermal expansion coefficient	α	10^{-6}/K	24 (20–100 °C)	21 (20–100 °C)	21 (20–100 °C)		21	20	20
Specific heat capacity	c_p	J/(g·K)	0.900 (20 °C) 0.939 (100 °C) 1.056 (400 °C)	0.91 (20 °C)	0.876 (20 °C)		0.880 (20 °C) 0.910 (100 °C) 1.075 (400 °C)	0.890 (20 °C)	0.900 (20 °C)
Specific electric resistivity	ρ_e	Ω·mm²/m	0.063–0.072	0.048–0.063	0.040–0.053		0.059–0.077	0.050–0.067	0.046–0.063
Thermal conductivity	λ	W/(K·m)	104–111 (20 °C) 147 (100 °C)	130–150 (20 °C)	140–170 (20 °C)		125–154 (20 °C) 161 (400 °C)	120–150 (20 °C)	130–160 (20 °C)

Properties		Unit	Most common magnesium high pressure die casting alloys			
DIN designation			MgAl9Zn1	MgAl5Mn	MgAl6Mn	MgAl4RE4
			EN-MC21120	EN-MC21220	EN-MC21230	
Alternative			AZ91	AM50	AM60	AE44
Melting point	T_m	°C	470–595		450–615	
Ultimate tensile strength	R_m	N/mm²	200–260	195–240	190–250	200–260

(continued)

Table 2 (continued)

Properties		Unit	Most common magnesium high pressure die casting alloys			
			MgAl9Zn1	MgAl5Mn	MgAl6Mn	MgAl4RE4
0.2% yield strength	$R_{p0.2}$	N/mm^2	140–170	100–125	120–150	100–140
Fracture elongation	A	%	1–6	6–14	4–14	5–11
Compression yield strength	$\sigma_{d0.2}$	N/mm^2	135	68		87
Young's modulus	E	kN/mm^2	43	43	45	45
Shear modulus	G	kN/mm^2	16	16		17
Poisson's ratio	ν	–	0.299	0.294		0.294
Impact bending toughness	@20 °C	J	6.5	18		14
Brinell hardness		HB	65–85	50–65	55–70	Min. 50
Density	ρ	g/cm^3	1.83	1.8	1.8	1.8
Thermal expansion coefficient	α	10^{-6}/K	27		26	
Specific heat capacity	c_p	J/(g·K)	1.05		1.02	
Specific electric resistivity	ρ_e	Ω·mm^2/m	0.141		0.111	
Thermal conductivity	λ	W/(K·m)	50.00–84.00		61	

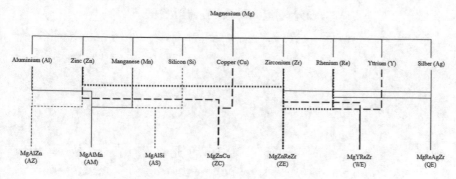

Fig. 12 Grouping of magnesium casting alloys according to DIN 1753 [10]

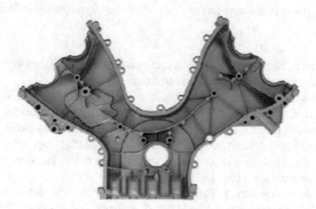

Fig. 13 Mg front cover (MgAl4RE4Mn) of a V8 engine [6]

Higher requirements regarding operating temperatures are covered by aluminum-free magnesium alloys WE43, WE54, QE22, Elektron21 or SC1. These alloys are used, for example, to cast helicopter gearboxes and engine blocks. Figure 12 shows the Magnesium casting alloys classification according to DIN 1753. Figure 13 shows an engine Magnesium front cover from a V8 engine.

4.4 Material Simulation

The material and component properties are characterized by their microstructure and its defects. The paved way created during the last years requires now to further refine the methods of prediction, so that in the future a congruence of simulation and reality will arise, considering the manufacturing process.

Thermodynamic methods allow predictions to be made in complex multi-component and multi-phase systems. If this method is extended to include kinetic

Solid solution with precipitates

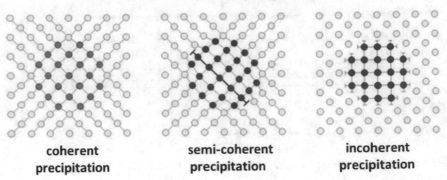

coherent   semi-coherent   incoherent
precipitation   precipitation   precipitation

Fig. 14 Development of coherence for different precipitation stages without consideration of voids [12]

processes such as nucleation, diffusion and dendrite arm change, a statement about phase constituents and compositions is obtained.

A further approach considers interface effects and enables the simulation of microstructure development. In order to stress this sensitive structure even more, the manufacturing process or the effect of the same should be mapped. The overall goal of all these developments is the prediction of the material properties, the possible defects and their location and also to accurately map the manufacturing processes within the virtual material development.

Unfortunately, simulation software like **JMatPro**, **Pandat**, **Thermo-Calc**, **MatCalc** [11] or others are not yet able to disentangle this complex situation. Most programs make their predictions based on thermodynamic equilibrium states and these states are not close to reality. Equilibrium states in the technical world are rather rare. Programs like **JMatPro** are already starting to integrate the diversity so that e.g. heat treatment states or mechanical properties at elevated temperature can be predicted. As stated before, this is not yet perfect and the expertise of specialized personnel is indispensable, but the interlinked prediction of material properties is clearly the goal. Figure 14 shows Development of coherence for different precipitation stages without consideration of voids.

4.5 Joining Technology

Driven by increasing costs for energy and raw materials especially by the European CO_2-emission targets, automotive industry faces the challenge to develop more lightweight and at the same time still rigid and crash-stable car bodies that are affordable for large-scale production. The implementation of weight-reduced constructions depends not only on the availability of lightweight materials and related forming technologies, but also on cost-efficient and reliable joining technologies suitable for

Table 3 Commonly used joint technology methods

Join methods
Resistance spot welding (steel-steel, Al-Al)
Laser beam welding (Al-Al, steel-steel)
Bonding
MIG welding (Al-Al)
Semi hollow punch-riveting
Grip punch riveting
Clinching
Roller Hemming
MAG welding (steel-steel)
Friction element welding
Flow drill screwing

multi-material design. The joining technology today makes a significant contribution to lightweight construction and the use of multi-material mix.

The development towards hybrid electric or battery electric drives is changing designs faster than any other innovation of the past decades. However, today's casting or die casting places special demands on joining technology.

Aluminum and magnesium casting alloys are more "brittle" than sheet materials or extrusions. For mixed joints, various welding lines have been or are being optimized or further developed e.g. CMT or the laser hybrid welding process, which have minimized the disadvantage of distortion due to heat input compared to the classic MIG/MAG process. Table 3 show the actual joint possibilities for hybrid material design in Body in White (BiW). In powertrain area the material hybrid design is not popular.

Thermo-mechanical joining processes combine the wishes of OEMs, which on the one hand include economic efficiency and on the other hand do not have to change existing production lines or only need to change them slightly. This process makes it very easy to permanently join lightweight materials such as aluminum and magnesium, but also to join carbon fiber reinforced polymer (CFRP) with the current body material steel. A resistance element, similar to the rivet used today, is inserted into the "brittle" partner, e.g. aluminum or magnesium die casting, and is then integrated into the BiW using the existing resistance element spot welding process.

The state-of-the-art self-piercing riveting process is also undergoing further development. This is achieved through innovation in the area of force initiation (preconditioning of the joining process, speed and method of insertion) as well as adaptation of the rivet shape and the associated die.

In order to support multi-material assemblies, friction stir welding and friction spot welding have recently become popular. Through the rotation of the pin, the downforce and forward pressure, the then necessary local heat is generated—the material becomes semi solid, in small areas liquid, and the permanent joint is formed.

The two materials can be joined using both butt and lap joints. It is a benefit if one of the two partners has a low melting point.

Another feature of die casting that must be taken into account, especially before bonding processes starts, is the state of the surface of the components that will be joined. The surfaces of cast components are often contaminated with residual organic substances which have to be washed off. These substances are created for example by the die-casting process (release agent) and the machining process (drilling-milling-cooling medium). The removal can be done by WBK (washing-pickling-preservation) or neutralized by pickling or surface lasers immediately before the bonding process. Bonding is well established and also belongs to the state-of-the-art processes. Up to now, both mechanical and thermal bonding have been used together with an adhesive. In order to harden, most adhesives require activation by heat (cathodic dip coating is often used at approx. 195 °C). However, the chemical industry is already developing adhesives that are insensitive against surface contamination of the joining partners and no longer require activation or curing energy. Figures 15, 16, 17 shows some typical joint processes used in the industry today.

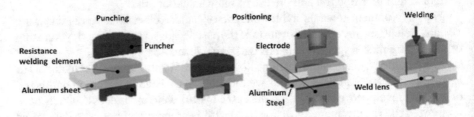

Fig. 15 Resistance element spot welding with prefabrication [13]

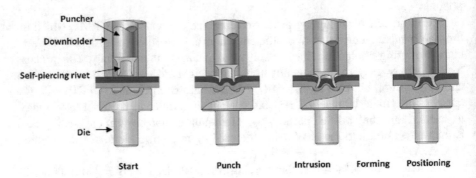

Fig. 16 Semi hollow punch rivet process [14]

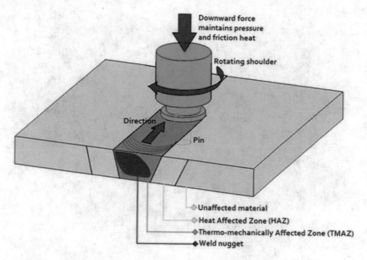

Fig. 17 Friction stir welding (FSW) principles of the process [15]

5 Additive Manufacturing

Observing the market of additive manufacturing for components in the last years, it becomes clear that this technology has evolved from prototypes manufacturing to a process used for series production. Meanwhile, there are many examples from aerospace and power generation technology where additive manufacturing substitutes components of traditional techniques such as precision casting, forging, milling or welded constructions.

By using larger, faster and more powerful machines, more and more components can be produced more cost-effectively by additive manufacturing than by traditional manufacturing methods. A critical issue from the design point of view is a 'mindset change' towards 'additive thinking'. It requires a holistic approach to additive manufacturing as an essential part of the value chain.

Nevertheless, in most cases (except for spare parts), a re-design of the components makes the additive process more efficient and attractive. Changing to additive design thinking is one of the most significant challenges that the technology will need to overcome in the coming years. Designers and engineers have to develop the necessary knowledge to introduce a design approach suitable for this technology.

In Powertrain there is lot of components produced in this technology in small series as prototypes. Figure 18 shows electric motor with integrated differential produced with additive manufacturing.

The additive manufacturing process is building up ultra-thin layers (0.03–0.1 mm) one over the other. Thus the realized geometry corresponds exactly to the representation of the Virtual Model. It is also advantageous to have only one design environment/job to avoid data conversion between CAD and 3D-Print machine. During the design of the component the following considerations should be taken into account:

Fig. 18 Electric motor in using additive manufacturing technology [16]

Thickness. For your model, make sure that layer thicknesses do not fall below the minimum. Keyword "slicer phase".

Walls thickness. Less than 1 mm wall thickness is possible but not very useful. At this thickness, the walls break easily. Better larger than 1.0 mm.

Level of detail. Fine details (up to 0.5 mm) are possible. Detail level means the distance between the surface of your model and the surface of your design detail, e.g. embossed and engraved texts. However, the feasibility depends on the orientation. As a recommendation can be used as embossed and engraved texts: Use Arial 20 pt with a height and depth of 0.4 mm.

Support material. Avoid support material whenever possible. Support material that is removed afterwards leaves areas that don't look very nice but can cause component faults. You can avoid support material by ensuring that your model is printed, beginning with the largest area. Walls or overhangs with angles less than 45° concerning the build platform usually need to be provided with supports; otherwise, they can lead to printing errors.

Surface quality and alignment. Angles less than 45° concerning the build platform tend to produce a poorer surface quality, while steep angles greater than 45° tend to form a smooth surface. Overhang structures (e.g., the underside of a table) tend to have poor surface quality.

Thermally-induced stress. The model is virtually equivalent to a layer-by-layer welding process. Due to solidification, thermally induced stresses occur as the molten powder cools. It is recommended to round or chamfer corners with a minimum radius of 3 mm by component design. For the same reason, it is also recommended to avoid

Table 4 Tolerance classification

Dimensions in mm	Casting standard DIN EN ISO 8062-3:2008-09	Tolerance in mm	IT-Grade ISO286
3	DCTG6	±0.2	IT14
3–30	DCTG6	±0.2	IT12
30–60	DCTG8	±0.5	IT13
60–120	DCTG8	±0.7	IT13
120–250	DCTG8	±0.9	IT13
250–400	DCTG8	±1.1	IT13

pointed corners. As a general rule, preference has to be given to organic shapes rather than angular designs.

Dimensional accuracy. For 3D-printed metal, a general accuracy of DCTG 8 described in DIN EN ISO 8062-3: 2008-09 for dimensions between 30 and 400 mm can be reached.

Table 4 shows the tolerance classification according to DIN EN ISO 8062-3:2008-9.

Removal of powder. The design must include holes at regular intervals to remove the unused powder trapped in the hollow body. These holes should have a diameter of at least 3 mm. For larger and complex hollow bodies, several holes with a diameter of 7 mm are required.

The longer and more complex you make the internal channels, the larger the minimum diameter you need to maintain. Therefore, in such cases, it is not always possible to guarantee a print result that corresponds to the theoretical CAD geometry.

The powder manufacturing sector, which previously had almost a monopoly share in the old pricing policy, is rapidly catching up in terms of material development and the number of suppliers. This will probably affect the pricing policy and the variety of available alloys in the future.

The development of additive technology will affect the field of the processes and the manufacturers of the machines. In particular, the possibility of modifying the material microstructure during the printing process has enormous potential. With additive manufacturing, it becomes possible to change locally the mechanical properties of the component at the desired locations under controlled conditions.

A good overview of additive machine manufacturers, covering the manufacturers' processes, including users, can be obtained from analyzing the market research of AM-power (based on ASTM/ISO 52900).

The following two diagrams provide an overview of additive technology in the area of plastic as well as metal. There are, of course, other 3D printing materials such as sand or ceramics, but applications in plastic and metal currently dominate the market. Tables 5 and 6 give an overview of the most important processes including raw material classification and use.

Table 5 Overview of the additive technology for polymers

Technology	Material form
Thermal powder bed fusion	Powder thermoplastic
Powder bed fusion	
Electrographic sheet lamination	
Selective powder deposition	
Pellet based material extrusion	Pellet thermoplastic
Continuous fiber thermoplastic deposition	Filament thermoplastic
Continuous fiber material extrusion	
Filament based material extrusion	
Continuous fiber set lamination	Sheet thermoplastic
Area-wise vat polymerization	Liquid thermoset
Vat polymerization	
Fiber alignment area-wise vat polymerization	
Material jetting	
Thermoset deposition	
Continuous fiber thermoset deposition	
Liquid vulcanization deposition	Liquid elastomer

Back in the 1980s, the first additive manufacturing process based on stereolithography was commercialized by Chuck Hull. Hull's company 3D-Systems is still one of the leading suppliers in the field of 3D printing today. Many suppliers are active in the areas of stereolithography and filament extrusion.

Today, processes from the field of laser-based powder bed fusion are still the most widely used technologies. EOS, SLM Solutions or Renishaw, among others, are active in this field.

Binder-based systems currently raise great expectations in terms of cost and production time reduction from suppliers such as HP or Desktop Metal. Binder systems target the high design freedom of established metal 3D-printing technologies, while at the same time, they claim to offer higher production speeds than powder bed processes.

In contrast, wire-based processes, such as those from Gefertec or Waam, do not focus on high resolution. These processes produce 'near-net-shape' raw parts whose surfaces are machined when high-resolution features are required.

Additive Manufacturing: Future Trends

The technology will find a permanent position among traditional manufacturing technologies in the next five years. However, the technology will not completely replace any of the conventional technologies. It will rather be applied where the

Table 6 Overview of the additive technology for metals [17]

Technology	
Liquid metal printing	Wire
Resistance welding	
Wire feed laser/electron beam/energy deposition	
Wire arc/plasma arc/energy deposition	
Coldspray	
Powder feed laser energy deposition	Powder
Electron beam powder bed fusion	
Laser beam powder bed fusion	
Metal selective laser sintering	
Mold short deposition	
Powder metallurgy jetting	
Binder jetting	
Metal lithography	
Metal pellet fused deposition	Pellet
Metal filament fused deposition modeling	Filament
Nanoparticle jetting	Dispersion
Friction deposition	Rods
Ultrasonic welding	Sheet

process can develop its full potential. In other words, there will be applications that will only be printed in the future, just as there are applications today that are only forged.

6 Filling and Solidification Simulation (FSS)

Computer-aided simulation of casting processes is used to simulate the flow response of melted material as well as the solidification during the cooling phase. The purpose of FSS is to identify porosities or other material defects of the casting part. Furthermore, the thermoelastic/thermoplastic stresses can be simulated as well as residual stresses and the distortion of the casting parts. The numerical simulation is based on the law of the conservation of mass and Navier–Stokes equation. The law of the conservation of mass is formulated for incompressible liquids. The Navier–Stokes equation mathematically describes the general motion of fluids caused by all acting forces, they represent the momentum theorem in a differential form for Newtonian liquids and are called impulse equations of real flow. In their simple form they are only valid for incompressible fluids and their instationary flows [18]. The FSS shows the flow profile with the flow rate and temperatures of the melt in the cavity. In the

FSS, the temperature, the rate of flow, the direction of the metal flow and the pressure in the metal cavity is simulated at any time and for each filled area of the cavity [12].

6.1 FSS of High-Pressure Die Casting

6.1.1 Characteristics of High Pressure Die Casting

The high-pressure die casting process consist of 3 phases: pre-filling, mold filling and final filing.

In the pre-filling phase, the melted metal is injected into the die cavity, which forces the metal flow through a horizontally mounted cylindrical shot sleeve. The pre-filling phase ends when molten metal enters the ingate.

In the mold-filling phase the melted material enters the cavity through the gating and fills the mold cavity.

In the last final-filling phase the piston of the HPDC machine applies the maximal pressure in order to complete the filling process and to push the air in the overflow.

In high-pressure die casting, the overflow, also known as air bean, has the task of diverting liquid metal, which contains air, separating agent vapors or oxides created by turbulence, from the mold cavity. The overflows are cutouts in the mold insert near the mold cavity, which are connected to the mold cavity by a thin gate. These are provided with ventilation channels that are intended to allow air and separating agent vapors to be extracted to the outside. If possible, overflows are placed in such a way that oxides and residues of the lubricant are washed out, and they should also ensure that certain areas of the mold are heated up. Figure 19 shows qualitative diagram of a typical piston movement pf a HPDC machine. One of the main challenges is setting

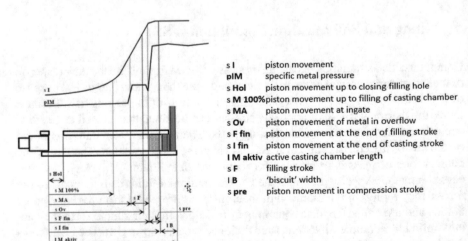

s I	piston movement
pIM	specific metal pressure
s **Hol**	piston movement up to closing filling hole
s **M 100%**	piston movement up to filling of casting chamber
s **MA**	piston movement at ingate
s **Ov**	piston movement of metal in overflow
s **F fin**	piston movement at the end of filling stroke
s **I fin**	piston movement at the end of casting stroke
I M aktiv	active casting chamber length
s **F**	filling stroke
I B	'biscuit' width
s **pre**	piston movement in compression stroke

Fig. 19 Piston movement diagram [4]

of the curve which are important for the quality of the component. These settings are performed today with the 'try an error' principle based on the experience. The use of machine learning (ML) methods could be used to optimize these curves using the data acquired during the Ramp-Up process [19].

Short Phase Description

Preliminary phase. During the preliminary phase (sMA), the metal is dammed up at low piston speeds (~0.05–0.3 m/s) and into the area of the gate guided (duration ~ 1–4 s). Due to the low speed, the cavity air is pressed out through the mould vents.

A non-optimal piston velocity could cause airwave pulsation between the mould and the piston crown wall, leading to air-gas shots in the melt that remains in the casting part in the form of air porosities.

The optimization of Preliminary phase leads to reduced air porosities in the melt. The parameters depend on the degree of filling, the position and geometry of ingate, the casting chamber length, piston diameter and temperature.

Mold-filling phase. During the preliminary phase (sFfin), the metal is injected with high pressure and high velocity (Al ~ 2–6 [m/s], Mg ~ 4–8 [m/s]) into the mould filling the mould cavity. Depending on the volume, the filling process takes about 10–400 [ms]. The metal solidification in this phase is prevented because of the short time of melt injection.

Post-filling phase. During this phase (IMactive), the metal begins to solidify, while the piston applies pressures about 200–1000 [bar] to compensate the metal shrinkage and compresses the gas pores on the rand of the component.

Switching point. The switching point determines the duration of the first phase when the slow piston speed is switched to the fast one.

If the switching point is delayed, there is a risk of cold flow and the mold filling time can become too long. In this case, the melt acceleration is not sufficient and causes a non-optimal filling process.

The quick switching point generates airwaves leading to air porosities in the melt.

The HPDC machine must be able to control the switching point with high accuracy to avoid any dynamic effects.

6.1.2 Numerical Simulation

From the simulation point of view, all casting processes can be divided into in three phases: the filling phase (preliminary filling, mold filling, post-filling), the solidification phase, and the cooling phase. Each of these phases can be described with a set of governing equations.

The filling phase is described with the continuity equation, momentum equation (Navier–Stokes) and energy equations. The solidification is described with energy equations and momentum equations. The target of solidification simulation is to calculate the temperature distribution in the solidification phase. The cooling phase

Table 7 List of most common foundry processes [12]

Process	Range of component mass	Time-scale of process	Material cast
Sand	100 g to 250 tonnes	Seconds to days	All metals
Investment	10 g to 100 kg	Minutes to hours	All metals
Resin shell	100 g to 100 kg	Minutes to hours	Fe, Cu
Permanent mould/gravity die	2–50 kg	Minutes	Primarily Al, Zn and Mg, some Cu
Low pressure die	5–25 kg	Minutes	Primarily Al and Mg
High pressure die	10 g to 20 kg	Seconds	Al, Mg, Zn
Squeeze casting	100 g to 20 kg	Minutes	Al

described with the energy equation can be expected using the equilibrium equation and Hooke's Law to simulate the solid's stress strain state [18].

Unfortunately, the casting simulation has not reached a confidence level as it had been reached by other simulation techniques like fluid dynamics or multi-body dynamics. The reason is the process complexity on the one hand but on the other hand primarily the time–space scale of the governing physical laws of casting process. Table 7 lists the main casting routes used in foundries and puts some of their estimations of scales [12].

The extremes of the processes cover about 8 orders of magnitude when talking about mass, about 5 orders of magnitude for filling and solidification times, and about 3 orders of magnitude in minimum length scale. If the modeler wishes to model grain structures in large castings, there is a scale factor of about 5–6 orders of magnitude between the geometry of the casting and the physical phenomenon to be modelled [12].

The main results of the simulation are the heat flux, thermal contraction, tempera-ture distributions at various times, velocity vectors, density and pressure changes, free surface movement, and areas of enclosed air. Since there are still no unique result evaluation methods, the simulation results have to be associated with defects like laminations, skin oxidation, porosity, distortion or trapped air. The usual approach is to evaluate simulation results and defects using praxis-based criteria.

Figure 20 show a filling process as typical result of FSS for a cylinder block

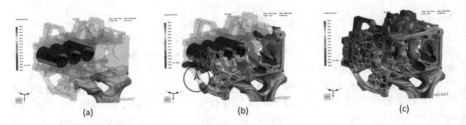

(a)   (b)   (c)

Fig. 20 FSS of cylinder block

produced in HPDC. The colors represent the temperature with a maximum value of 600 °C.

Figure 20a shows the position and temperature distribution of the melted metal by the end of the pre-filling phase. The front of melted metal passed the ingate position of the cavity. The blue areas represent a drop of temperature, which should not exceed a lower limit during the casting phase. Figure 20b represents material and temperature distribution in 0.7694 s. The area A shows the closed air, which could be critical for porosities. Figure 20c shows the last phase of filling and the beginning of solidification. The temperature distribution on this phase is very important for the mechanical material properties of the component.

The most common porosity types of defects that appear in aluminum castings processed with high-pressure die casting technology are gas porosity, shrinkage porosity and leaker.

Gas porosity can be described as trapped air in the casting, which can be caused by several sources. It can be caused by poor melt quality, poor shot end control, poor venting and overflow function or bad gating and runner design. The shrinkage porosities can be the cause of possible crack initiation in the casting, which can have several different causes, mainly being thick walls of the casting. This defect is caused by metal reducing its volume during solidification and an inability to feed shrinkage with more metal before solidification. Hot spots can also cause shrinkage porosity to be concentrated in a specific zone.

Leaker porosities represent large air volumes in the material having several different reasons.

The simulation of porosities is one of the main tasks of FSS and is evaluated using the percentage of air in the element volume. In praxis there are some guidelines used to evaluate the porosities but they depend on the simulation software and mesh of the simulation model as well as the component itself. Unfortunately, there is no standard method to evaluate the porosities and in most of the cases it is very important to perform a model calibration using some porosities measurement on the component using destructive or non-destructive methods.

As the local porosities influences the mechanical properties of the component it is very important to take the material model for the life cycle evaluation of the component into account. Unfortunately, there is no standard methodology at the moment but this will be an important issue for the next development of material life cycles modeled for casting components [20].

Figure 21 shows results of FSS simulation and a typical porosity evaluation of a 4-cylinder engine block.

Inhomogeneous temperature distribution and porosities lead very often to different local mechanical properties of components.

Fig. 21 Porosity evaluation of a 4-cylinder engine block [21]

6.2 FSS of Iron Sand Casting

The filling and solidification simulation includes the simulation of the casting parameters such as flow profile, flow rate and cooling down of the melted iron. The main focus during the evaluation of the results is put on the identification of local gas porosities and their elimination through the modification of the casting's gating system. The casting process is mathematically described by the continuity equation, the Navier–Stokes equation, the single-phase flow of the liquid metal and the function of the volume fraction. The homogeneous solidification and cooling are central design criteria for the casting system. In order to achieve the required material quality in terms of mechanical properties, the last solidification area should take place in the gating to avoid any air porosities. This is a challenge for the design of the gating system (Fig. 22).

6.3 Numerical Simulation and Software

Although numerical simulation of casting process is not widely used in the casting industry, there is commercial software on the market that has been established during the last decades offering various modules and material databases. The most known on the market are: **AutoCAST, EKKCapcast, MAGMASoft, NOVACAST, ProCAST, Flow-3DCast, SOLIDCast**.

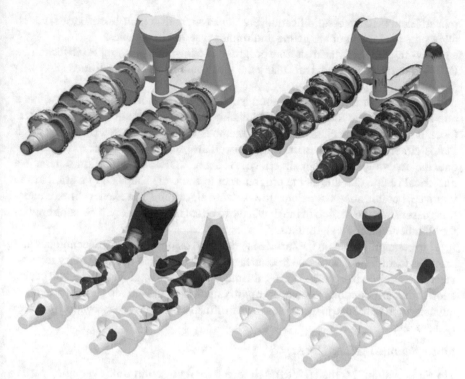

Fig. 22 Solidification simulation of cast iron crankshaft [22]

The most common numerical methods implemented in casting simulation software are

1. Finite Difference Method (FDM) implemented in MAGMASoft, Flow-3DCast, and SOLIDCast,
2. Finite Volume Method (FVM) implemented in NOVACAST
3. Finite Element Method (FEM) implemented in EKKCapcast, ProCAST
4. Vector Element Method (VEM) implemented in AutoCAST.

Finite Difference Method (FDM)

Finite-difference methods (FDM) are a class of numerical techniques for solving differential equations by approximating derivatives with finite differences and Taylor-Series expansion. The material properties are assumed to stay at a constant level in spatial domain. To achieve high accuracy, it is necessary that the spatial domain is divided into maximal number of controlled volumes resulting in long simulation times.

MAGMASoft solves heat and mass transfer on a rectangular grid on the surface of the model. Besides the filling behavior, it provides information about other casting related features like pre-solidification, entrapped air, melt velocity distribution, which allows the evaluation of the gating and runner system effectiveness.

Flow-3D calculates the melt front progression using the Volume of Fluid technique (VOF) and model intricate parts using the Fractional Area/Volume Ratios (FAVOR) [23].

The volume of fluid (VOF) method is a free-surface modelling technique, i.e. a numerical technique for tracking and locating the free surface (or fluid–fluid interface). It belongs to the class of Eulerian methods which are characterized by a mesh that is either stationary or is moving in a certain prescribed manner to accommodate the evolving shape of the interface. As such, VOF is an advection scheme—a numerical recipe that allows the programmer to track the shape and position of the interface, but it is not a standalone flow-solving algorithm. The Navier–Stokes equations describing the motion of the flow have to be solved separately. The same applies for all other advection algorithms.

The rectangular grid on the surface is generated using well-known techniques and provides features like improved free surface flow accuracy, reduced memory requirement and quick convergence. The disadvantaged of the rectangular surface grid generation is the introduction of geometry approximation on the curved geometry such as radius, which introduce inaccuracies for the diffusion and mass conservation problem in these areas.

Finite Volume Method (FVM)

The Finite Volume Method (FVM) is a method for representing and evaluating partial PDE in the form of algebraic equations. In the FVM, volume integrals are converted to surface integrals using the divergence theorem. These terms are evaluated as fluxes at the surfaces of each finite volume. An advantage of the FVM is that it is easily formulated to allow unstructured meshes. The method is used in many computational fluid dynamics packages. *Finite Volume* refers to the small volume surrounding each node point on a mesh.

Although the idea of domain discretization is the same compered to FDV, the surface integral method of FVM is more efficient treating the Neumann Boundary Condition (the condition specifies the values of the derivative applied at the boundary of the domain) as well as the discontinuities because of poor requirements of singularity and smoothness.

NOVACAST generates cubic elements on the volume domain and surface shell element on the surface domain. Using this technique, the simulation is fast and enough accurate. The height/width of metal front, which is calculated during the filling, provides a filling of necessary fraction of the shell instead of filling cell by cell as observed in FDM.

ProCAST and EKKCapcast are based on the Finite Element Method.

Vector Element Method (VEM)

In the VEM, originally called node finite element analysis, the unknown field values are assigned to the element vertexes where the vector field quantities are described with their components on the vertexes. The domain discretization is done using node elements which are used for the vertex calculation. In this case, two/three scalars must be assigned to each vertex for the two-/three-dimensional case, respectively. One of the difficulties of the method is the continuity requirements for the approximated vector fields which cannot be easily fulfilled and the application of the boundary conditions are rather problematic. To overcome these problems, are introduced the so-called edge elements [24].

The approach in casting simulation based on the calculation of the highest thermal gradient at any point of the domain which is given by the vector sum of flux vectors in all directions [25]. The domain is divided in pyramidal sectors from the considered point. The flux vector for heat or cooling (which is proportional to volume) and the surface area is calculated for each sector. The calculation continues along the direction of the result and flux vector unless the vector becomes zero. The location of the hot spot is the last location and the feed metal path considering to be the path of the calculation. From the numerical point of view, the method is simple compared to other methods but delivers reliable and robust results. VEM requires a small memory and is fast enough. AUTOCAST is based on VEM.

Material Model

One of the main challenges of casting simulation is the material model. Most of the software available on the market provides some material database with standard alloys based on aluminum, copper, magnesium and zinc, and also superalloys such as nickel, chromium and titanium-based alloys as well as material database for ductile iron and steel. The most popular software packages like MAGMASoft and ProCast provide most commonly used mold and core materials. Most software packages have an open database system for user-defined material.

Discussion and Conclusions

The casting technology is a cost-effective method to produce powertrain components. Nevertheless, in the near future, the casting industry has to deal with challenges related to electrification in the transport sector and digitalization. The electrification itself introduces new parts such as battery housings, electric motor housings, and high-performance electronics housings. Due to a high diversity of powertrain types for passenger cars being forecasted in the coming years, especially due to hybridization, several variants of powertrains and smaller production numbers on the market are to be expected. Since the traditional casting technology was optimized for higher quantities, requiring more flexibility becomes a challenge for the optimization of the casting processes.

The casting itself is a high energy demanding process with a significant impact on CO_2 emissions. An essential contribution to CO_2 reduction can be achieved with the help of digitalization and the use of Machine Learning methods for the process optimization regarding energy consumption and minimization of scrap. This will intensify the research activities in this area in the coming years. Another focus will be on increasing energy efficiency, which includes both process optimization and evaluation of energy sources.

References

1. Just, M.: Basic Casting Production Methods and Metallurgy, Compendium Training Document 2010. GF Casting Solutions, Schaffhausen (2010)
2. Kalpakjian, S., Schmid, S.R., Werner, E.: Werkstofftechnik. Pearson Studium (2017)
3. Precision Casting and Additive Manufacturing. Business Unit Presentation. GF Casting Solutions (2021)
4. Grundlagen der Gießereitechnik. Presentation. VDG. www.vdg.de (2005)
5. Database, GF Casting Solutions Materials Database (2020)
6. Handout Product Samples. GF Casting Solutions (2020)
7. Kallas, M.K.: Multi-directional unibody casting machine for a vehicle frame and associated methods. Patent Appl. Pub. No. US 2019/0217380 A1. Pub. Date: July 18, 2019
8. List of CEN technical committees. Wikipedia: https://en.wikipedia.org/wiki/List_of_CEN_tec hnical_committees
9. Hossinger, J.: The Tables From One to Ten of Heat Treatment. GF Casting Solutions, Schaffhausen (2020)
10. Kammer, C.: Standardization of magnesium alloys. In: Magnesium Pocketbook, pp. 141–154. Aluminium-Verlag, Düsseldorf (2000)
11. JMatPro (www.matplus.de), Pandat (www.computherm.com), Thermo-Calc (www.thermo calc.com), MatCalc (www.matcalc-engineering.com)
12. Jolly, M.R., Gebelin, J.-Ch.: Casting simulation: how well do reality and virtual casting match? A state of the art review. Int. J. Cast Metals Res. (2002)
13. Schwingfestigkeit thermisch-mechanisch gefügter Verbindungen für Mischbauanwendungen mit ultrahochfesten Stählen. Schlussbericht zu IGF-Vorhaben Nr. 18344 N, 16.01.2017
14. Pudar, M., Vallant, R.: Fügen von Magnesium im Karosseriebau. Magna Steyr and TU Graz, Join-Ex 2012 Vienna, 10+11 October, TU Graz Homepage. https://online.tugraz.at/tug_online/voe_main2.getvolltext?pCurrPk=72385
15. Taheri, H., Kilpatrick, M., Norvalls, M., Harper, W.J., Koester, L.W., Bigelow, T., Bond, L.J.: Investigation of nondestructive testing methods for friction stir welding. Metals **9**, 624 (2019). https://doi.org/10.3390/met9060624
16. Schäfer, P.: Porsche zeigt additiv gefertigtes E-Antriebgehäuse. Additive Fertigung. Springer Professional, 12-01-2021. https://www.springerprofessional.de/en/additive-fertigung/elektr ofahrzeuge/porsche-zeigt-additiv-gefertigtes-e-antriebgehaeuse/18733248
17. AM-Power Homepage: https://am-power.de
18. Nogowizin, B.: Theorie und Praxis des Druckgusses. Schiele & Schön (2010)
19. Reducing Costs and Environmental Impact Through Innovation. GF Casting Solutions. https://www.gfcs.com/en/technology-space/artificial-intelligence-in-manufacturing.html
20. Vanderesse, N., Maire, É., Chabod, A., Buffière, J.-Y.: Microtomographic study and finite element analysis of the porosity harmfulness in a cast aluminium alloy. Int. J. Fatigue **33**(12), 1514–1525 (2011)
21. Menne, R.J., Weiss, U., Brohmer, A., Egner-Walter, A., Weber, M., Oelling, P.: Implementation of casting process simulation for increased engine performance and reduced development time and costs—selected examples from FORD R&D engine projects. In: 28th International Vienna Motor Symposium, pp. 92–203
22. Papadimitriou, I., Track, K.: Lightweight potential of crankshafts with hollow design. MTZ Worldwide **79**, 42–45 (2018)
23. Flow-3D: Modeling techniques. https://www.flow3d.com/
24. Takahashi, N., Nakata, T., Fujiwara, K., Imai, T.: Investigation of effectiveness of edge elements. IEEE Trans. Magnetics **28**(2), 1619–1622 (1992)
25. Amin, L.D., Patel, S., Mishra, P., Joshi, D.: Rapid development of industrial castings using computer simulation. Indian Foundry J. **60**(8), 40–42 (2014)

Author Index

419

Printed in the United States
by Baker & Taylor Publisher Services